Chemistry for the 21st Century
Edited by E. Keinan, I. Schechter

Chemistry for the 21st Century

Edited by
Ehud Keinan, Israel Schechter

⊗WILEY-VCH

Weinheim – New-York – Chichester – Brisbane – Singapore – Toronto

MYL

The Editors of this Volume

Prof. Dr. E. Keinan
Department of Chemistry
Technion – Israel Institute of Technology
Haifa 32 000
Israel

The Scripps Research Institute
10550 North Torrey Pines Road, MB20
La Jolla
Ca 92037
USA

Prof. Dr. I. Schechter
Department of Chemistry
Technion – Israel Institute of Technology
Haifa 32 000
Israel

Library of Congress Card No.:
applied for

British Library Cataloguing-in-Publication Data
A catalogue record for this book is
available from the British Library.

**Die Deutsche Bibliothek – CIP Cataloguing-
in-Publication Data**
A catalogue record for this publication is
available from Die Deutsche Bibliothek

Printed in the Federal Republic of Germany

Printed on acid-free paper

Composition   Kühn & Weyh, Freiburg
Printing   Strauss Offsetdruck GmbH,
Mörlenbach
Bookbinding   Wilh. Osswald & Co.,
Neustadt (Weinstraße)

ISBN   3-527-30235-2

Contents

Preface

"Creative thinking in science is mainly concerned with searching for fresh meaningful combinations of old pieces of information."

W. I. B. Beveridge

"Novelties come from previously unseen association of old material. To create is to re-combine."

François Jacob

On May 11–12 1998, the Israeli scientific community staged a very special event celebrating both the 20th anniversary of the Wolf Prize and the 50th anniversary of the State of Israel. The scientific program of that event included many lectures by Wolf Prize Laureates in all five scientific disciplines of the Prize: Agriculture, Chemistry, Mathematics, Medicine and Physics. This program, which was sponsored by Novartis AG, highlighted a revolutionary trend that characterizes science at the turn of the 20th century – the unification and integration of science.

Historically, divisions of the sciences evolved with the establishment of European universities during the Late Middle Ages. The medieval perception made clear distinctions between theology and physics, and even between terrestrial and celestial physics. Furthermore, it separated the mathematical sciences from natural philosophy and metaphysics. Institutional divisions separated the elevated sciences (which were based on philosophical principles) from the intellectually inferior arts (which were based on the various domains of craftsmanship). Interestingly, mathematics was considered at that time to be a craft and therefore unfit for explaining natural phenomena.

During the Renaissance and particularly during the Scientific Revolution of the 17th century, attempts were continuously made to shake this old perception. Many outstanding pioneers, including Copernicus, Kepler, Galileo, Bacon, Harvey, Descartes, Boyle, Newton, and Hooke, creatively endeavored to establish new theoretical and methodological links between previously separated scientific domains. Galileo took mathematics seriously and connected it with metaphysics; Bacon called the philosophers to come out of their libraries to the real world and learn from other disciplines. Promulgating the new "experimental philosophy", Boyle probably established the most sophisticated laboratory in Europe. Craftsmen, meanwhile, started reading philosophical books and made their own contributions to science.

Interestingly, and probably not surprisingly, most of these geniuses operated outside the academic divisions of the university. Yet, the establishment of the modern universities at the beginning of the 19th century was again accompanied by the ordering, classification and partitioning of science. Understandably, the structure of universities around the world reflects their need to maintain comprehensive teaching programs and cover the entire base of scientific knowledge. Thus, the universities' commitment to provide scientific education has led to compartmentalization, hierarchy and even the formation of barriers between well-defined scientific disciplines, such as mathematics, physics, chemistry, biology and medicine, each of which is further divided into subcategories and subsections.

Now, at the turn of the 20th century we are witnessing a profound movement to change the traditional scientific arrangement that was so well preserved by our universities for almost 200 years. The current scientific revolution is reminiscent of the one that occurred at the beginning of the 17th century. As the contemporary explosion of information can cause further specialization and partitioning of scientific disciplines, integration of science sounds counterintuitive. Yet, the increased societal demand for novel technologies challenges the scientific community and pushes towards collaboration and cross-fertilization between seemingly remote scientific domains. There is now an interesting tendency to establish new academic as well as industrial institutions that break the walls between traditional disciplines and encourage creativity, brainstorming and heterogeneous teamwork in research.

Originally, we planned to publish a single volume compilation of scientific contributions that covers a wide range of active disciplines, entitled "Science for the 21st century". Publisher's preferences, however, lead us to arrange the material in the form of a three-volume set. While this first volume is dedicated to the chemical sciences, the second volume will focus on the life sciences and the third will deal with physics and the mathematical sciences.

The astounding history of science and technology of the 20th century testifies for the highly unpredictable nature of scientific developments. It is essentially impossible, and would be quite arrogant on our part, to make any prediction of which scientific disciplines will prevail in the future and therefore merit greater attention today. Consequently, this compilation represents a broad-scope sampling, but certainly not comprehensive coverage, of all scientific domains. All chapters are written by scientists who are well known in their fields and include their scientific achievements and a personal view of the perspectives and prospects in these fields. We hope that readers will have the opportunity to encounter new ideas and will be exposed to subjects well beyond that of their own scientific disciplines. However, our most important task is to provide the readers, now and in the future, with the "taste" and "flavor" of the science at the turn of the millennium.

August 2000

Ehud Keinan
Israel Schechter

List of Contributors

J. L. Atwood
Department of Chemistry
University of Missouri-Columbia
Room 123 Chemistry Building
Columbia, MO 65211
USA

Rodney J. Bartlett
Quantum Theory Project
PO Box 118435
University of Florida
Gainesville, FL 32611-8435
USA

William A. Bonner
Deptment of Chemistry
Stanford University
Stanford, CA 94305-5080
USA

Robert H. Burris
Department of Biochemistry
University of Wisconsin
433 Babcock Drive
Madison, WI 53706-1544
USA

Marvin L. Cohen
Department of Physics
University of California at Berkeley
Berkeley, California 94720
USA

T. Daniel Crawford
Center for Computational Quantum Chemistry
School of Chemical Sciences
University of Georgia
Athens, GA 30602
USA

S. J. Danishefsky
Department of Chemistry
Columbia University
Havemeyer Hall, Box 255
New York, NY 10027
USA

R. B. van Dover
Bell Laboratories
Lucent Technologies
Murray Hill, NJ 07974
USA

Yong Duan
Department of Pharmaceutical Chemistry
University of San Francisco
513 Parnassus
San Francisco, CA 94143
USA

G. Ertl
Abteilung Physikalische Chemie
Fritz-Haber-Institut der MPG
Faradayweg 4 - 6
14195 Berlin
Germany

Patrick Frank
Deptment of Chemistry
Stanford University
Stanford, CA 94305-5080
USA

Christina R. Harris
The Sloan-Kettering Institute for Cancer Research
Laboratory for Bioorganic Chemistry
1275 York Avenue
New York, NY 10021
USA

Stephen C. Harvey
Department of Biochemistry and Molecular
Genetics
University of Alabama at Birmingham
Birmingham
Birmingham, Alabama 94143
USA

Rollin A. King
Center for Computational Quantum Chemistry
School of Chemical Sciences
University of Georgia
Athens, GA 30602
USA

P. A. Kollman
Department of Pharmaceutical Chemistry
University of San Francisco
513 Parnassus
San Francisco, CA 94143
USA

Scott D. Kuduk
The Sloan-Kettering Institute for Cancer Research
Laboratory for Bioorganic Chemistry
1275 York Avenue
New York, NY 10021
USA

J.-M. Lehn
Laboratoire de Chimie Supramoléculaire
ISIS, Université Louis Pasteur
4, rue Blaise Pascal
67000 Strasbourg
France

Matthew L. Leininger
Center for Computational Quantum Chemistry
School of Chemical Sciences
University of Georgia
Athens, GA 30602
USA

R. D. Levine
The Fritz Haber Research Center
for Molecular Dynamics
The Hebrew University
Jerusalem 91904
Israel

Leonhard R. MacGillivray
Steacie Institute for Molecular Sciences
National Research Council of Canada
Ottawa
Ontario, K1A 0R6
Canada

L. A. Paquette
Evans Chemical Laboratories
The Ohio State University
100 West 18th Avenue
Columbus, OH 43210
USA

H. F. Schaefer III
Center for Computational Quantum Chemistry
School of Chemical Sciences
University of Georgia
Athens, GA 30602
USA

L. F. Schneemeyer
Bell Laboratories
Lucent Technologies
Murray Hill, NJ 07974
USA

Morris Schnitzer
Eastern Cereal and Oilseed Research Centre
Agriculture and Agri-Fodd Canada
Ottawa, ON, K1A 0C6
Canada

Solomon H. Snyder
Department of Neuroscience
Johns Hopkins University School of Medicine
725 North Wolfe Street
Baltimore, Maryland 21 205
USA

Edward F. Valeev
Center for Computational Quantum Chemistry
School of Chemical Sciences
University of Georgia
Athens, GA 30602
USA

Steven S. Wesolowsy
Center for Computational Quantum Chemistry
School of Chemical Sciences
University of Georgia
Athens, GA 30602
USA

R. N. Zare
Deptment of Chemistry
Stanford University
Stanford, CA 94305-5080
USA

1

Some Reflections on Chemistry – Molecular, Supramolecular and Beyond*

Jean Marie Lehn

1.1
From Structure to Information. The Challenge of Instructed Chemistry

In chemistry, as in other areas, the language of information is extending that of constitution and structure as the field develops towards more and more complex architectures and behaviors. And supramolecular chemistry is paving the way towards comprehending chemistry as an *information science*. In the one hundred years since 1894, molecular recognition has evolved from Emil Fischer's "Lock and Key" image of the age of mechanics towards the *information paradigm* of the age of electronics and communication. This change in paradigm will profoundly influence our perception of chemistry, how we think about it, how we perform it. Instructed chemistry extends from selectivity in the synthesis and reactivity of molecular structures to the organization and function of complex supramolecular entities. The latter rely on sets of instructed components capable of performing on mixtures specific operations that will lead to the desired substances and properties by the action of built-in self-processes.

Supramolecular chemistry has started and developed as defined by its basic object: the chemistry of the species generated by non-covalent interactions. Through recognition and self-processes it has led to the concepts of (passive and active) information and of programmed systems, becoming progressively the chemistry of molecular information, its storage at the molecular level, its retrieval, transfer and processing at the supramolecular level.

*) The present text is an adapted version of
   Chapter 10 in : J.-M. Lehn, Supramolecular
   Chemistry – Concepts and Perspectives, 1995,
   VCH, Weinheim.

The outlook of supramolecular chemistry is towards a general *science of informed matter*, bringing forward in chemistry the third component of the basic trilogy matter–energy–information.

Chemical systems may store information either in an *analogical* fashion, in the structural features (size, shape, nature and disposition of interaction sites, etc.) of a molecule or a supermolecule, or in a *digital* fashion, in the different states or connectivities of a chemical entity. The evaluation of the information content of a *molecular recognition* process based on structural sensing in receptor/substrate pairs requires an assessment of the relevant molecular characteristics. Recognition is not an absolute but a relative notion. It results from the structural (and eventually dynamical) information stored in the partners and is defined by the fidelity of its reading, which rests on the difference in free energy of interaction between states, represented by different receptor/substrate combinations. It is thus not a yes/no process but is relative to a threshold level separating states and making them distinct. It depends on free energy and consequently on temperature. The parameter kT could be a possible reference quantity against which to evaluate threshold values, differences between states and reading accuracy. Both analogical and digital processing of chemical information depend on such factors.

Digital storage and retrieval of chemical information is found in the nucleic acids where the basic digital operation is a two state 2/3 process (2 versus 3 hydrogen bonds in A:T and G:C base pairs respectively) corresponding to the usual 0/1 commutation of electronic computers. It may also be envisaged for multisite receptors or multiredox systems possessing distinct states of site occupation or of oxidation.

A system of intriguing potential is represented by recently described inorganic superstructures where metal ions are arranged in grid-like arrays. These resemble grids based on quantum dots that are of much interest in microelectronics. They may be considered to consist of *ion dots* of still smaller size than quantum dots and do not necessitate microfabrication but form spontaneously by self-assembly. Such architectures may foreshadow multistate digital supramolecular chips for information storage in and retrieval from inscribed patterns that might be light or electrically addressable. For instance, the use of techniques of STM or AFM type to induce redox changes at specific locations in a single unit would then correspond to a sort of *single electronics* at ion dots. The grid-type arrangements also pose the intriguing question of performing matrix algebra operations on these inorganic superstructures. Extension into three dimensions through stacks of grids would lead to layered arrays in foliated spaces.

Molecular and supramolecular devices incorporated into ultra-micro-circuits represent potential hardware components of eventual systems that might qualify as molecular computers, whose highly integrated architecture and operation would not be of the von Neumann type. On the biological side, the fabrication of components for sensory and motor protheses could be considered. All these entities may result from the self-assembly of suitably instructed subunits so that computing via self-assembly may be envisaged.

One may note that folding and structure generation in biological macromolecules could occur through parallel search with formation of local arrangements which then interact with each other giving larger structured areas which will in turn interact, and so on, in a convergent fashion until the final superstructure is reached. For instance, protein folding, as a self-organizing process, would present characteristics of parallel computing.

Entities resulting from self-assembly and self-organization of a number of components may undergo *self-correction* and *adaptation*. This might also explain why large multisite protein architectures are formed by the association of several smaller protein subunits rather than from a single long polypeptide.

Beyond *programmed chemical systems*, the next step in complexity consists in the design of chemical *"learning" systems*, systems that are not just instructed but can be trained, that possess *self-modification* ability and *adaptability* in response to external stimuli. This opens perspectives towards systems that would undergo *evolution*, i.e. progressive change of internal structure under the pressure of environmental factors. It implies also the passage from closed systems to *open systems* that are connected spatially and temporally to their surroundings.

1.2
Steps Towards Complexity

The progression from elementary particles to the nucleus, the atom, the molecule, the supermolecule and the supramolecular assembly represents steps up the ladder of complexity. Particles interact to form atoms, atoms to form molecules, molecules to form supermolecules and supramolecular assemblies, etc. At each level novel features appear that did not exist at a lower one. Thus a major line of development of chemistry is towards complex systems and the emergence of complexity.

Very active research has been devoted to the development of complexity measures that would allow the quantitative characterization of a complex system. In the present context, complexity is not just described by the number of states, the multiplicity of a system, as defined in information science, or by the characteristics of the graphs representing a molecule or an assembly of molecules, or by structural complexity. Complexity implies and results from multiple components *and* interactions between them with integration, i.e. long range correlation, coupling and feedback. It is interaction between components that makes the whole more than the sum of the parts and leads to collective properties. Thus, the complexity of an organized system involves three basic features:

Complexity = (Multiplicity) (Interaction) (Integration) = MI2

The species and properties defining a given level of complexity result from and may be explained on the basis of the species belonging to the level below and of their multibody interaction, e.g. supramolecular entities in terms of molecules, cells in terms of supramolecular entities, tissues in terms of cells, organisms in terms of tissues and so on up to the complexity of behavior of societies and ecosystems. For example, in the self-assembly of a virus shell, local information in the subunits is sufficient to "tell" the proteins where to bind in order to generate the final polyproteinic association, thus going up a step in complexity from the molecular unit to the supramolecular architecture. Ultimately one will have to go ever deeper and wider so as to link the structures and functions from the atom to the organism, along a hierarchy of levels defining the architecture of complexity.

The novel features that appear at each level of complexity and characterize it do not and even *cannot* conceptually *exist* at the level below but may be explained in terms of MI2, from the simplest particle to the highly complex multibody, multi-interactive societies of living organisms. Such an attitude is not reductionist, it is not a reduction of a level to the lower one(s) but an *integration*, connecting a level to the other ones by integrating species and interactions to describe and explain increasing complexity of behavior.

A simple but telling illustration is for instance the boiling point of a liquid. A single molecule of water has no boiling point, the concept of boiling itself does not, cannot even, exist for it. Only for a population of interacting water molecules is there such a thing as a boiling point, or a freezing point, or any other *collective property*.

A corollary is the question of how many individuals it takes to form a collectivity and to display the collective properties: how many molecules of water to have a boiling point, how many atoms to form a metal, how many components to display a phase transition? Or, how do boiling point, metallic properties, phase transition etc. depend on and vary with the number of components and the nature of their interaction(s)? In principle any finite number of components leads to a collective behavior that is only an approximation, however close it may well be, an asymptotic approach to the "true" value of a given property for an infinite number of units.

The path from the simple to the complex in behavioral space corresponds to that from the single to the collective and from the individual to the society in population space. A sum of individuals becomes a collectivity when there are interactions and at each level novel interactions appear leading to higher complexity. With respect to molecular chemistry, one may consider supramolecular chemistry as a sort of molecular sociology.

The global behavior of condensed matter may be described by phenomenological physical laws. An understanding of these macroscopic events will ultimately require their explanation in terms of the underlying molecular and supramolecular features, i.e. in terms of the chemical nature of the microscopic components and of their interactions. For instance, how is viscosity or a phase change related to the constituting molecules? How do individual partners synergetically cooperate to produce macroscopic spatial, temporal or functional features, inducing a transition from chaos to order through self-organisation? How does turbulent flow vary with the type of multibody interactions between them? Or how is structuration in a energy flow determined by molecular features of the components and their supramolecular interactions?

There is here a very exciting and fundamental field of investigation for supramolecular chemistry concerning the emergence of order and complexity, the passage from the microscopic to the macroscopic, from the isolated to the collective, with the aim of providing an etiological explanation of the phenomenological description. It requires bridging of the gap between and integration of the points of view of the physicist and chemical physicist on one hand and of the structural and synthetic chemist on the other.

Thus, the horizon of supramolecular chemistry lies on the road towards complexity, from the single molecule towards collective properties of adaptive multibody systems of interacting components.

Thomas Mann himself set the goal (!) when he wrote: "*Irgendwann musste die Teilung zu "Einheiten" führen, die zwar zusammengesetzt, aber noch nicht organisiert, zwischen Leben und Nichtleben vermittelten, Molekülgruppen, den Übergang bildend zwischen Lebensordnung und bloßer Chemie*". ("*At some point the partition must have led to units which, although assembled but not yet organized, mediated between life and non-life, groups of molecules forming the transition between life order and mere chemistry*").

A challenge is set, that of showing that there is no such thing as "just" chemistry, but that through the control over supramolecular structures, functions and organization, a bridge is built and continuity is provided between the animate and the inanimate, between life and non-life.

1.3
Chemistry and Biology, Creativity and Art

The highest level of complexity is that expressed in that highest form of matter, living matter, life, which itself culminates in the brain, the plasticity of the neural system, epigenesis, consciousness and thought.

Chemistry and notably supramolecular chemistry entertain a double relationship with biology. Numerous studies are concerned with substances and processes of biological or *biomimetic* nature. There has been a profound evolution by which the chemist appropriates and diverts the power of the natural chemical processes of biology to the goals of chemistry, for instance, in the use of enzymes as reagents, the generation of catalytic antibodies, the control of gene expression, the development of molecular diversity techniques, etc. Conversely, the scrutinization of biological processes by chemists has provided understanding on a precise molecular basis and ways for acting on them by means of suitably designed substances. Thus, the cultures of chemistry and biology are intimately linked and are coming closer and closer together.

On the other hand, the challenge for chemistry lies in the development of *abiotic*, non-natural systems, figments of the imagination of the chemist, displaying desired structural features and carrying out functions *other* than those present in biology with (at least) comparable efficiency and selectivity. Subject only to the constraints one chooses to impose and not to those of the living organism, abiotic chemistry is free to invent new substances and processes. The field of chemistry is indeed broader than that of the systems actually realized in nature.

The emphasis on molecular information in supramolecular chemistry finds its response in the operation of biological systems.

The future path of chemistry will be shaped by both inside and outside forces. Its evolution towards increasing diversity and towards increasing complexity also takes biological phenomena as points of reference. The specificity of chemistry may be stressed by comparing biology and chemistry with respect to two basic parameters, *complexity* and *breadth or diversity*. Biology is of extreme complexity; however, the substances on which it is based belong to defined classes, and, although tremendously rich, are nevertheless limited in variety of types. Chemistry, on the other hand, is still of low complexity compared to biology, but its breadth, the diversity of its substances, is infinite, being limited only by the imagination of the chemist in endlessly combining and recomposing the basic bricks of chemical architectures.

The chemist finds illustration, inspiration and stimulation as well as confidence and reassurance in natural processes, since they are proof that such highly complex systems can indeed be achieved on the basis of molecular components. One might say that science, notably chemistry, relies on the biological world through an existence axiom: the mere fact that biological systems and, in particular, we human beings exist, demonstrates the fantastic complexity of structure and function that the molecular world can present; it shows that such a complexity can indeed exist despite our *present* inability to understand how it operates and how it has come about. So to say, if we did not exist we would be unable to imagine ourselves! And the molecular world of biology is only one of all the possible worlds of the universe of chemistry that await to be created at the hands of the chemist.

With respect to the *frontiers of life* itself three basic questions may be asked: How? Where? Why?

The first concerns the origin of life on earth as we know it, of our biological world. The second considers the possibility of extraterrestrial life, within or beyond the solar system. The third question wonders why life has taken the form we know, and it has as corollary the question whether other forms of life can (and do) exist: is there "artificial life"? It also implies that one might try to set the stage and implement the steps that would allow, in a distant future, the creation of artificial forms of life.

Such an enterprise, which one cannot (and should not) at the present stage outline in detail except for initial steps, rests on the presupposition that there may be more than one expression of the processes characterizing life. It thus invites us to explore the *"frontiers of other forms of life"* and of the *chemical evolution* of living worlds.

Questions have been addressed about which one may speculate, let one's imagination wander, perhaps even set paths for future investigations. However, where the answers lie is not clear at present, and future chemical research towards ever more complex systems will uncover new modes of thinking and new ways of acting that we at present do not know about and may even be unable to imagine.

The perspectives are definitely very wide and extend far into the new millennium. It will be necessary to distinguish the daring and visionary from the utopian and illusory! On the other hand, we may feel as if we are progressing in a countryside of high mountains: the peaks, the goals are visible and identifiable or may become so as progress is made, but we do not yet know how to reach them. We may find landslides, rockfalls, deep crevices, tumultuous streams along the way, we may have to turn around and try again, but we must be confident that we will eventually get there. We will need the courage to match the risks, the persistence to fill in the abyss of our ignorances and the ambition to meet the challenges, remembering that *"Who sits at the bottom of a well to contemplate the sky will find it small" (Han Yu, 768–824).*

But to chemistry, the skies are wide open, for if it is a science, it is also an art. By the beauty of its objects, of course, but also in its very essence, by its ability to invent the future and to endlessly recreate itself.

Like the artist, the chemist engraves into matter the products of creative imagination. The stone, the sounds, the words do not contain the works that the sculptor, the composer, the writer express from them. Similarly, the chemist creates original molecules, new materials and novel properties from the elements provided by nature, indeed entire new worlds, that did not exist before they were shaped at the hands of the chemist, like matter is shaped by the hand of the artist.

Indeed, chemistry possesses this creative power; as stated by *Marcelin Berthelot*: "*La chimie crée son objet.*" ("*Chemistry creates its object*".) It does not merely fabricate objects, but creates its own object. It does not preexist, but is invented as progress is made. It is not just waiting to be discovered, but is to be created.

The essence of chemical science finds its full expression in the words of that epitome of the artist-scientist *Leonardo da Vinci*: "*Where nature finishes producing its own species, man begins, using natural things and, with the help of this nature, creating an infinity of species ...*"

The essence of chemistry is not only to discover but to invent and, above all, *to create*. The book of chemistry is not only to be read but to be written! The score of chemistry is not only to be played but to be composed!

2

Chemical Synthesis and Biological Studies of the Epothilones – Microtubule Stabilizing Agents with Enhanced Activity Against Multidrug-Resistant Cell Lines and Tumors.

Christina R. Harris, Scott D. Kuduk, and Samuel J. Danishefsky

2.1
Introduction

Epothilones A (**1**, EpoA) and B (**2**, EpoB), along with other minor related constituents (Fig. 2.1), are cytotoxic macrolide natural products isolated from the myxobacterium *Sorangium cellulosum*. [1, 2] These compounds have been determined to stabilize microtubules [3] in a mechanism reminiscent of the clinically used paclitaxel (Taxol™). [4–7] Although the 16-membered ring structures of the epothilones bear little structural resemblance to the taxanes, both classes of molecules share a common modality of action. Both the epothilones and the taxanes function by stabilizing cellular microtubule assemblies, thus leading to the arrest of cell division and, eventually, apoptosis. Other naturally occurring, non-taxoid molecules, including the discodermolides [8] and eleutherobins [9] (Fig. 2.1) have recently been discovered that operate by this same mechanism. Surprisingly, in light of the structural diversity

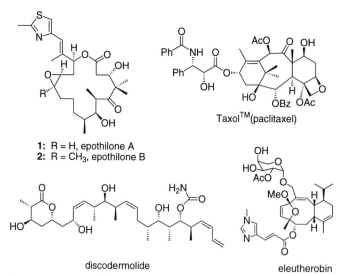

1: R = H, epothilone A
2: R = CH₃, epothilone B

Taxol™(paclitaxel)

discodermolide

eleutherobin

Fig. 2.1 Structures of the naturally occurring mictrotubule stabilizing compounds paclitaxel, epothilones A and B, discodermolide, and eleutherobin.

apparent in each of these molecules (Fig. 2.1), they each share the ability to stabilize microtubules through a paclitaxel-like mechanism.

Perhaps the most intriguing discovery unearthed during the recent investigations of the epothilones is that EpoB is notably more potent (both *in vitro* and *in vivo*) than paclitaxel in inhibiting cell growth. [3] Furthermore, while the cellular mechanism of each agent is apparently the same, the epothilones differ in their ability to retain activity against multidrug-resistant (MDR) cell lines and tumors where paclitaxel (Taxol™) and other major chemotherapeutic agents fail.

Recognizing that the epothilones, or suitably modified derivatives, might find status as cancer chemotherapy agents, we embarked upon a multidisciplinary pursuit toward this end. Included in our objective was the goal of total synthesis of the epothilones and their analogs, which has been accomplished both by us [10] and by others. [11] While the epothilones are available by fermentation, [2, 4] the more active epothilone B is particularly scarce, and thus we sought chemical synthesis as a means for our laboratory to gain a more practical access to this series of compounds and their analogs for *in vitro* and *in vivo* analysis.

The first part of this report includes a comprehensive account of our efforts and successes in this regard. In the latter discourse, we shall diverge to report our findings regarding the biological activity of these remarkable compounds and their potential usefulness in a clinical setting.

2.2
Total Synthesis of Epothilones

In contrast to the structurally complex taxanes, the relatively simple structure of the epothilones was regarded as better suited for the rapid development of structure activity comparisons (SAR). The initial goal of our synthetic program was that of devising a realizable total synthesis of the epothilones (and their analogs) that would provide purely synthetic material for preliminary biological testing, both *in vitro* and *in vivo*. From this examination it was anticipated that preliminary structure-activity relationships (SAR) would be obtained through totally synthetic epothilones. Hopefully from such studies might arise epothilones with increased chemotherapeutic value. In addition to considerations of potency, we were particularly interested in the MDR reversing ability of the epothilones in comparison to their MDR susceptible taxoid counterparts. [7]

2.3
First Generation Syntheses of Epothilones A and B

Although the epothilones are structurally less complex that the taxanes, we foresaw several synthetic issues concerning this venture that would require careful attention. The structures of the epothilones invite retrosynthetic dissection into two regions. The polypropionate domain of the epothilones constitutes carbons C1–C8, while the

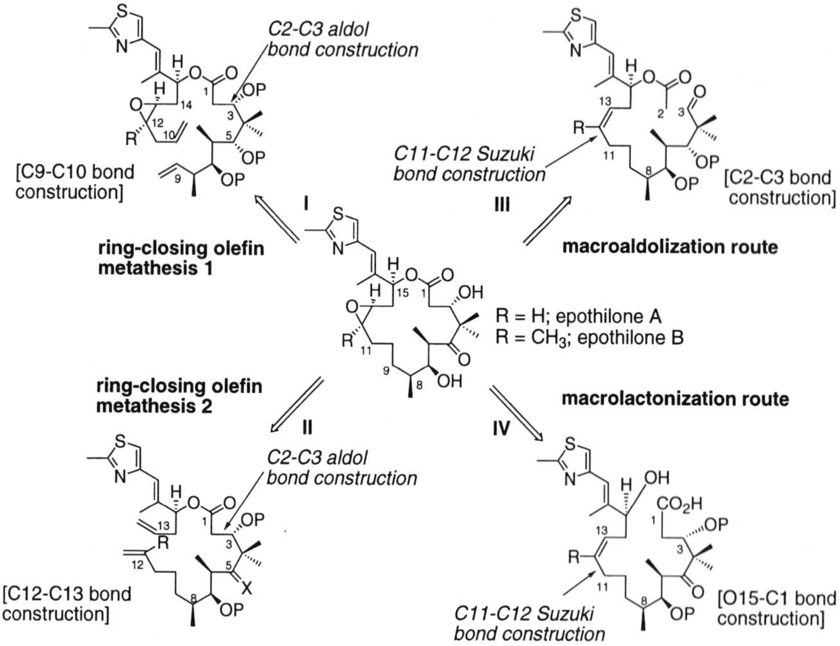

Scheme 2.1 Bond construction strategies for total synthesis of epothilone A and B.

O-alkyl sector comprises C12 through C15 (Scheme 2.1). These regions are joined through a third achiral "hinge" region (C9–C11) which consists of three contiguous methylene groups. The acyl sector is adorned with four discontiguous stereocenters enjoined between two potentially labile β-hydroxy carbonyl moieties. The *O*-alkyl sector contains the C12–C13 *cis*-epoxide moiety insulated by a single methylene group from C15, which bears an allylic alcohol and an α,β-unsaturated thiazole linkage.

Scheme 2.1 outlines the synthetic strategies pursued in our laboratory towards the total synthesis of the epothilones. Although the structural issues addressed above were not insurmountable, several obstacles were encountered during the course of our pursuit. In nearly each instance, the two sectors containing the stereochemical information were prepared independently and then assembled in a late stage merger.

2.4
First Generation Synthesis of the Acyl Domain

Synthesis of the C1–C11 domain of the epothilones at first appeared quite forbidding because of the need to gain facile control over the stereocenters at C3, C6, C7, and C8. Indeed, this region has been the scene of many variations in strategy and synthetic design. [12] Our first generation synthesis of the C1–C11 acyl domain, although somewhat arduous, accomplished the central goal of stereospecificity in the control of the stereocenters placed at C6, C7, and C8. A rigid cyclic template, e.g. **7**, obtained via a Lewis acid-catalyzed diene-aldehyde cyclocondensation (LACDAC) between aldehyde **5** and diene **6** (Scheme 2.2), [13] was anticipated to allow translation of absolute stereochemistry to C6, C7, and C8 of epothilone.

For the initial synthetic endeavor, [14] the first generation ring-closing metathesis (RCM) (Scheme 2.1, path I), required the goal system **13**. Later this segment would be coupled with an appropriate acyl fragment (see below). Starting with known aldehyde **3** (Scheme 2.2, Path A), [15] elongation with phosphorane **4** led to **5** in 83 % yield. [13a] Using chemistry developed in these laboratories during the 1980s, [16] aldehyde **5** functioned as a "heterodienophile" in the context of a Lewis acid-catalyzed diene-aldehyde cyclocondensation (LACDAC) reaction with butadiene **6** to afford, after Luche reduction, [17] the racemic dihydropyran **7** in 65 % yield. [18] Lipase-mediated kinetic resolution of racemic **7** afforded the desired acetylated pyran **9**, which was deprotected and re-protected as the *para*-methoxybenzyl ether. [19] Reaction of **10** with 3,3-dimethyldioxirane (DMDO) afforded the epoxide, which on oxidative solvolysis with sodium metaperiodate gave a 92 % yield of aldehyde **11**. Allylation of the resultant aldehyde afforded a 96 % yield of a 4:1 mixture of diastereomers at the future C12. Finally, mesylation followed by deprotection of the *para*-methoxybenzyl group and cyclization of the resultant hydroxy mesylate with lithium hexamethyldisilazide produced the desired goal system, **13**.

An alternative route to **10** via the LACDAC condensation was also investigated (Scheme 2.2, Path B) starting with known lactaldehyde derivative **14** [20]. In the event, reaction with **6** gave rise to a 93 % yield of a dihydropyrone **15**. Following ketone reduction and protection as the TIPS ether, the benzyloxymethyl (BOM) group was removed by dissolving metal reduction to produce **16**. Dess-Martin oxidation gave the ketone, which was condensed with the phosphine oxide **18** in a Horner reaction, [21] thereby producing optically pure **7**, which was then converted to intermediate **10**.

The construction of the acyl fragment is shown in Scheme 2.3. Realizing that a carboxylic acid coupling to allylic alcohol **13** would be ideal in the metathesis approach, it was envisioned that an appropriate acid might arise after a two-carbon extension of aldehyde **28**. Here again we sought to address this problem via a dihydropyrone using the LACDAC reaction. In the event, titanium-mediated cyclocondensation of the known optically pure β-benzyloxyisobutyraldehyde **19** [22] with diene **20**, [23] previously developed in our laboratory, [24] afforded dihydropyrone **21** in 87 % yield (see Scheme 2.3). Following LAH reduction, glycal **22** was subjected under modified Conia-Simmons-Smith conditions to afford cyclopropano derivative **23**. Fragmentation of this cyclopropane was accomplished through the agency of *N*-iodosuccinimide in methanol to afford methyl glycoside **24**. Subsequent reductive de-iodination and protection as the triphenylsilyl ether provided the desired TIPS-protected methyl glycoside, **25**. The pyran ring cleavage was mediated by 1,3-propanedithiol and titanium (IV) chloride to afford the dithioacetal **26**. [25] Finally, following protection of the alcohol, olefin formation, and removal of the dithioacetal, aldehyde **28** was available for coupling with the left hand sector.

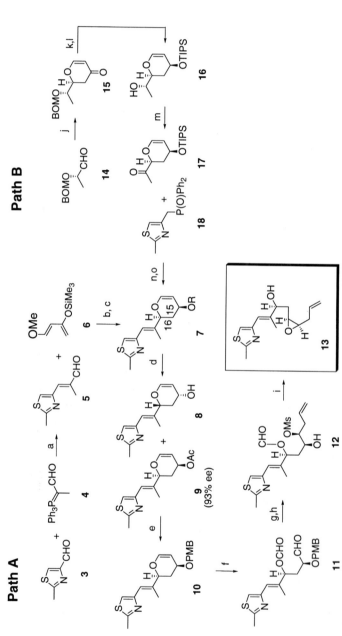

Scheme 2.2 (a) C_6H_6, reflux (83 %); (b) *trans* 1-methoxy-3-trimethylsilyloxy-1,3-butadiene (**6**), BF$_3$ · OEt$_2$, CH$_2$Cl$_2$; then CSA (65 %); (c) NaBH$_4$, CeCl$_3$ · 7H$_2$O, MeOH, 0 °C to rt (99 %); (d) Lipase-30, vinyl acetate, DME, rt, (−)-**9** (45 %; 93 % ee); (e) (*i*) K$_2$CO$_3$, MeOH, rt; (*ii*) PMBCl, NaH, DMF, 0 °C to rt (97 % overall); (f) 3,3-dimethyldioxirane, K$_2$CO$_3$, CH$_2$Cl$_2$, 0 °C; then NaIO$_4$, H$_2$O/THF (92 %); (g) allyl triphenylstannane, SnCl$_4$, CH$_2$Cl$_2$, −78 °C (98 % of **12** + epimer (4:1)); (h) (*i*) MsCl, Et$_3$N, CH$_2$Cl$_2$, 0 °C; (*ii*) DDQ, CH$_2$Cl$_2$/H$_2$O (20:1), 0 °C to rt (93 % overall); (*i*) (*i*) LiN(SiMe$_3$)$_2$, THF, −78 to 0 °C; (*ii*) K$_2$CO$_3$, MeOH/H$_2$O (78 % of the *cis* epoxide); (j) *trans*-1-methoxy-3-trimethylsilyloxy-1,3-butadiene (**6**), MgBr$_2$ · OEt$_2$, THF, −10 °C; then AcOH, H$_2$O (93 %); (k) (*i*) NaBH$_4$, CeCl$_3$ · 7H$_2$O, MeOH, 0 °C to rt (87 % overall); (l) Na, NH$_3$ (l), THF, −78 °C; then MeOH, −78 to 25 °C (92 %); (m) Dess–Martin periodinane, pyridine, CH$_2$Cl$_2$, rt (98 %); (n) **18**, *n*-BuLi, THF, −78 °C; then **17**, THF, −78 °C to rt (87 %); (o) (*i*) *n*-Bu$_4$NF, THF, rt; (*ii*) PMBCl, NaH, DMF, 0 °C to rt (97 % overall); BOM = CH$_2$OCH$_2$Ph; TIPS = Si(*i*-Pr)$_3$; PMB = *p*-OMeC$_6$H$_4$CH$_2$; Ms = SO$_2$CH$_3$.

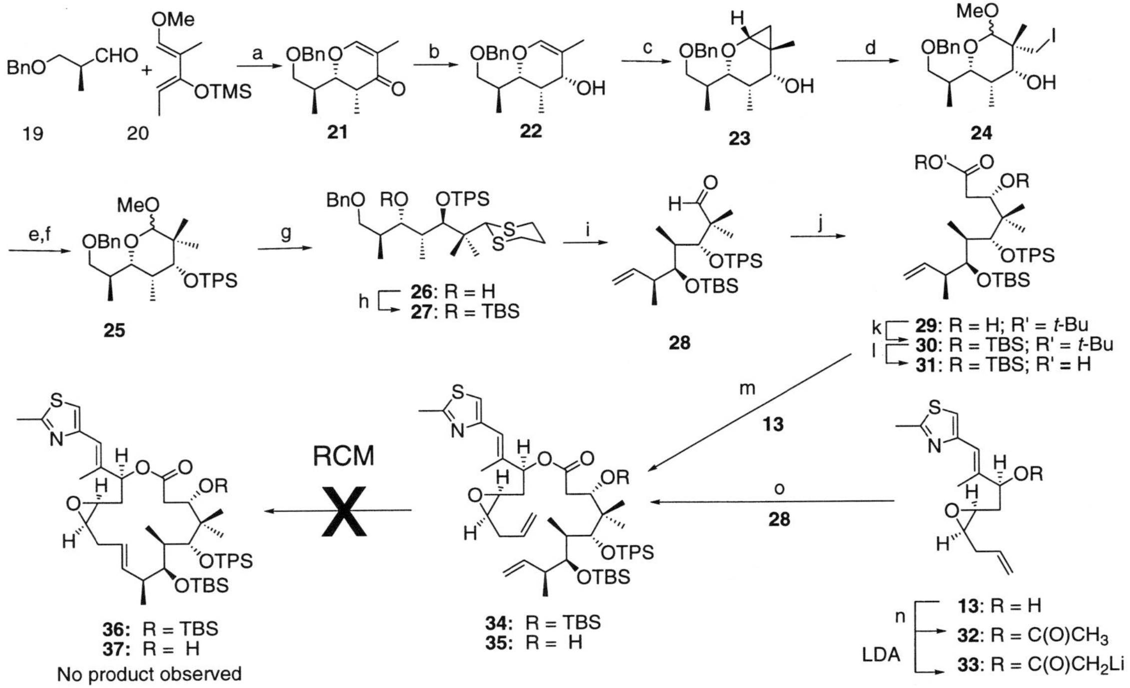

Scheme 2.3 (a) TiCl₄, CH₂Cl₂, –78 °C; then CSA, PhH, rt (87%); (b) LiAlH₄, Et₂O, –78 °C (91%); (c) Et₂Zn, CH₂Cl₂, Et₂O, rt (93%); (d) NIS (7 equiv.), MeOH, rt; (e) *n*-Bu₃SnH, AIBN (cat.), PhH, reflux (80% from **23**); (f) Ph₃SiCl, imid., DMF, rt (97%); (g) 1,3-propanedithiol, TiCl₄, CH₂Cl₂, –78 to –40 °C (78%); (h) *t*-BuMe₂SiOTf, 2,6-lutidine, CH₂Cl₂, 0 °C (98%); (i) (*l*) 2,3-dichloro-5,6-dicyano-1,4-benzoquinone (DDQ), CH₂Cl₂/H₂O (19:1), rt (89%); (*ii*) (COCl)₂, DMSO, CH₂Cl₂, –78 °C; then Et₃N, –78 °C to rt (90%); (*iii*) CH₃PPh₃Br,

NaN(SiMe₃)₂, PhCH₃, 0 °C to rt (76%); (*iv*) PhI(OCOCF₃)₂, CH₂Cl₂/CH₃CN/H₂O, rt (85%); (*l*) *t*-BuOC(O)CH₂Li, THF, 0 °C (90%; ca. 2.5:1 mixture of C3 epimers in favor of **29**); (k) *t*-BuMe₂SiOTf, 2,6-lutidine, CH₂Cl₂, rt; (l) TESOTf, 2,6-lutidine, CH₂Cl₂, rt (90% overall); (m) **31**, EDC, CH₂Cl₂, 4-DMAP, rt; then **20** (78%); (n) Ac₂O, Et₃N, 4-DMAP, CH₂Cl₂, rt (94%); (o) **28** + **32**, LDA, THF, –78 °C (2.6:1 mixture of C3 epimers in favor of **35**; 85%); Bn = CH₂Ph; TMS = SiMe₃; TPS = SiPh₃; TBS = Si(*t*-Bu)Me₂.

2.5
Investigation of C9–C10 Bond Construction Through Ring Closing Metathesis.

Prior to the actual metathesis event, coupling of **13** and **28** via an ester linkage was required (Scheme 2.3). Two methods were employed in this connection. The first involved the aforementioned two-carbon expansion of aldehyde **28**. Thus, condensation of **28** with Rathke anion (lithiated *tert*-butyl acetate) generated a mixture of diastereomeric alcohols; the major product was shown to have the requisite 3*S* configuration. TBS protection of ester **29** and subsequent ester hydrolysis generated the desired acid, **31**, which could be further esterified with alcohol **13** in 78 % yield.

Our second approach involved condensation of the lithium enolate of acetate **32** with aldehyde **28**. In the event, the aldol reaction afforded an 85 % yield of a ca. 5:1 mixture of C3 epimers with the desired diastereomer (**35**) comprising the major product.

Initial efforts in the ring-closing metathesis approach were attempted with substrates **34** and **35**. However, after employing a variety of catalysts and experimental conditions, no cyclized systems (**36** or **37**) were obtained. Other substrates were prepared to further probe this unexpected failure; however, no observable reaction was realized. Model systems later suggested that the dense functionality between C3 and C8 was the culprit for lack of macrocycle formation. Eventually a second generation C12–C13 RCM (not shown here) approach was developed [26] which yielded mixtures of C12–C13 *Z/E* isomers that were used in early SAR studies. [26b] However, since the separation of products was so difficult, we did not seriously pursue this route for total synthesis.

2.6
B-Alkyl Suzuki Strategy

With the apparent failure of the ring-closing olefin metathesis approach, we sought to investigate the potential for forming the C11–C12 bond via a modified Suzuki coupling strategy (Path III and IV, Scheme 2.1). Indeed, the successful synthesis of both epothilones A and B, first accomplished by our laboratories, was achieved through the *B*-alkyl Suzuki strategy. The concept for this route is generalized in Scheme 2.4.

Thus, we envisioned that the epothilones could be assembled through a merger of the *O*-acyl and polypropionate domain through a *B*-alkyl Suzuki coupling of an appropriate vinyl iodide, **42**, and olefin **44** or **48**. Scheme 2.4 outlines our synthesis of the requisite vinyl iodide moiety. Thus, DIBAL reduction of the appropriate thiazole ester **38** and subsequent Wittig olefination afforded enal **39**, which could be allylated under either Brown or Keck conditions. Subsequent dihydroxylation and oxidative cleavage afforded aldehyde **41**, which could then be converted to the desired vinyl iodide **42** through a Wittig reaction with the known phosphorane.

With the appropriate vinyl iodide in hand, we sought to prepare the polypropionate moiety suitable for eventual Suzuki merger. Thus, we envisioned merger of

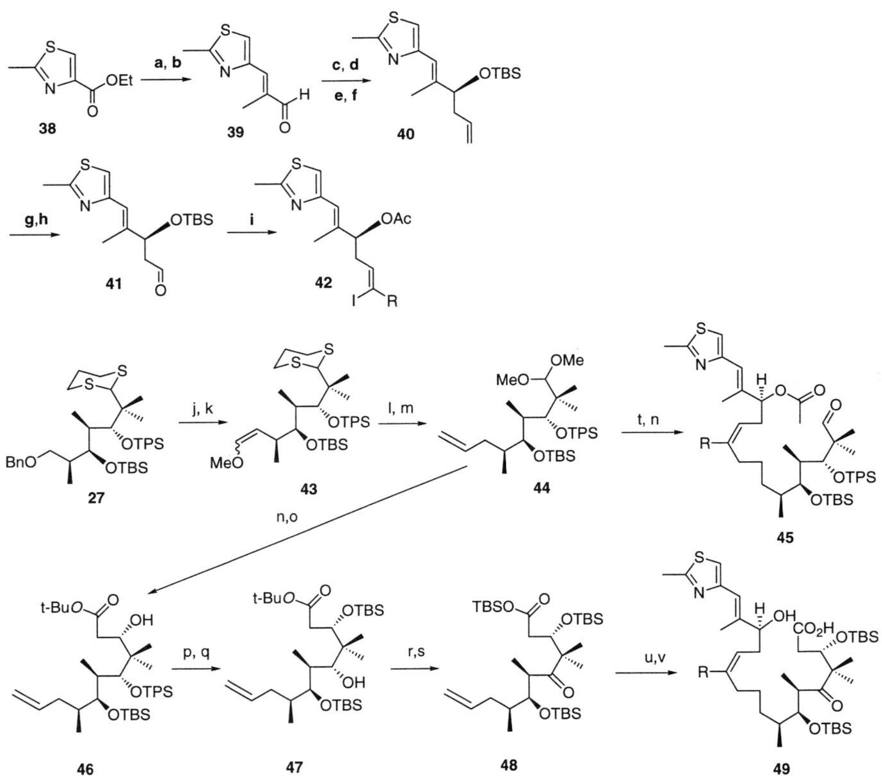

Scheme 2.4 (a) DIBAL, −78 °C, (90%); (b) Ph₃P=C(Me)CHO, (80%); (c) Keck allylation, 5 days, 50% conv., (85–90% BORSM); (d) TBSOTf, 2,6-lutidine, −78 °C, (90%); (e) Brown, Ipc₂BCH₂CH=CH₂, (85%); (f) TBSCl, imidazole, DMF, 87% (g) OsO₄, NMO, (95%); (h) Pb(OAc)₄, Na₂CO₃, (85%); (i) (i) CH₃CH₂PPh₃I, BuLi, I₂, NaHMDS (60%); (ii) HF-pyridine, THF, (98%); (iii) Ac₂O, DMAP, pyridine, (100%); (i) (i) DDQ, CH₂Cl₂/H₂O, (89%); (ii) (COCl)₂, DMSO, CH₂Cl₂, −78 °C; then Et₃N, −78 °C to 0 °C, (90%); (k) MeOCH₂PPh₃Cl, t-BuOK, THF, 0 °C to rt, (86%); (l) (i) p-TsOH, dioxane/H₂O, 0 °C to rt, (99%);

(ii) CH₂PPh₃Br, NaHMDS, PhCH₃, 0 °C to rt, (76%); (m) PhI(OCOCF₃)₂, MeOH/THF, rt, 0.25 h, (92%); (n) p-TsOH, dioxane/H₂O (5:1), 50 °C; (o) t-butyl acetate, LDA, THF, −78 °C; (p) HF-pyridine, pyridine, THF, rt, (98%); (q) TBSOTf, 2,6-lutidine, CH₂Cl₂, −30 °C, (96%); (r) Dess-Martin periodinane, CH₂Cl₂, rt, (89%); (s) TBSOTf, 2,6-lutidine, CH₂Cl₂, rt, (95%); (t) 9-BBN, THF, rt; then **42**, PdCl₂(dppf)₂, Cs₂CO₃, Ph₃As, H₂O/DMF, rt, 75%; (u) 9-BBN, THF, rt; then **42**, PdCl₂(dppf)₂, Cs₂CO₃, Ph₃As, H₂O/DMF, rt, 56%; (v) K₂CO₃, MeOH/H₂O, 84%; Bn = CH₂Ph; TPS = SiPh₃; TBS = Si(t-Bu)Me₂.

these two segments to proceed through one of either two polypropionate segments **44** or **48**. The two different polypropionate variations would permit formation of the macrocycle via either macroaldolization or macrolactonization. Thus, beginning with thioacetal **27**, a one-carbon homologation sequence was achieved through a series of manipulations to afford the desired dimethyl acetal, **44**. In the event, hydroboration of the alkene **44** and direct subjection of the resultant borane to the vinyl iodide, **42**, under Suzuki conditions afforded the desired macroaldolization precursor, **45**.

The macrolactonization candidate substrate was prepared in a similar manner. Thus, the dimethyl acetal, **44**, was further elaborated by liberating the protected aldehyde and subsequent treatment with the Rathke anion. Although the aldol reaction in this series afforded a 2:1 mixture of diastereomers, the undesired product could be successfully converted to the desired β-hydroxy ester through an oxidation-reduction sequence. The C5 protecting group was removed (**47**) and the desired C5 ketone (**48**) was generated after two subsequent steps. Finally, hydroboration of the highly elaborate olefinic ester, **48**, was successfully accomplished through the action of 9-BBN to afford the desired borane, which was coupled with vinyl iodide, **42**, to afford the desired hydroxy acid, **49**, for macrolactonization.

2.7
Macrolactonization and Macroaldolization Approaches

Having successfully achieved the C11–C12 bond formation through Suzuki coupling, we then sought to close the macrocyclic ring. The first would be via a straightfoward macrolactonization route (Scheme 2.5). Compound **49**, prepared as described above, underwent smooth macrolactonization under Yamaguchi conditions [27] to afford the desired macrolactone, **50**, in 88 % yield. An alternative cyclization approach involved the macroaldolization of compound **45**. The approach was possible because the *gem*-dimethyl substitution at C4 blocks the possibility of enolization of the aldehyde function. In the event, deprotonation of compound **45** (see Scheme 2.5) with KHMDS in THF at −78 °C did indeed stereoselectively close the macrocycle, resulting in the preferential formation (6:1) of the desired (*S*)-C3 alcohol **53**. The undesired epimer **54** could be recycled through ketone **55** via an oxidation-reduction sequence, allowing a very efficient access to **53**. Deprotection of **50** to desoxyepothilone A **56** (dEpoA) was achieved with HF-pyridine. Completion of the synthesis was accomplished by introduction of the C12–C13 epoxide with DMDO under carefully controlled conditions to afford a 49 % yield (=16:1 mixture of diastereomers in favor of the desired) of epothilone A, **1**. Similar routes were also employed for the synthesis of epothilone B (**2**). [28]

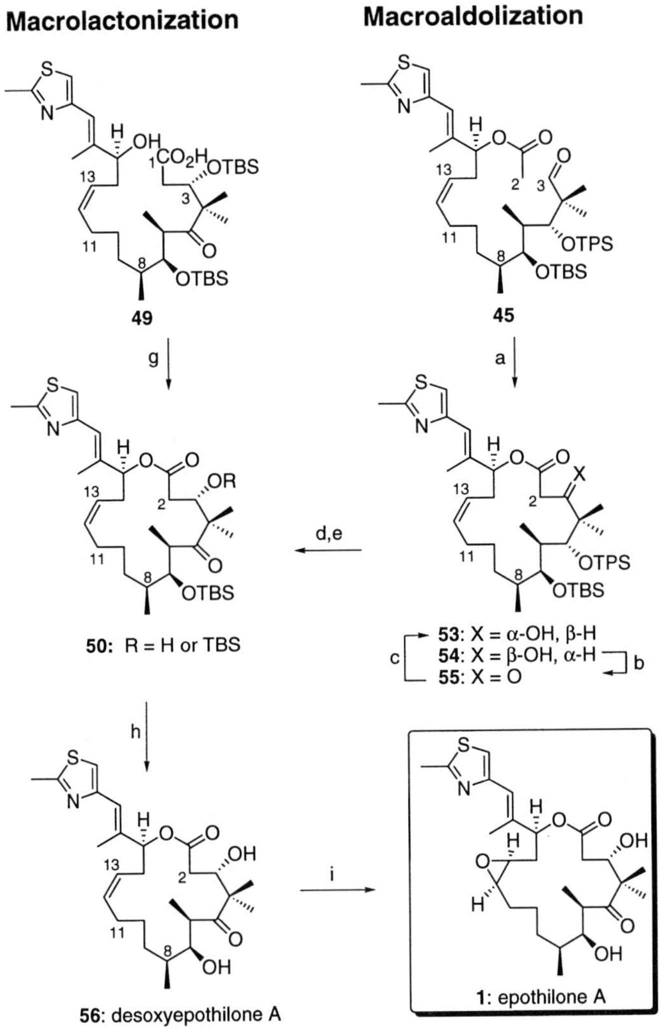

Macrolactonization

49

g ↓

50: R = H or TBS

h ↓

56: desoxyepothilone A

i →

Macroaldolization

45

a ↓

c [**53:** X = α-OH, β-H
54: X = β-OH, α-H ⎤ b
55: X = O ⎦

d,e ←

1: epothilone A

Scheme 2.5 (a) KHMDS, THF, –78°C, 0.001M (51%; 6:1 a/b); (b) Dess-Martin periodinane, CH₂Cl₂, rt; (c) NaBH₄, MeOH, THF, –78°C to rt (80% for two steps); (d) HF · pyridine, pyridine, THF, rt (99%); (e) TBSOTf, 2,6-lutidine, CH₂Cl₂, –30°C (93%); (f) Dess-Martin periodinane, CH₂Cl₂, rt (84%); (g) 2,4,6-trichlorobenzoyl chloride, TEA, 4-DMAP, toluene, rt (88%); (h) HF · pyridine, THF, rt (99%); (i) 3,3-dimethyl-dioxirane, CH₂Cl₂, –35°C (49%; ≈16:1 mixture of diastereomers in favor of **1**); TPS = SiPh₃; TBS = Si(*t*-Bu)Me₂.

2.8
A New and More Efficient Synthesis of Epothilone B

Our initial synthetic ventures in the epothilone field provided satisfactory amounts of fully synthetic compound for *in vitro* cytotoxicity and tubulin binding studies. However, it was only through a tenacious effort that suitable quantities of epothilone B could be prepared for *in vivo* studies against xenografts of human tumors in nude mice. While our initial *in vivo* studies suggested some potentially serious toxicity problems with the parent agent, experiments with 12,13-desoxyepothilone B (57) proved to be more promising (see below). We were pleased to discover that although the *in vivo* and *in vitro* potency of 57 is significantly lower that those of 2, the 12,13-desoxy system dramatically outperforms paclitaxel *in vivo*. Our interest in this field was fueled further when our preliminary *in vivo* studies established that the desoxy compounds were well tolerated and virtually curative against a variety of xenograft tumors. Moreover, in drug-resistant xenografts, compound 57 was determined to be clearly superior to paclitaxel. [29] These findings augmented the need for a practical chemical synthesis of 57 which would be able to support a serious and substantial discovery research program directed toward the clinical development of this ostensibly useful compound. Toward this end, we considered a major departure from our original academic syntheses. The governing concepts for our new generation synthesis are summarized below.

Scheme 2.6 provides an overall view of our strategy towards solving this problem. As depicted, our late generation synthesis embraces three key discoveries that were crucial to its success. We anticipated that the difficult C1–C11 polypropionate domain could be assembled through a double stereodifferentiating aldol condensation of the C5–C6 Z-metalloenolate system **B** and chiral aldehyde **C**. Two potentially serious problems are apparent upon examination of this strategy. First was the condition that the aldol reaction must afford the requisite *syn* connectivity between the emerging stereocenters at C6–C7 (by *uk* addition) concomitant with the necessary *anti* relationship relative to the resident chirality at C8 (by *lk* diastereoface addition). Secondly, it would be necessary to steer the required aldol condensation to C6 in preference to the more readily enolizable center at C2.

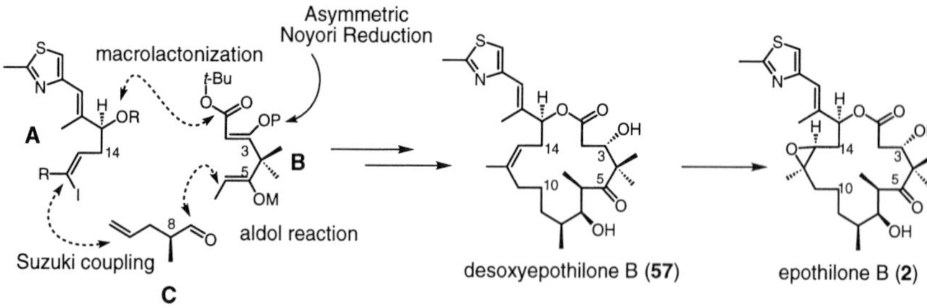

Scheme 2.6 Late generation synthesis of desoxy epothilone B and epothilone B. The key steps in this total synthesis are a stereospecific aldol reaction, B-alkyl Suzuki coupling, and stereoselective Noyori reaction.

2.9
Dianion Equivalents Corresponding to the Polypropionate Domain of Epothilone B

Three tactical approaches were surveyed in the evolution of our program. As outlined in Scheme 2.7, initially the aldol reaction (Path A) was performed directly between aldehyde 63 and the dianion derived from tricarbonyl 58. In this way, it was indeed possible to generate the Z-lithium enolate of 58 as shown in Scheme 2.7 which underwent successful aldol condensation. However, the resultant C7 β-hydroxyl functionality tended to cyclize to the C3 carbonyl group, thereby affording a rather unmanageable mixture of hydroxy ketone 59a and lactol 59b products. Lactol formation could be reversed following treatment of the crude aldol product under the conditions shown (Scheme 2.7); however, under these conditions an inseparable 4:1 mixture of diastereomeric products, 60 (a or b): 61 (a or b) [30], was obtained. This avenue was further impeded when it became apparent that neither the acetate nor TES groups were compatible with the remainder of the synthesis.

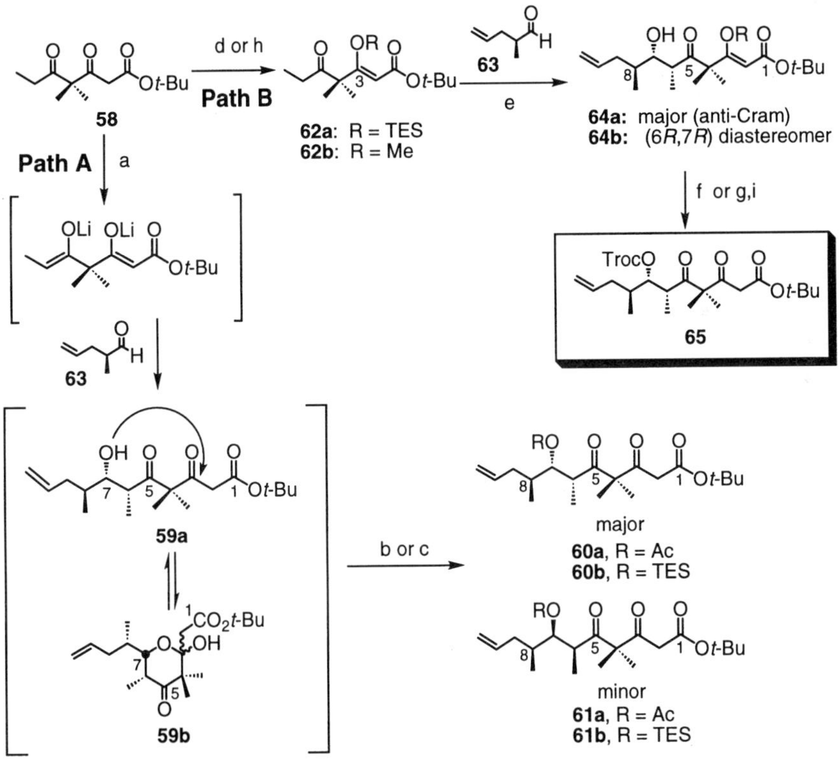

Scheme 2.7 (a) 2.2 equiv LDA, THF, −50 °C; (b) TESCl, imidazole, DMF, 12 h, (55 %, 2 steps), 2.3:1 ds; (c) Ac$_2$O, DMAP, Et$_3$N, CH$_2$Cl$_2$, 12 h, (58 %, 2 steps), 2–3:1 ds; (d) NaH, TESOTf, −50 °C, (86 %); (e) LDA, THF, −120 °C, (75 %, 5.4:1 ds); (f) TrocCl, pyridine, CH$_2$Cl$_2$; then 0.5 M HCl/MeOH (86 %, 2 steps); (g) TrocCl, pyridine, CH$_2$Cl$_2$, (86 %); (h) TMSCHN$_2$, Hunig's Base, CH$_3$CN, MeOH, (81 %); (i) p-TSA, acetone, rt, (90 %).

Unfortunately, more durable blocking groups could not be introduced onto the C7 hydroxyl center by trapping of the desired hydroxy ketone **59a**.

Although our initial attempts in this direction proved unavailing, our efforts were not completely futile in that we were able to confirm that the critical aldol reaction with S-aldehyde **63** did indeed provide the desired C6-C7 *syn* and C7-C8 *anti* relationship (by anti-Felkin Ahn addition) as the major diastereomer.

Realizing that simply engaging the C3 carbonyl group of the nucleophile in another functional arrangement might alleviate the problem of undesired lactolization, we sought to engage the C3 carbonyl group as an enol ether (Path B, Scheme 2.7). In this way, unwanted cyclization would not occur, and the C7 hydroxyl group could be readily protected with a more robust moiety. As reported in our recent disclosure, [31] the triethylsilyl (TES) enol ether (**62a**) is readily prepared from tricarbonyl system **58**. Moreover, the resultant C3-enol, C6-enolate ether successfully underwent an aldol coupling with aldehyde **63**. Without the difficulties encountered by undesired hemiacetal formation, the C7 hydroxyl moiety could be readily protected as the trichloroethoxycarbonate ester (Troc). We were gratified to discover that the aldol reaction between enol ether **62a** and aldehyde **63** gave rise to a 1:5.5 mixture of C7-C8 *syn* and desired C7-C8 *anti* isomers. These diastereomers could be separated by flash column chromatography. Unfortunately, our initial success in this regard did not translate well in attempted scale-ups; the aldol reaction between **62a** and aldehyde **63** proved to be troublesome. We soon discovered that the resultant C3 TES enol ether in **64a** (R = TES) was prone to decomposition under the very basic conditions of the aldol reaction and was also quite sensitive to silica gel chromatography.

Finally, we sought to modify the C3 protecting group to provide a more stable moiety that would survive both the basic aldol reaction conditions and silica gel chromatography. To this end, we investigated the usefulness of a methyl enol ether linkage. [32] Happily, we discovered that the requisite methyl enol ether, **62b**, [33] could be readily prepared from trimethylsilyl diazomethane, TMSCHN$_2$, with Hunig's base in high yield and as a single olefin isomer (Path B, Scheme 2.7). Fortunately, the aldol reaction performed on the methyl enol ether also afforded the same diastereoselectivity as the TES enol ether. Likewise, the diastereomers (**64a**, R = Me, and its C8 epimer) obtained in the aldol reaction could be readily separated by flash column chromatography on silica gel. Without the difficulties encountered by decomposition, the C7 hydroxyl moiety could be readily protected as the trichloroethoxycarbonate ester (Troc) when performed on a larger scale.

2.10
B-Alkyl Suzuki Merger

A second key component of our practical total synthesis of epothilone B includes a B-alkyl Suzuki merger, Scheme 2.8, which successfully occurs between the previously described vinyl iodide [34] **66** and tricarbonyl **65**. With the coupling step accomplished, the resultant TBS protecting group could be hydrolyzed to afford the requisite C15-hydroxy ester **67**.

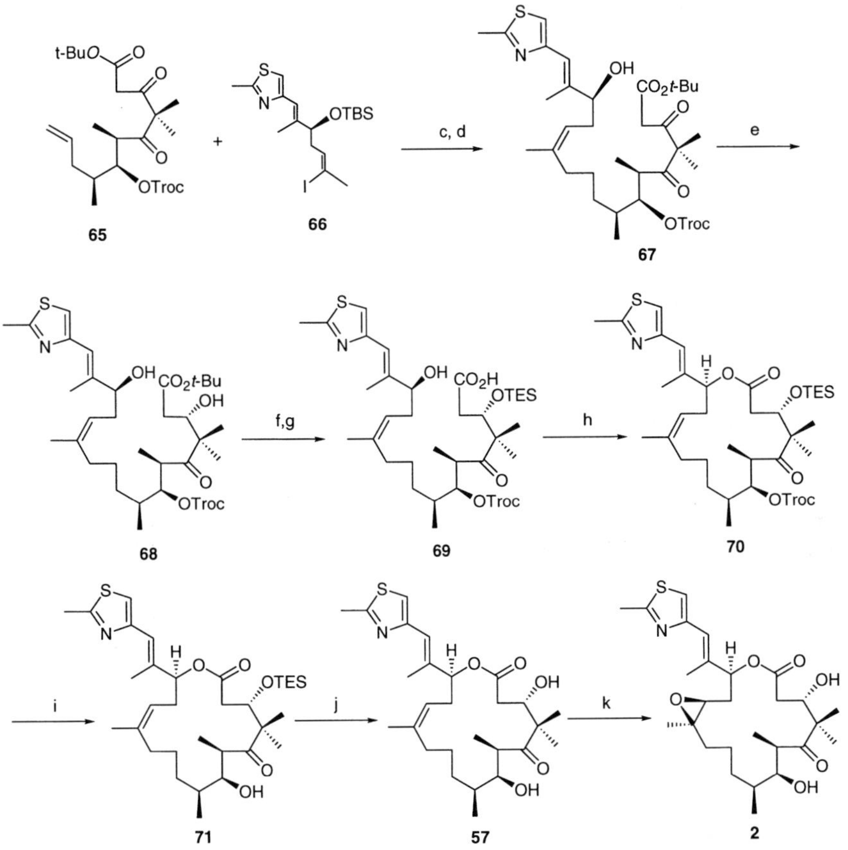

Scheme 2.8 (a) TrocCl, pyridine, CH₂Cl₂, 0 °C to rt; (b) *p*-TsOH, (87–88 %, 2 steps) (c) 9-BBN; then **66**, Pd(dppf)₂Cl₂, Ph₃As, Cs₂CO₃, H₂O,DMF, rt, 2 h, (60 %); (d) 0.4 N HCl/MeOH, (96 %); (e) (*R*)-(BINAP)RuCl₂, H₂, 1200 psi, MeOH, HCl, 25 °C, 7 h, (88 %); (f) TESOTf, 2,6-lutidine, CH₂Cl₂, –78 °C to rt; (g) then 0.1N HCl/MeOH (77 %, 2 steps) (h) 2,4,6-trichloro-benzoylchloride, TEA, 4-DMAP, PhCH₃, (78 %); (i) SmI₂, cat. NiI₂, –78 °C, THF, (95 %); (j) HF-pyridine, THF, (98 %); (k) DMDO, CH₂Cl₂ (80 %).

2.11
Stereoselective Noyori Reduction

After extensive developmental studies, [35] the final crucial element in our most recent synthesis of epothilone B involves an asymmetric catalytic reduction of the C3 ketone of **67** proceeding via a modified Noyori procedure (Scheme 2.8, **67→68**). In the event, Noyori reduction of ketone **67** afforded the desired diol **68** with excellent diastereoselectivity (>95:5). The ability to successfully control the desired C3 stereochemistry of the late stage intermediate **68** permitted us to introduce the C1-C7 fragment into the synthesis as an achiral building block.

With these critical issues addressed, no significant obstacles remained in the total synthesis, and the conversion of **68** to dEpoB (**57**), Scheme 2.8, was effected using previously developed methods. Thus, the diol **68** was successfully hydrolyzed to the requisite hydroxy acid **69**, and subsequent macrolactonization afforded the fully protected macrolactone **70**. Samarium(II) iodide proved singularly effective in accomplishing the 2,2,2-trichloroethoxy carbonate (Troc) deprotection of **71** at C7. [36] Finally, standard HF-pyridine deprotection of the C3 triethylsilyl group afforded the desired dEpoB **57**, which could be epoxidized to afford the natural product epothilone B (**2**) with a high degree of chemo- and stereoselectivity.

2.12
Discovery of a Remarkable Long-Range Effect on the Double Diastereoface Selectivity in an Aldol Condensation

Traditional models for diastereoface selectivity were first advanced by Cram and later by Felkin for predicting the stereochemical outcome of aldol reactions occurring between an enolate and a chiral aldehyde. [37] During our investigations directed toward a practical synthesis of dEpoB, we were pleased to discover an unanticipated bias in the relative diastereoface selectivity observed in the aldol condensation between the Z-lithium enolate **B** and aldehyde **C**, Scheme 2.6. The aldol reaction proceeds with the expected simple diastereoselectivity with the major product displaying the C6-C7 *syn* relationship shown in Scheme 2.7 (by *ul* addition); however, the C7-C8 relationship of the principal product was *anti* (by *lk* addition). [38] Thus, the observed *syn:anti* relationship between C6-C7:C7-C8 in the aldol reaction between the Z-lithium enolate of **62** and aldehyde **63** was wholly unanticipated. These fortuitous results prompted us to investigate the cause for this unanticipated but fortunate occurrence.

Superficially, the relative face selectivity exhibited in the aldol condensation seems to be contrary to the predicted models for double stereodifferentiation encompassed in the Felkin rules. [39] Our results were reconcilable when viewed in the context of a perception originally suggested by Roush to account for attrition in *anti* selectivity with certain aldehydes. [40] The Roush model proposes that in the reacting Curtin-Hammett conformer, the R group (larger group) of the aldehyde is distanced from the R′ group of the enolate (**I**) to avoid an unfavorable, potentially serious *syn*-pentane interaction (formula **I** in Table 2.1). Thus, the basis of the Roush model focuses on minimization of steric hindrance between the largest functions of the enolate and the α-branched aldehyde in the reacting ensemble. Consequently, the observed anti-Felkin Ahn selectivity displayed in these reactions is "typical" for such α-methyl aldehydes. When the R group is approximately equal to the methyl center in its effective A-value, the selectivity for the *anti*-product must correspondingly deteriorate.

Our discoveries using aldehyde **63** (Table 2.1, formula **II**) and related congeners were unique in that our substrate aldehydes lack the typical resident protected alcohol derivative that is usually involved in fashioning *anti* diastereoface selectivity. [41]

Rather, the conformational bias in our substrates is apparently dependent on a very particular relationship between the formyl moiety and unsaturated in the pendant side chain. As a result of our studies, we speculate that the presence of unsaturation at C4-C5 in the aldehyde moiety provides a subtle stabilizing nonbonded interaction between the unsaturation in the aldehyde and the carbonyl of the enolate (Table 2.1, **25**). [42] This interaction can be envisioned to stabilize the transition state leading to the observed major anti-Felkin diastereomer **75a**.

Table 2.1 outlines the results of several aldol reactions between the enolate **74** and various other aldehydes. Our exhaustive study identified the significance of the tether length between the formyl group and the site of unsaturation. Thus, reduction of the double bond of the side chain led to a sharply diminished selectivity, affording a 1.3:1 mixture of diastereomeric products (entry b). Likewise, lengthening of the tether beyond that found in entry a led to a 2:1 ratio of diastereomers (entry c). By contrast, shortening the tether (entry d) gave strong *syn* diastereoface selectivity consistent with previous findings with this particular aldehyde. The results of entry e, in which similar steric factors are virtually equivalent (propyl versus 2-propenyl at the branching site) demonstrate a small but clear preference for the C7-C8 *anti* product, presumably reflecting the special effect of the olefin-aldehyde interaction. The Roche aldehyde (entry f), a substrate well known for its tendency to favor the anti-Felkin Ahn adduct [44] on the basis of presumed chelation control, performs as expected to afford a 4:1 mixture of *anti:syn* diastereomeric products (entry f).

We next contemplated whether the unsaturation site could be encompassed in the context of a properly positioned benzo linkage. We were intrigued to discover that excellent diastereoface selectivity was obtained in the aldol condensation of the Z-lithium enolate with the benzyl-substituted formyl moiety, entry g.

Realizing that functional group substitution about the aromatic ring would dramatically affect the donating ability of the ring, we examined the effects of *para*-positioned functional groups on the resultant C7-C8 relationship (entries k-n, formula **III**, Table 2.1). Some minor slippage in the *anti:syn* ratio is seen in the electron deficient *para*-bromo substrate (entry k). The benchmark ratio seen in entry g is restored with the electron-rich *para*-methoxy substrate (entry l), while a small improvement was realized with the *para*-dimethylamino derivative (entry m). By contrast, in the strongly electron-deficient *para*-nitrophenyl substrate (entry n), the C7-C8 *anti* selectivity is severely abrogated. Clearly, the "aryl effect" is closely coupled to the electron-donating ability of the ring.

By contrast with the reconcilable data observed with *para*-substituted substrates, a range of *ortho* substituents (entries h-j, Table 2.1) all resulted in significant weakening of the C7-C8 *anti* selectivity. We take these data to suggest that *ortho* substitution results in some steric inhibition of the rotamer in which the faces of the aromatic ring and formyl group are parallel (see structure **III**).

To interpret these experiments, we suggest that our data point to a stabilizing through-space interaction of a donor olefinic linkage with the formyl function as the likely source of preference of conformers **II** and **III** leading to the sense of attack anticipated by the Roush model. [7]

Tab. 2.1 This depicts our rationalization of the observed relative face selectivity and the effects of substitution on the pendant alkyl chain of the aldehyde in the observed ratio of diastereomers obtained in the aldol condensation of the Z-lithium enolate **74** and substituted α-methyl aldehydes.

C_1C_2 syn, C_2C_3 anti C_1C_2 syn, C_2C_3 anti C_1C_2 syn, C_2C_3 anti

74 → THF, −120°C → **Felkin 75a** + **Anti-Felkin 75b**

Entry	Aldehyde R =	Ratio (C7:C8) (syn:anti)		Entry	Aldehyde R =	Ratio (C7:C8) (syn:anti)	
a	CH₃	1	5.5	h	Br / CH₃	1	2.1
b	CH₃	1	1.3	i	OMe / CH₃	1	1.2
c	CH₃	1	2.0	j	Me / CH₃	1	2.8
d	Ph / CH₃	11	1.0	k	Br / CH₃	1	4.2
e		1	2.0	l	MeO / CH₃	1	5.0
f	BnO / CH₃	1	4.0	m	Me₂N / CH₃	1	5.4
g	CH₃	1	5.0	n	O₂N / CH₃	1	1.2

2.13
Preparation of Other Epothilone Analogs

The epothilone analogs in Table 2.2 were prepared according to a modified route of the total synthesis described in Scheme 2.8. For example, analog **80**, which bears a fused phenyl ring at C12-C13 was prepared as outlined in Scheme 2.9. Once again, the scheme centered on a *B*-alkyl Suzuki coupling. Thus, zinc-mediated nucleophilic addition of 2-iodobenzylbromide **(76)** with aldehyde **39** afforded racemic **77** in 67% yield. Oxidation of **77** followed by asymmetric reduction to **79** afforded only modest enantioselectivities. The highest enantiomeric excess observed for the reduction of **77** by Corey methods was 60%. While far from ideal, this enantioselectivity margin was sufficient for our purposes. The intermediate was used in accordance with the previous protocols in Scheme 2.8 to obtain analog **80**.

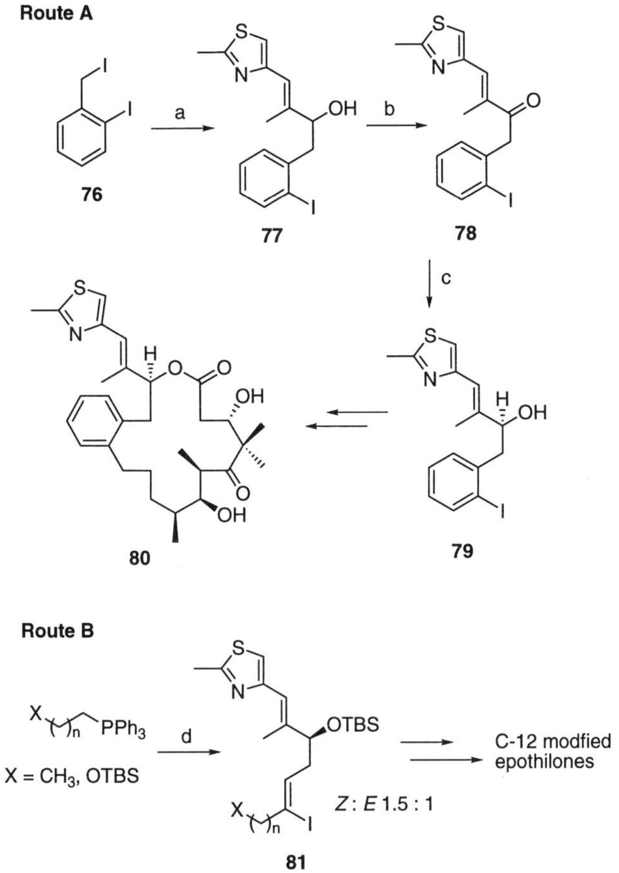

Route A

76

77

78

79

80

Route B

X = CH$_3$, OTBS

Z : *E* 1.5 : 1

C-12 modfied epothilones

81

Scheme 2.9 (a) Zn, CuCN, LiCl, THF, then BF$_3$, **3**, (67%); (b) TPAP, NMO, CH$_2$Cl$_2$, (89%); (c) CBS reduction, 60% ee, (90%); (d) *(i)* *n*-buLi, THF, I$_2$, –30°C *(ii)* NaHMDS, then **81**, (75–95%).

Similarly, access to the C12-modified epothilone analogs [45] was secured by modifying the alkyl group geminal to the vinyl iodide as depicted in Scheme 2.9 below. Thus, Wittig olefinations of the appropriate phosphonium salts with aldehyde **41** afforded good yields of the requisite vinyl iodides **81**. Although *Z:E* mixtures were obtained, these diastereomers were separable with the desired *Z* isomer predominant. These intermediates were quite amenable to our total synthesis described above to furnish C12-modified analogs.

2.14
Biological Evaluation of Epothilones

The current front line therapeutic agent against a variety of solid forms of cancer including ovarian, breast, colon, lung, and liver neoplasms is Taxol™ (paclitaxel), which is a member of the taxane family of natural products. [46] Acquired resistance to paclitaxel and other commonly used cancer chemotherapy agents may be mediated by a number of mechanisms including overexpression of the energy-dependent drug transport protein P-glycoprotein (PgP). [47] Broad-spectrum resistance to structurally and mechanistically diverse anticancer agents constitutes the multidrug-resistance (MDR) phenotype. Although certain MDR reversal agents appear promising when co-administered with the anti-cancer agent, [48] a search for paclitaxel analogs with improved performance *in vitro* and *in vivo* has met with limited success. [49, 50]

As previously discussed, the epothilones are naturally occurring, cytotoxic molecules that function with a paclitaxel-like mechanism. Despite their structural dissimilarity, both families of agents lead to the arrest of cell division and cell death by stabilizing cellular microtubule assemblies. The epothilones, however, differ in their ability to retain activity against multidrug-resistant (MDR) cell lines and tumors where paclitaxel fails. The goal of our ongoing research program in this field is to focus on the relationship between epothilone and paclitaxel in the context of tumors with the MDR resistance phenomena. Our long-term efforts encompass the discovery and development of novel epothilone-like compounds with improved pharmacological and clinical efficacy.

2.15
SAR Analysis of Epothilones: The Zone Approach

Our research initially began with an in-depth SAR analysis of the epothilones. The chemistry described above allowed us to prepare epothilone analogs in sufficient quantity for initial *in vitro* cytotoxicity screening and tubulin binding studies. The goal here was to determine which areas or zones of the molecule are tolerant or intolerant of structural change. Table 2.2 and Fig. 2.2 demonstrate a selected range of structures prepared in this laboratory and the *in vitro* cytotoxicity of each analog when evaluated against both sensitive CCRF-CEM and resistant CCRF-CEM/paclitaxel cell lines. CCRF-CEM is a human T-cell acute lymphoblastic leukemia cell line. CCRF-CEM/paclitaxel cell line was developed in our laboratory after continuous exposure of CCRF-CEM cells with increasing and sublethal (IC_{50}). In order to present a more clear illustration of the epothilone SAR, we have chosen to divide the structure of epothilone into an acyl sector (shown arbitrarily as carbons 1–8), an *O*-alkyl sector (carbons 9–15), and a pendant aryl sector projecting from C15.

First we consider the acyl sector of the epothilones, which proved to be intolerant of modification. For example, inversion of stereochemistry at C3 (*S* to *R*), or reduction at C5 results in serious arrest of activity. Analogs with functionality at C3, C5, C6, C7 and C8 removed demonstrate both diminished tubulin-binding activity and cytotoxicity (structures not shown). Deletion of the single methyl group at C8 has a highly pronounced deleterious effect on activity. Removal of the "C9" methylene group resulting in a 15-membered macrolide, **87**, results in a major loss of activity in tubulin polymerization/depolymerization assays.

Secondly, we considered the *O*-alkyl sector. We first reported that epoB **2** is more potent than epoA **1** in both cytotoxicity and tubulin-binding ability. Since substitution of a methyl group at C12 (e.g., epoB vs epoA) imparts improved activity, we examined the limits of substitution at C12. Substitution at C12 with either ethyl, propyl, or hexyl is well tolerated (see compounds **84**, **85** and **86**).[15] We further examined the "side chain" substitution while being particularly attentive to the effects of more polar alcohol or ethylidene acetal functionality. These more polar functional groups had a profound effect on the susceptibility of the compound to succumbing to multidrug-resistance. The highly polar alcohol functionality provided analogs (compounds **88–90**) that were MDR substrates and were presumably exported out of the cell more efficiently by the drug efflux transport protein, PgP. The alkyl chain in the alcohol series significantly effects the multidrug resistance susceptibility. Thus, analogs containing a shorter pendant alkyl chain demonstrated the most susceptibility to the MDR phenomenon. However, the ethylidene acetals, **91** and **92**, are poor MDR substrates, and the potency of the analog is related to the number of carbons between C12 and the acetal linkage. Thus, the 3-carbon acetal **92** is less potent that the 2-carbon acetal **91**. In fact, analog **91** is both MDR resistant and 2-fold more potent against the sensitive tumor CCRF-CEM than lead compound dEpoB (**57**).

Continuing with the *O*-alkyl sector, the *Z*-desoxycompounds **56–57** (dEpoA and dEpoB) and **82–83** retain activity comparable to their natural epoxy-containing counterparts, **1** and **2**, respectively. Again, the pre-"epoB" compound **57** (with the C12 methyl) is more active than the pre-"epoA" compound **56**. Even the *E*-desoxy compounds **93–95** maintain significant biological activity, although reduced relative to the parent *Z*-series (**57** and **94** for example). Surprisingly enough, upon inversion of stereochemistry at C15 (see compound **96**), activity is still maintained, albeit in attenuated form. A very interesting analog is **80**, in which C12/C13 is confined within an aromatic ring. Intriguingly, preliminary evaluation of this compound demonstrated considerable tubulin binding, but poor cyctotoxicity was observed *in vitro*. Conceivably this result can be attributed to metabolic oxidation of the aromatic ring. Further study of this analog is currently under active investigation. In general, the

O-alkyl sector is remarkably tolerant of modification with basic maintenance of *in vitro* function.

The last sector examined was the "aryl" zone, [14] which proved to be quite tolerant to permutation. For example, replacement of thiazole by oxazole (**97**) is well tolerated and is equipotent with **57** (depoB) in both tubulin polymerization (Fig. 2.3) and cytotoxicity assays. The more drastic substitution of phenyl in place of the thiazolyl moiety retained 60% of the activity of **2** in the tubulin polymerization assay, although there was a slight loss of cytotoxicity. Thiazole deletion, **98**, exhibited very low cytotoxicity ($IC_{50} > 10.0$ mM) and shows almost no activity in tubulin binding assays. The olefinic spacer element connecting the aromatic rings with C15 was examined via compounds **99** and **100**, which demonstrated a major loss of cytotoxicity. These analogs illustrate that "Zone 3" requires the olefinic spacer linking the lactone at C15 to an aromatic subsection which is then substantially amenable to modification. While this zone tolerates modification, the aryl functionality may not be removed.

Although we did not prepare a massive collection of compounds, the SAR data obtained from our pool of analogs was more than adequate. It also allowed for the selection of candidates for more advanced *in vitro* and *in vivo* studies. Nicolaou et al. have reported an exhaustive SAR study, and the same general conclusions have been drawn. [51] In summary, it is clear that the polypropionate sector constitutes a "hot-spot" region with great sensitivity to structural change. In comparison, the *O*-alkyl and aryl sectors exhibit regions of significant tolerance, both in the cytotoxicity assays and in tubulin binding assays.

2.16
In Vitro Analysis Comparison to Paclitaxel and Related Agents

Having developed significant SAR data for our totally synthetic epothilones, our next scientific efforts examined the cell growth inhibitory effects in two rodent and three human tumor cell lines and, more importantly, their drug resistant sublines. While paclitaxel showed strong cross-resistance to the sublines also resistant to adriamycin, vinblastine or actinomycin D, most epothilones demonstrate little or no cross-resistance. Thus, EpoB proved to be only 6.1-fold resistant and dEpoB 1.8-fold resistant whereas clinically used vinblastine and paclitaxel (Taxol™) were 527-fold resistant and 1971-fold resistant against multidrug resistant CCRF-CEM/VBL100 cells, respectively. In fact, EpoB is ca. 3400 times more active than paclitaxel against the resistant human leukemic cell line CCRF-CEM/VBL in cell-culture cytotoxicity studies. In the DC-3F cell line (hamster lung tumor cells resistant to actinomycin D), there was a 13 000-fold resistance to actinomycin D, and a 388-fold resistance to paclitaxel. By way of comparison, these cells were only 28-fold resistant to EpoB and showed no observable resistance to dEpoB. In general, among the compounds and cell lines tested in our laboratory, dEpoB exhibited the least cross-resistance to several drug-resistant cell lines.

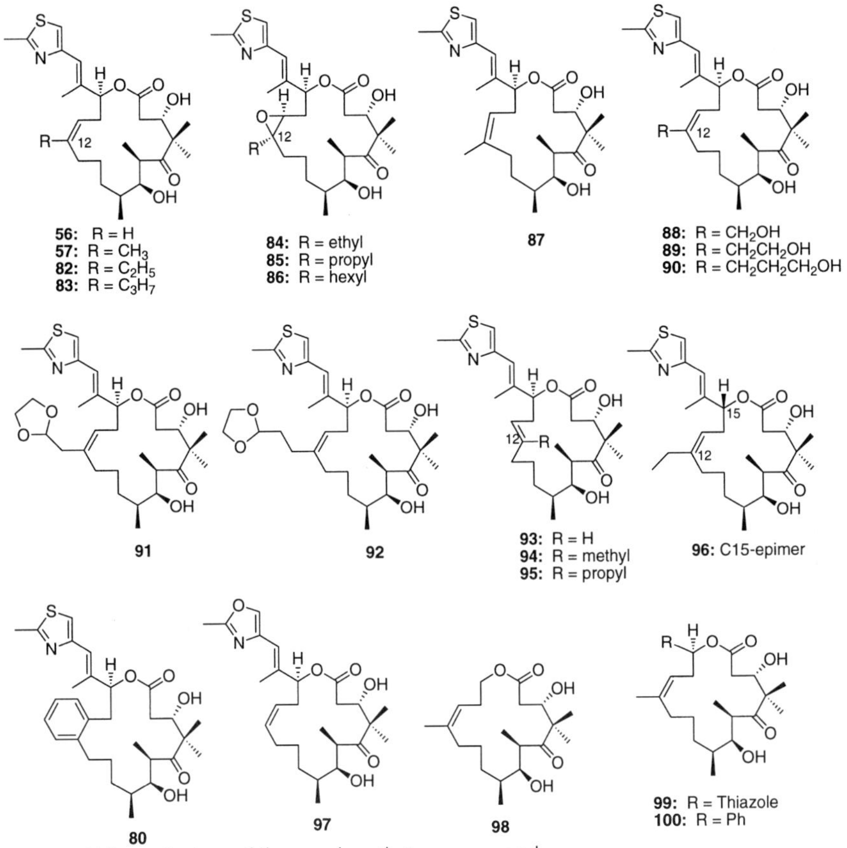

Fig. 2.2 Fully synthetic epothilone analogs that were prepared for SAR studies.

Tab. 2.2 Structure-activity relationships for epothilone and various epothilone analogs.

Compound	CCRF-CEM IC_{50} (nM)[a]	CCRF-CEM/VBL IC_{50} (nM)[a]	Tubulin binding[b]
Paclitaxel	2.1	>4000	95
1 Epothilone A	3.0	200	99
2 Epothilone B	0.2	1.0	100
56 (depoA)	22	12	97
57 (depoB)	9.0	17	98
80	2050	4300	33
82	21	77	95
83	39	67	96
84	1.0	7.0	93
85	4.0	6.0	92

Tab. 2.2 (Continued)

Compound	CCRF-CEM IC$_{50}$ (nM)^a	CCRF-CEM/VBL IC$_{50}$ (nM)^a	Tubulin binding^b
86	27	49	72
87	>10 000	>10 000	1
88	49	> 2000	n.d.
89	32	1033	n.d.
90	9.5	167	n.d.
91	4.3	20	72
92	80	409	n.d.
93	52	35	54
94	90	262	85
95	90	254	89
96	55	197	84
98	>10 000	>10 000	6
99	> 3000	1200	n.d.
100	1800	> 5000	n.d.

a) Relative efficacy of epothilone compounds against drug-sensitive and resistant CCRF-CEM cell lines. Formation of microtubules in the presence of 10 µM of the tested compounds.
b) Microtubules formed in the presence of 10 µM epothilone B is defined as 100 %.

2.17
In Vivo Analysis: Comparisons to Paclitaxel

With these promising results in hand, a series of *in vivo* studies were perfomed with selected epothilone analogs. The syntheses described above provided more than ample quantities of material for the tasks at hand. Our initial *in vivo* efforts were directed at comparing EpoB (**2**), dEpoB (**57**), and paclitaxel. *In vivo* studies, using i.p. administration, indicated that the parent, EpoB, resulted in substantially greater reduction in tumor size compared to paclitaxel. However, EpoB was highly toxic to mice and showed little therapeutic potential when compared with our eventual lead compound, dEpoB. Clearly, the C12-C13 epoxide must be responsible for the undesired toxicity. Fortunately, our current lead compound dEpoB still retains a very high level of efficacy against tumor xenografts without concomitant toxicity. The therapeutic effects of the various drugs were evaluated in athymic mice bearing human mammary adenocarcinoma MX-1 xenografts. dEpoB exhibited far superior therapeutic effects *in vivo* compared with the more potent and toxic EpoB. More significantly, dEpoB (25–40 mg/kg, Q2D×5, i.p.) showed far superior therapeutic effect and lower toxicity than paclitaxel, doxorubicin, camptothecin or vinblastine (at maximal tolerated doses) in parallel experiments.

Other *in vivo* experiments were carried out with dEpoB. For example, with mammary adenocarcinoma xenografts resistant to adriamycin, MCF-7/Adr, therapeutic effects obtained with dEpoB were superior to those obtained with paclitaxel when i.p. regimens were used. For ovarian adenocarcinoma xenografts, SK-OV3, dEpoB (i.p.) and paclitaxel (i.v.) gave similar therapeutic effects. In nude mice bearing a human mammary carcinoma xenograft (MX-1), marked tumor regression and cures were obtained with dEpoB.

Subsequently, we expanded on these efforts by studying the effects of various formulations, routes, and schedules of i.v. administration. We discovered that slow infusion (6 h, Q2D, 30 mg/kg × 5 doses) of dEpoB in nude mice bearing human xenograft tumors with a Cremophor-ethanol vehicle was the most beneficial in terms of efficacy and decreased toxicity.

While dEpoB performed similarly to paclitaxel in sensitive tumor xenografts (MX-1 human mammary and HT-29 colon tumor), clearly superior effects of dEpoB were observed against MDR tumors under these slow infusion conditions. Thus, dEpoB (6 h, Q2D, 30 mg/kg × 5 doses, i.v.) demonstrated a full curative effect when administered to nude mice bearing the resistant human lymphoblastic T-cell leukemia,

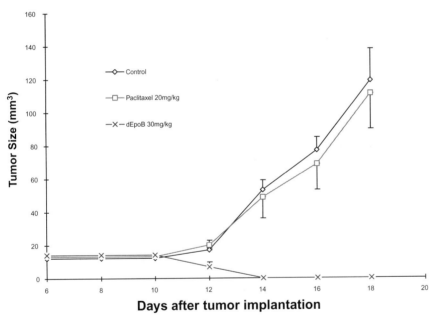

Fig. 2.3 Therapeutic effect of dEpoB and paclitaxel (Taxol[TM]) in nude mice bearing CCRF-CEM/paclitaxel xenografts (57-fold resistant to paclitaxel) following Q2D × 5, i.v. 6 h infusion. Human T-cell lymphoblastic leukemia (CCRF-CEM/paclitaxel) cells resistant to paclitaxel (10^7 cells) were inoculated subcutaneously into nude mice on day 0. Intravenous infusion was given on days 6, 8, 10, 12 and 14. <>, control with vehicle only; ω, paclitaxel 20 mg/kg; and X, dEpoB 30 mg/kg. The average tumor volume of the control group on days 12, 14, 16, 18, and 24 was 20 ± 3, 53 ± 6, 77 ± 8, 119 ± 22, and 415 ± 62 mm^3 respectively (mean ± SEM; n=3). The vehicle for 6 h i.v. infusion was 100 μl (Cremophor: ethanol, 1:1) + 3 ml saline.

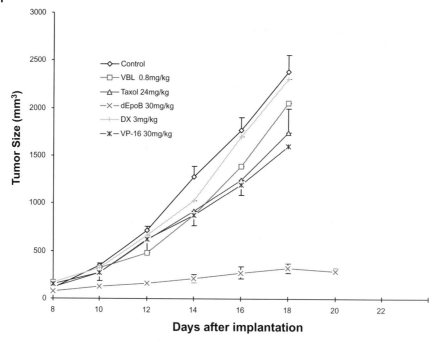

Fig. 2.4 Therapeutic effect of dEpoB and paclitaxel (Taxol™) in nude mice bearing MCF-7/Adr xenografts following Q2D × 5, i.v. treatment. Adriamycin (DX)-resistant human mammary adenocarcinoma (MCF-7/Adr) tissue 50 mg was implanted subcutaneously into nude mice on day 0. Six-hour i.v. infusions for control, dEpoB and paclitaxel and i.v. injection for VBL, DX, and VP-16 were given on days 8, 10, 12, 14 and 16. ◇, control with vehicle only; □, VBL , 0.8 mg/kg; Δ, paclitaxel, 24 mg/kg; +, DX, 3 mg/kg; X (*top*), VP-16, 30 mg/kg; and X (*bottom*), dEpoB, 30 mg/kg. The average tumor volume of the control group on day 14, 16 and 18 was 1281 ± 145, 1767 ±161 and 2381 ± 203 mm³ (mean ± SEM; *n*=5), respectively. The vehicle for i.v. 6 h infusion was 100 ul (Cremophor:ethanol, 1:1) + 3.5 ml saline. The *vertical bars* are the standard errors of means for the control, dEpoB, and paclitaxel.

CCRF-CEM/paclitaxel, that was 57-fold resistant to paclitaxel. Paclitaxel given under exactly the same conditions showed no significant therapeutic effects (Fig. 2.3). Likewise, in nude mice implanted with the paclitaxel resistant human mammary adenocarcinoma MCF-7/Adr xenograft, dEpoB markedly reduced the established tumors (average tumor size reduction, 89 % versus 27 %, dEpoB versus paclitaxel) and suppressed tumor growth, although a full cure was not achieved. The beneficial effects of dEpoB relative to other commonly used anti-cancer agents were demonstrated when nude mice bearing the refractory MCF-7/Adr tumor xenografts were challenged with chemotherapeutic agents such as adriamycin, vinblastine, and etoposide. In these experiments, Adr demonstrated a lack of therapeutic effect even at the nearly lethal dose, and both VBL and VP-16 showed little therapeutic effect (Fig. 2.4).

Based upon the promising profile of dEpoB, we sought to perform *in vivo* studies with other desoxy analogs such as **83** and **91** that demonstrated favorable *in vitro* efficacy. Our hope was that modification at C12 (propyl, 2-carbon acetal) would lead to improved pharmacological effects similar to that of dEpoB (C12 methyl) relative to dEpoA (C12 proton) where placement of a methyl group at C12 improves the activity. However, our supposition proved incorrect, and in fact substitution was deleterious. Thus, analogs **83** and **91** showed markedly higher toxicity than the lead compound dEpoB when administered to nude mice bearing human xenografts.

2.18
Conclusions

Although dEpoB is naturally occurring (epothilone D), it is isolated as a minor fermentation product compared to the principal constituents, epothilones A and B. Unfortunately, the chemical conversion of EpoB to dEpoB does not appear to be a promising source. Thus chemical synthesis was required for the discovery of the biological attributes of this promising compound. Fortunately, through these rigorous biological examinations and structure activity profiles, we have designated dEpoB as our leading candidate for clinical application. Further development of dEpoB as a potential clinical candidate continues to remain the main focus of our research.

2.19
Acknowledgements

This research was supported by the National Institutes of Health (grant numbers: S.J.D. (CA-28824 and CA-08748). Postdoctoral Fellowship support is gratefully acknowledged by: C.R.H. (American Cancer Society, PF-98–173-001) and S.D.K. (US Army Breast Cancer Research Fellowship, DAMD 17–98–1–1854). The authors wish to acknowledge the work on the epothilone project of their colleagues whose names are listed in the references.

References

1 GERTH, K.; BEDORF, N.; HÖFLE, G.; IRSCHIK, H.; REICHENBACH, H. *J. Antibiot.* **1996**, *49*, 560

2 HÖFLE, G.; BEDORF, N.; STEINMETZ, H.; SCHOMBURG, D.; GERTH, K.; REICHENBACH, H. *Angew. Chem. Int. Ed. Engl.* **1996**, *35*, 1567.

3 BOLLAG, D. M.; McQUENEY, P. A.; ZHU, J.; HENSENS, O.; KOUPAL, L.; LIESCH, J.; GOETZ, M.; LAZARIDES, E.; WOODS, C. M. *Cancer Res.* **1995**, *55*, 2325.

4 ROWINSKY, E.K.; EISENHAUER, E.A.; CHAUDHRY, V.; ARBUCK. S.G.; DONEHOWER, R.C.; *Seminars Oncol.* **1993**, *20*, 1–15.

5 CORTEZ, J.E.; PAZDUR, R. *J. Clin. Oncol.* **1995**, *13*, 2643–2655.

6 ROSE, W.C. *Anti-Cancer Drugs* **1992**, *3*, 311–321.

7 CASS, C. E.; JANOWSKA-WIECZOREK, A.; LYNCH, M. A.; SHEININ, H.; HINDENBURG, A. A.; BECK, W. T. *Cancer Res.* **1989**, *49*, 5798–5804.

8 GUNASEKERA, S.P.; GUNASAKERA, M.; LONG-LEY, R.E.; SCHULTE, G.K. *J. Org. Chem.* **1990**, *55*, 4912–4915.

9 LINDEL, T.; JENSEN, P.R.; FENICAL, W.; LONG, B.H.; CASAZZA, A.M.; CARBONI, J.; FAIRCHILD, C.R. *J. Am. Chem. Soc.* **1997**, *119*, 8744–8745.

10 SU, D.-S.; BALOG, A.; MENG, D.; BERTINATO, P.; DANISHEFSKY, S.J.; ZHENG, Y.-H.; CHOU, T.-C.; HE, L.; HORWITZ, S.B. *Angew. Chem. Int. Ed. Engl.* **1997**, *36*, 2093–2096.

11 NICOLAOU, K. C.; HE, Y.; VOURLOUMIS, D.; VALLBERG, H.; YANG, Z. *Angew. Chem. Int. Ed. Engl.* **1998**, *37*, 2014–2045 and references therein.

12 HARRIS, C. R.; BALOG, A.; SAVIN, K.; DANISHEFSKY, S. J.; CHOU, T.-C.; ZHANG, X.-G. Actualites de Chimie Therapeutique, Vingt-cinqieme serie, Paul Ehrlich Lecture, Societe de Chimie Therapeutique **1999**, Elsevier, Paris, New York, pp. 187–205.

13 (a) MENG, D.; SORENSEN, E. J.; BERTINATO, P.; DANISHEFSKY, S. J. *J. Org. Chem.* **1996**, *61*, 7998; (b) BERTINATO, P.; SORENSEN, E. J.; MENG, D.; DANISHEFSKY, S. J. *J. Org. Chem.* **1996**, *61*, 8000.

14 For an in-depth description of our first generation synthetic effort, see: MENG, D.; BERTINATO, P.; BALOG, A.; SU, D.-S.; KAMENECKA, T.; SORENSEN, E.J.; DANISHEFSKY, S.J. *J. Am. Chem. Soc.* **1997**, *119*, 10073.

15 (a) SHAFIEE, A.; MAZLOUMI, A.; COHEN, V. I. *J. Heterocycl. Chem.* **1979**, *16*, 1563; (b) SCHAFIEE, A.; SHAHOCINI, S. *J. Heterocycl. Chem.* **1989**, *26*, 1627.

16 (a) DANISHEFSKY, S. J. *Aldrichim. Acta* **1986**, *19*, 59; (b) DANISHEFSKY, S. J. *Chemtracts* **1989**, *2*, 273.

17 LUCHE, J.-L. *J. Am. Chem. Soc.* **1978**, *100*, 2226.

18 DANISHEFSKY, S. J.; KITAHARA, T. *J. Am. Chem. Soc.* **1974**, *96*, 7807.

19 (a) BERKOWITZ, D. B.; DANISHEFSKY, S. J. *Tetrahedron Lett.* **1991**, *32*, 5497; (b) BERKOWITZ, D. B.; DANISHEFSKY, S. J.; SCHULTE, G. K. *J. Am. Chem. Soc.* **1992**, *114*, 4518.

20 HEATHCOCK, C. H.; YOUNG, S. D.; HAGEN, J. P.; PIRRUNG, M. C.; WHITE, C. T.; VANDERVEER, D. *J. Org. Chem.* **1980**, *45*, 3846.

21 For examples, see: (a) LYTHGOE, B.; NAMBU-DIRY, M. E. N.; RUSTON, S.; TIDESWELL, J.; WRIGHT, P. W. *Tetrahedron Lett.* **1975**, 3863; (b) LYTHGOE, B. *Chem. Soc. Rev.* **1981**, 449; (c) TOH, H. T.; OKAMURA, W. H. *J. Org. Chem.* **1983**, *48*, 1414; (d) BAGGIOLINI, E. G.; IACO-BELLI, J.A.; HENNESSY, B. M.; BATCHO, A. D.; SERENO, J. F.; USKOKOVIC, M. R. *J. Org. Chem.* **1986**, *51*, 3098.

22 (a) COHEN, N.; EICHEL, W. F.; LOPRESTI, R. J.; NEUKOM, C.; SAUCY, G. *J. Org. Chem.* **1976**, *41*, 3505; (b) NAGAOKA, H.; KISHI, Y. *Tetrahedron* **1981**, *37*, 3873; (c) MEYERS, A. I.; BABIAK, K. A.; CAMPBELL, H. L.; COMINS, D. L.; FLEMING, M. P.; HENNING, R.; HEUSCHMANN, M.; HUDSPETH, J. P.; KANE, J. M.; REIDER, P. J.; ROLAND, D. M.; SHIMIZU, K.; TOMIOKA, K.; WALKUP, R. D. *J. Am. Chem. Soc.* **1983**, *105*, 5015; (d) ROUSH, W. R.; PALKOWITZ, A. D.; ANDO, K. *J. Am. Chem. Soc.* **1990**, *112*, 6348.

23 DANISHEFSKY, S. J.; YAN, C.-F.; SINGH, R. K.; GAMMILL, R. B.; McCURRY, P. M., JR.; FRITSCH, N.; CLARDY, J. *J. Am. Chem. Soc.* **1979**, *101*, 7001.

24 (a) DANISHEFSKY, S. J.; PEARSON, W. H.; HARVEY, D. F.; MARING, C. J.; SPRINGER, J. P. *J. Am. Chem. Soc.* **1985**, *107*, 1256; (b) DANISHEFSKY, S. J.; MYLES, D. C.; HARVEY, D. F. *J. Am. Chem. Soc.* **1987**, *109*, 862.

25 For an example, see: EGBERTSON, M.; DANISHEFSKY, S. J. *J. Org. Chem.* **1989**, *54*, 11.

26 (a) BERTINATO, P.; SORENSEN, E. J.; DANISHEFSKY, S. J. *J. Org. Chem.* **1996**, *61*, 8000. (b) MENG, D.; SU. D.-S.; BALOG, A.; BERTINATO, P.; SORENSEN, E. J.; DANISHEFSKY, S. J.; ZHENG, Y.-H.; CHOU, T.-C.; HE, L.; HORWITZ, S. B. *J. Am. Chem. Soc.* **1997**, *119*, 2733.

27 (a) YAMAGUCHI, M.; INANAGA, J.; HIRATA, K.; SAEKI, H.; KATSUKI, T. *Bull. Chem. Soc. Jpn.* **1979**, *52*, 1989. (b) MULZER, J.; MARESKI, P. A.; BUSCHMANN, J.; LUGER, P. *Synthesis*, **1992**, 215.

28 SU, D.-S.; MENG, D.; BERTINATO, P.; BALOG, A.; SORENSEN, E. J.; DANISHEFSKY, S. J.; ZHENG, Y.-H.; CHOU, T.-C.; HE, L.; HORWITZ, S. B.; *Angew. Chem. Int. Ed. Engl.* **1997**, *36*, 757.

29 CHOU, T.-C.; ZHANG, X.-G.; HARRIS, C. R.; KUDUK, S. D.; BALOG, A.; SAVIN, K.; DANISHEFSKY, S. J. *Proc. Nat. Acad. Sci.* **1998**, *95*, 15798.

30 The stereochemistry of the diastereomeric products isolated in this reaction were confirmed to be those arising from incomplete relative face selectivity of the aldol reaction (Felkin:Anti-Felkin). Thus the stereochemistry was proven by comparing the ^{1}H NMR spectrum of the major diastereomer with a compound of known stereochemistry that had been independently synthesized.

31 BALOG , A; HARRIS, C.; SAVIN, K.; ZHANG, X.G.; CHOU, T.-C.; DANISHEFSKY, S.J. *Angew. Chem. Int. Ed. Engl.* **1998**, 37, 2675–2678.

32 HARRIS, C. R.; KUDUK, S. D.; SAVIN, K.; BALOG, A.; DANISHEFSKY, S.J. *Tetrahedron Lett.* **1999**, *40*, 2263.

33 AOYAMA, T.; TERASAWA, S.; SUDO, K.; SHIORI, T. *Chem. Pharm. Bull.* **1984**, *32*, 3759.

34 SU, D.-S.; MENG, D.; BERTINATO, P.; BALOG, A.; SORENSEN, E. J.; DANISHEFSKY, S. J.; ZHENG, Y.-H.; CHOU, T.-C.; HE, L.; HORWITZ, S.B. *Angew. Chem., Int. Ed. Engl.* **1997**, *36*, 757

35 The Noyori reduction of various diketo esters in this series was very dependent upon the amount of acid present in the reaction. Without the presence of a stoichiometric amount of acid, the rate of reduction as well as the selectivity in the reduction dropped off. At higher pressures, the chemoselectivity of the reduction was poor resulting in the reduction of both alkene groups. Further, the carbonyl at C5 was never reduced under these reaction conditions but was absolutely necessary for the reduction of the C3 carbonyl. When C5 was in the alcohol oxidation state, no reduction was seen. A. Balog, unpublished results.

36 MACHROUHI, F.; NAMY, J.-L.; KAGAN, H. B. *Synlett.*, **1996**, 633

37 LODGE, E.P.; HEATHCOCK, C.H. *J. Am. Chem. Soc.* **1987**, *109*, 2819 AND REFERENCES THEREIN

38 HEATHCOCK, C. H. In "Asymmetric Synthesis," J. D. MORRISON, Ed., 1984, Academic Press, Inc.: New York, Vol. 3, pp. 111–213.

39 (a) MUKAIYAMA, T. *Org. React.* **1982**, *28*, 203. (b) MASAMUNE, S.; CHOY, W.; PETERSEN, J. S.; SITA, L. R. *Angew. Chem. Int. Ed. Engl.* **1985**, *24*, 1

40 ROUSH, W. R. *J. Org. Chem.* **1991**, *56*, 4151–4157.

41 HEATHCOCK, C. H.; BUSE, C. T.; KLESCHICK, W. A.; PIRRUNG, M. C.; SOHN, J. E.; LAMPE, J. *J. Org. Chem.* **1980**, *45*, 1066.

42 HARRIS, C. R.; KUDUK, S. D.; BALOG, A.; SAVIN, K.; DANISHEFSKY, S.J. *Tetrahedron Lett.* **1999**, *40*, 2267.

43 GHERA, E.; KLEINMAN, V.; HASSNER, A. *J. Org. Chem.* **1999**, *64*, 8.

44 REETZ, M.T. *Angew. Chem. Int. Ed. Engl.* **1984**, *23*, 556.

45 CHOU, T.-C.; ZHANG, X.G.; HARRIS, C. R.; KUDUK, S. D.; BALOG, A.; SAVIN, K.; DANISHEFSKY, S.J. *Proc. Natl. Acad. Sci. USA* **1998**, *95*, 15798.

46 LANDINO, L.M.; MacDONALD, T.L. in: *The Chemistry and Pharmacology of Taxol and Its Derivatives.* FAVIN, V. (ed) Elsevier, New York (1995) Chapter 7, p.301

47 GIANNAKAKOU, P.; SACKETT, D. L.; KANG, Y.-K.; ZHAN, Z.; BUTERS, J. T.; FOJO, T.; PORU-CHYNSKY, M. S. *J. Biol. Chem.* **1997**, *272*, 17118–17125, and references therein.

48 OJIMA, I.; BOUNAUD, P.-Y.; BERNACKI, R. J. *Chemtech* **1998**, *28*, 31–36.

49 ROSE, W.C. *Anti-Cancer Drugs* **1992**, *3*, 311–321.

50 ROWINSKY, E.K.; EISENHAUER, E.A.; CHAUDHRY, V.; ARBUCK. S.G.; DONEHOWER, R.C. *Seminars Oncol.* **1993**, *20*,1–15.

51 NICOLAOU, K.C.; ROSCHANGAR, F.; VOURLOU-MIS, D. *Angew. Chem. Int. Ed. Engl.* **1998**, *37*, 2015.

3
The Spirotetrahydrofuran Motif: its Role in Enhancing Ligation in Belted Ionophores, Biasing Cyclohexane Conformation, and Restricting Nucleoside/Nucleotide Conformation

Leo A. Paquette

3.1
Introduction

The capacity of oxygen-containing molecular arrays to serve as host molecules is significantly enhanced if the ligating atoms are suitably preorganized for effective coordination. Reduction of the populations of conformations unfavorable to chelation in favor of those spatial arrangements conducive to multipoint binding interactions has served as an important principle of rational design. [1] Despite the fact that many naturally-occurring ionophores utilize cyclic ether fragments as ligand arrays for ionic recognition and transport, the significantly enhanced intrinsic basicity of a cyclic ether relative to an acyclic ether variant [2–4] is often taken for granted or not explicitly addressed.

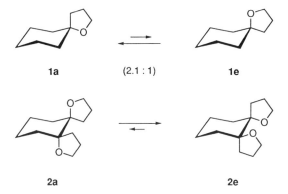

Beyond these fundamental considerations are steric and electronic factors whose magnitude can gain primary significance as the number of oxygen atoms is increased appreciably. Consider the following. The A values for $-OCH_3$ (0.55–0.75 kcal/mol) and $-CH_2CH_3$ (1.79 kcal/mol) [5] reveal both substituents to prefer equatorial occupancy in a monosubstituted cyclohexane, more so the ethyl group because of its greater steric bulk. [6] The direction and magnitude of this stereochemical bias, which is well mirrored in **1**, [7] is not seen in **2**, where the significantly more stable conformer has the electronegative hetero atoms projected in a gauche orientation. [8] This phenomenon also operates generally in 1,2-disubstituted ethanes [9] and *trans*-1,2-dialkoxycyclohexanes, [10] the positioning of the nonbonded electrons being attributed to stabilizing attractive forces [11,12] and/or hyperconjugative control stemming from an avoidance of unfavorable antiperiplanar acceptor-acceptor interactions. [13]

Heteroatom-directed π-facial diastereoselection during Diels-Alder cycloaddition to dispiro[4.0.4.4]tetradeca-11,13-dienes such as **3** provides additional important insight from the reactivity perspective. [14] Where ordinary dienophiles such as *N*-phenylmaleimide are concerned, conventional concertedness operates, and capture *syn* to the oxygen atoms materializes exclusively (see **4**) in order to skirt steric interactions with the more spatially-demanding methylene groups. The obvious contra-steric behavior exhibited by *N*-methyltriazolinedione can be attributed to the repulsive interactions than come into play between the nonbonded electron pairs on the oxygens resident in diene **3** and on the nitrogen atoms of the azo linkage in the reactive dienophile. The mechanistic details surrounding the formation of **5** likely involve initial formation of an aziridinium imide intermediate with subsequent rearrangement via a 1,4-zwitterion to the observed urazole.

The presence of one or more spirotetrahydrofuran rings in a molecule can provide for operation of unprecedented rearrangement pathways. A case in point is the remarkable ease with which **6** can be reversibly transformed into **7** under acidic con-

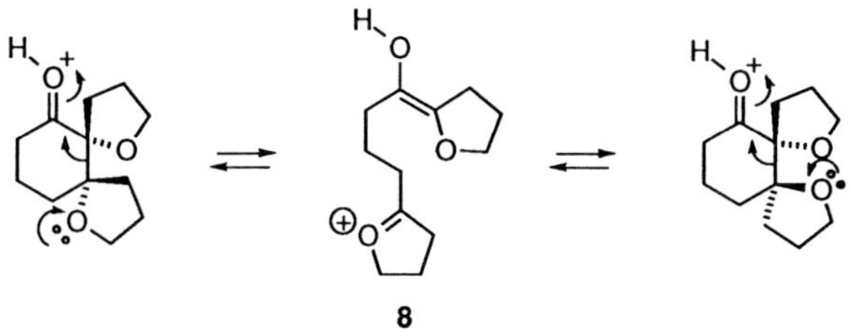

Scheme 3.1

ditions. [15] The unusual feature of this epimerization is its operation within a framework constructed of two vicinal quaternary centers. The interconversion operates by way of a push-pull fragmentation involving electron flow from the β-oxido atom to the protonated carbonyl oxygen (Scheme 3–1). Since **6** and **7** are chiral while **8** is not, both optically enriched ketones should experience racemization concurrent with isomerization and indeed do so. [15]

Dienes such as **3** exhibit an even greater sensitivity to acid than do **6** and **7**. [16] Since formation of the [4.4.4]propelladiene **10** competes very favorably with conversion to **9**, the transition state for aromatization is not particularly advantaged (Scheme 3–2). To our amazement, tetracyanoethylene has been found to be capable of promoting the identical chemical transformations. [17]

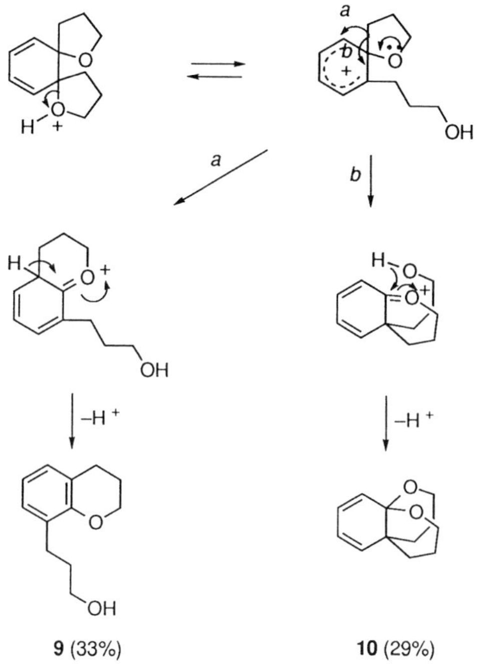

9 (33%)   **10 (29%)**

Scheme 3.2

For alcohols that feature a neighboring quaternary center, it is generally assumed that this adjoining site will not be affected stereochemically during oxidation to the carbonyl derivative. Although this expectation has been borne out by a myriad of examples and is considered secure by most investigators, **11** constitutes a *bona fide* example in which stereochemical integrity is indeed compromised. [18] Its oxidation with various types of reagents results in competitive desymmetrization to provide mixtures of **12** and **13** by a mechanism consisting of β-elimination of an oxygen atom with scission of a neighboring heterocyclic ring, and ultimate reconstitution of the latter.

The recent availability of poly(spirotetrahydrofuranyl) systems has brought to light several of their interesting properties. The above examples were discovered en route to specific targets rather than representing the end-products of thoughtful design. In the sequel, the latter criterion has been applied in a conscious effort to garner new insight into the value of the spirotetrahydrofuran motif. The three case studies documented here hopefully provide some sense of the diverse applications for which these building blocks qualify, and usher in new thinking in the next millennium.

3.2
syn-1,3,5-Orientation on a Cyclohexane Core

Several years ago, we suggested that the enhanced binding efficiency of tetrahydrofurans might be very suitably exploited by proper assimilation of these building blocks into belted spirocyclic platforms. [19] The heightened level of structural preorganization resident in *cis,cis*-trispiro ether **14**, for example, was expected to lend itself admirably to an enhanced capability for metal ion chelation. However, binding-site convergence and the resultant cumulative noncovalent interactions can be realized only if the **14ax** conformation is readily adopted. The complex interplay of steric and electronic effects expected to control the conformational bias in this ligand system was not conducive to a reliable prediction.

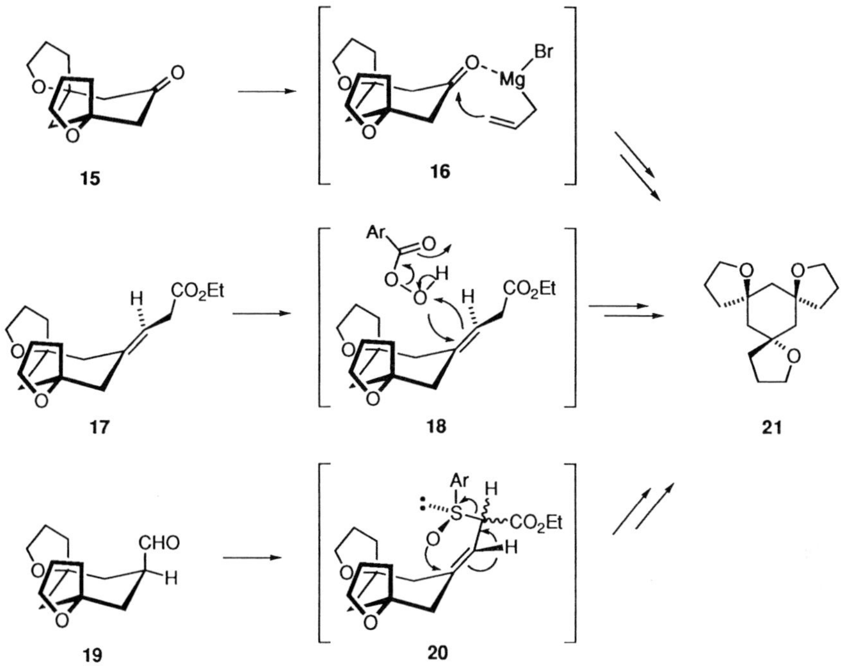

Scheme 3.3

The synthetic route to **14** began with 3,5-dimethoxybenzoic acid and progressed uneventfully after several laboratory steps to the dispirocyclohexanone **15**, α,β-unsaturated ester **17**, and aldehyde **19**. All three intermediates were found to share in common the capacity for equilibrating between their two chair conformers. [20] Notwithstanding the facility with which the equatorial and axial C–O bonds interconvert, all three compounds undergo highly stereoselective reactions, thereby implicating the involvement of a specific conformer during product formation (Scheme 3–3). The transition states depicted in **16**, **18**, and **20** are believed to be operational during Grignard additions to **15**, the peracid oxidation of **17**, and an SPAC reaction [21] on aldehyde **19**. Since the nucleophilic addition in **16** proceeds by equatorial attack and the oxygenation reactions in **18** and **20** involve axial delivery, the interesting dilemma that unfolds is the fact that all three processes give rise uniquely to the unwanted isomer **21**.

The presence of the two spirotetrahydrofuran rings in these reactive intermediates is obviously not conducive to the stereocontrolled synthesis of **14**. This problem was resolved by making recourse to **22**, a precursor to **15**, the silyl-protected (3-hydroxypropyl) substituent which was viewed to be sufficiently bulky to guarantee its equatorial occupancy in the lowest energy conformation (Scheme 3–4). Furthermore, treatment of **22** with Grignard reagent should be met with rapid deprotonation of the hydroxyl and formation of chelate **23**. At this point, customary equatorial

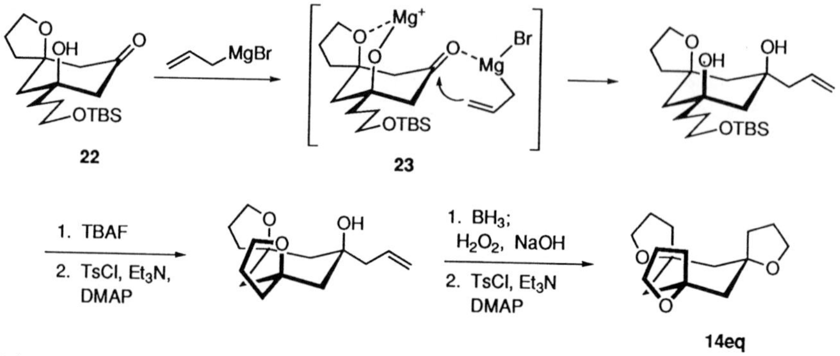

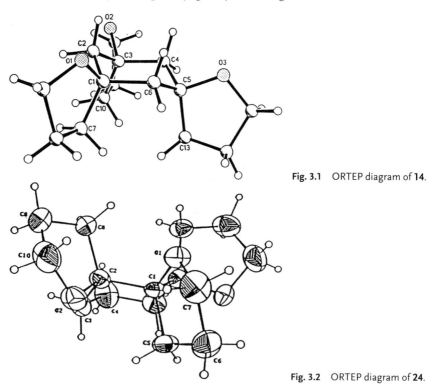

Scheme 3.4

entry of the nucleophile operates and fixes the triad of oxygens in the all-cis relationship present in **14**. [20,22]

The C_{3v} symmetry of **14** was reflected in its five-line ^{13}C NMR spectrum. X-ray crystallographic analysis revealed further that the all-equatorial conformation is adopted in the solid state (Fig. 3–1). This bias persists as well in solution and may be a consequence of more favorable dipole-dipole interactions. The parallelism between **14** and its isomer **24**, whose ground-state conformer also projects all three C–O bonds in the equatorial plane (Fig. 3–2), is striking.[8]

Fig. 3.1 ORTEP diagram of **14**.

Fig. 3.2 ORTEP diagram of **24**.

Despite the favored population of **14eq**, this trispiro ether binds strongly to Li^+ and Na^+ ions, as well as to $CH_3NH_3^+$. [20,22] All three complexes provided crystals enabling direct corroboration of their three-dimensional structures as **25–27**. The major differences between **25** and **26** is the increased distance of Na^+ (1.554 Å) relative to Li^+ (1.279 Å) above the plane defined by the three oxygen centers. The association constants (K_a) for the complexation of **14** in solution show the fit to be best for lithium ion. The selectivity exhibited toward Li^+ is impressively high ($K_a(Li^+)/K_a(Na^+) = 7.9 \times 10^7/2.5 \times 10^6 = 32$, ($K_a(Li^+)/K_a(K^+) = 7.9 \times 10^7/3.3 \times 10^4 = 2400$) and rivaled only by highly preorganized spherands [23] and smaller cryptands. [24]

24

25, M = Li
26, M = Na

27

It will be recognized that **14** is capable of coordination uniquely in a *monofacial* sense. On this basis, this substrate represents a logical starting point for the development of a class of *bifacial* ligands featuring extended three-dimensional features not previously examined in the context of crown ethers. Since the potential application of such systems in many areas of investigation can be envisioned, our attention was subsequently directed to the acquisition of such compounds.

3.3
Maximally Substituted Hexa(spirotetrahydrofuranyl)-cyclohexanes

Motivated by the possible bifacial complexation capability of **28**, a general synthetic strategy was developed which has also allowed access to four additional members of this family of sterically crowded heterocycles (**29–32**). [25–27] The starting materials are cyclohexanones **6** and **7**, which are readily available by 2-fold ring expansion of cyclobutanone by means of an oxonium ion-activated protocol. [28] The stereodefined elaboration of vicinal trispiro homologs can be accomplished in two reliable, complementary ways. [29] The first alternative begins with Claisen-Schmidt condensation involving an aromatic aldehyde, relies on sterically enforced nucleophilic attack by the Normant reagent, [30] and is generally trans-selective in regard to the adjacent C–O bond as required of **28** (Scheme 3–5). The second option, which is not illustrated, proceeds by means of α-oxygenation and O-silylation in order to take advantage of the large OTBS group to achieve net cis "capping" of the ketone carbonyl. The iterative sequence that leads from **7** to **33** and subsequently to **34** exemplifies the manner in which progression around the cyclohexane core can eventually deliver the vicinal tetraspiro cyclohexanone of choice. Beyond this point, any attempt at enolate generation is met with immediate and irreversible β-elimination with cleavage of the first tetrahydrofuran established (see 34 → 35).

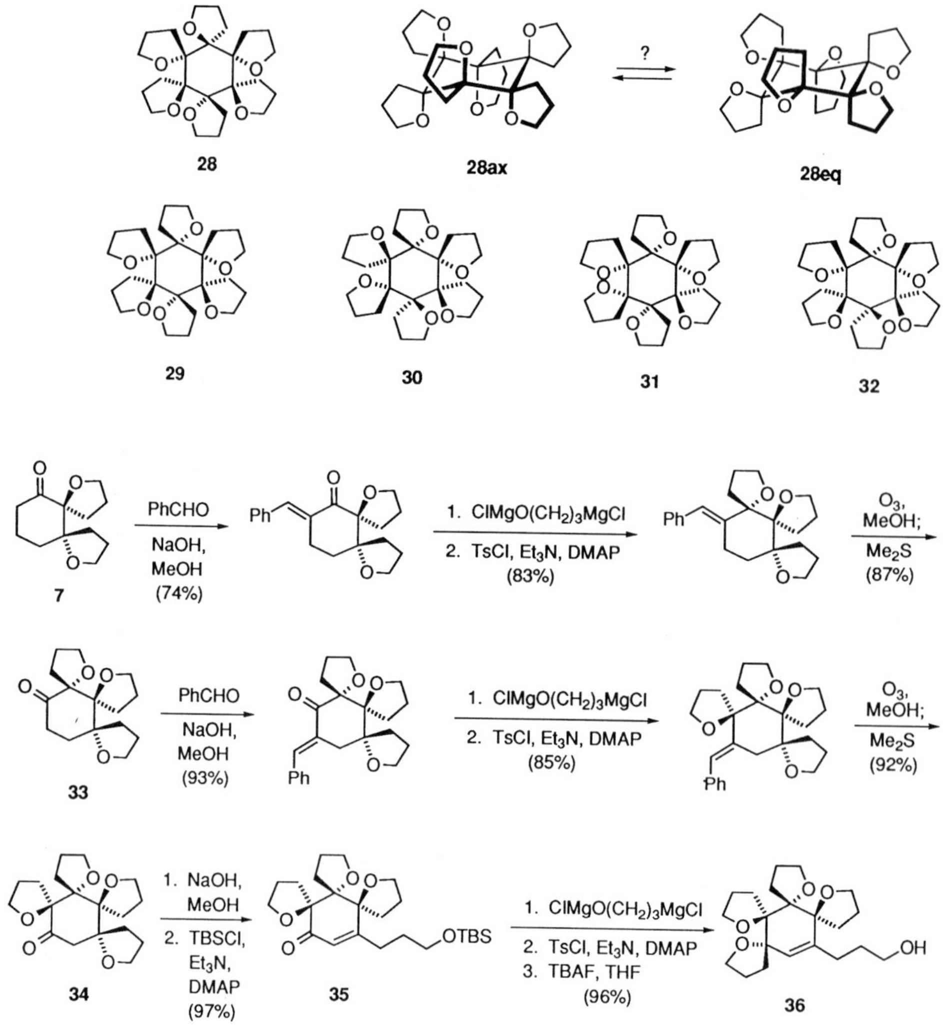

28   28ax   28eq

29   30   31   32

Scheme 3.5

Reinstallation of the fractured spiro ring can be accomplished either directly at the enone level or, more interestingly, after the ketone carbonyl has been capped as in 36. When 36 and stereoisomers thereof are individually subjected to intramolecular oxymercuration, the stereochemical course of the cyclization is dictated by the approach of Hg^{2+} to that surface of the double bond where coordination to a proximal axially oriented ether oxygen can operate. [31] Where 37 and 40 are concerned,

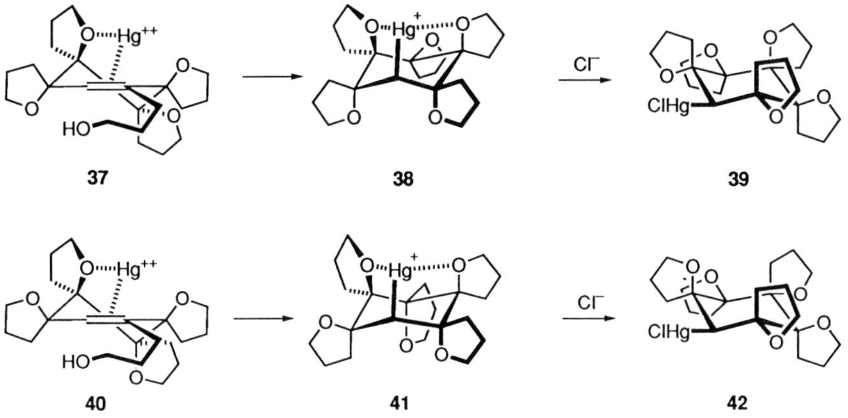

Scheme 3.6

the contribution of the energetically favorable pair of diaxial 1,3 Hg/O interactions sets the stage for the kinetically controlled generation of **38** and **41**, respectively (Scheme 3–6). X-ray analysis of the derived chloromercurials has established that the ultimate conformation adopted by these compounds positions as many ether oxygens equatorial as possible.

As matters would have it, the mercuricyclization of **36** can advance comparably on both surfaces of the π bond, with the result that **43** and **44** are formed competitively (Scheme 3–7). [31] These diastereomeric products diffracted well and provided the important structural information displayed in Figs. 3–3 and 3–4. [32] In **43**, all six heteroatoms are projected equatorially. This is not so for **44** where the axial C–Hg and C–O bonds generated along the reaction coordinate are preserved. The latter phenomenon is made possible because the methylene groups from four tetrahydrofuran rings are able to remain axially disposed.

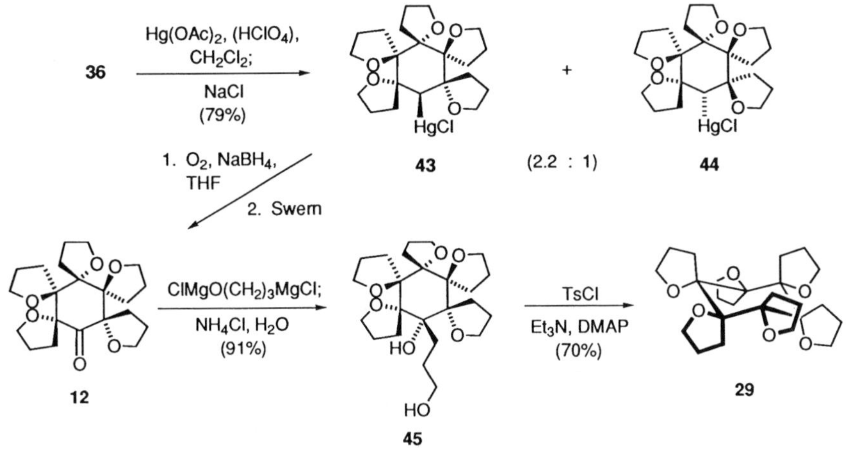

Scheme 3.7

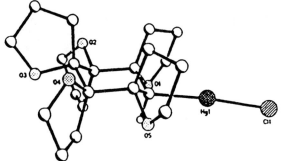

Fig. 3.3   ORTEP diagram of **43**.

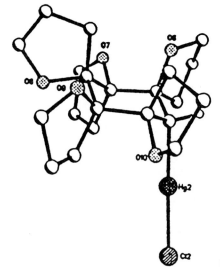

Fig. 3.4   ORTEP diagram of **44**.

Following the oxidative demercuration of **43** and subsequent conversion of the cyclohexanol to **12**, the time had come to implement the final capping maneuver. Admixture of **12** with the Normant reagent was met with efficient 1,2-addition and formation of diol **45**, from which **29** was crafted. [26] The conformation of this hexaspiro system was defined as shown by crystallographic methods. Since the **12** → **45** conversion installs three contiguous oxygen centers in syn fashion, an alternative protocol that would serve to elaborate the carbon-oxygen bond last was mandated. The manner in which this was accomplished is outlined in Scheme 3–8. [25,27] The successful tactic for inverting stereochemistry at the carbinol site in **45** involved dehydration to give **46**, epoxidation of the double bond predominantly from the β-surface, oxidation to the aldehyde level, and base-promoted β-elimination. This series of steps delivered **47**, reduction of which led to the formation of **48** and subsequently the highly crystalline, D_{3d}-symmetric **28**. Not unexpectedly, crystallographic analysis confirmed that its six oxygen atoms are projected equatorially. Disappointingly, however, **28eq** exhibits no capacity for conformational ring inversion to **28ax**

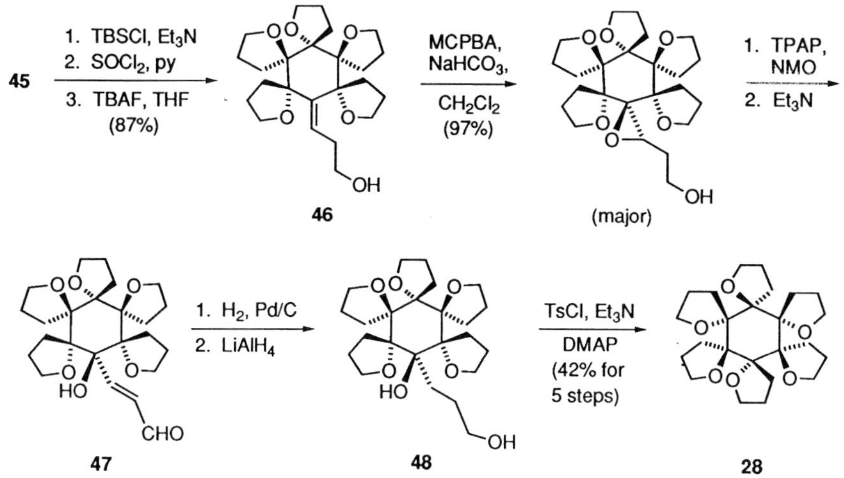

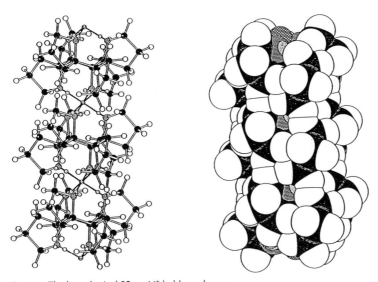

Scheme 3.8

Fig. 3.5 The hypothetical **28ax** · Li$^+$ ladder polymer.

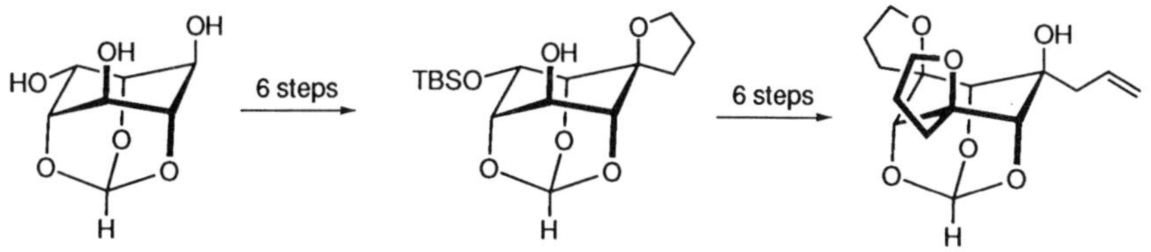

Scheme 3.9

even under extreme conditions and understandably is incapable of binding usefully to alkali metal ions. The barrier to chair-to-chair interconversion present in **28** may be the highest yet observed. [33]

Obviously, the characteristics of **28** are not conducive to bifacial complexation such that ladder polymers consisting of **28ax** and alkali metal ions as in the hypothetical 3 × **28ax** · 4Li$^+$ complex shown in Fig. 3–5 in two formats. It is now clear that goals related to this phenomenon can be realized only if the strong gauche preference exhibited by two neighboring C–O bonds is overridden. Attempts to run this gauntlet are in progress. One thrust has begun with the protected *myo*-inositol **49** [34] (Scheme 3–9). The ortho ester subunit serves to lock the oxygen centers at positions 1, 3, and 5 of the cyclohexane ring in an all-axial arrangement while enabling stereocontrolled transformations to be carried out at the other three skipped sites. A small number of synthetic steps has made possible the stepwise introduction of one spirotetrahydrofuran ring as in **50** and subsequently a second. [35] The ensuing introduction of an allyl group proceeds from the exterior to set the sixth C–O bond axially. As indicated clearly by the ORTEP diagram of **51** (Fig. 3–6), this intermediate represents, as does **49**, a polyoxygenated cyclohexane having all six C–O bonds projected above and below the equatorial plane.

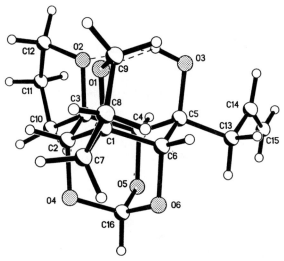

Fig. 3.6   ORTEP diagram of **51**.

What does the future hold? With specific regard to **51**, one may ask whether ligation of an oxygen triad to Li$^+$ will be adequate to hold conformation while chemical transformations are performed on the opposite surface? If not, can systematic dismantling of the ortho ester bridge provide opportunities for chemical change without incurring chair-to-chair interconversion? Since no precedence exists in these areas, the future holds considerable excitement.

3.4

Spirocyclic Restriction of Nucleosides/Nucleotides

The systematic variation of nucleotide components is widely credited as being an important source of improved understanding of the conformational and supramolecular binding properties of this remarkable class of molecules. Potential therapeutic applications as inhibitors of protein expression remain high provided that these analogs display strong and sequence-specific binding to single-stranded RNA or double-stranded DNA, and exhibit resistance to enzymatic degradation. In recent years, structural modifications featuring restrictions in conformational flexibility are being accorded increased attention. [36–38] In the context of this theme, our attention has been focused on the preparation of the diastereomeric pairs of spironucleosides defined by **52–55**. For reasons to be detailed below, members of this subset are expected to attract considerable attention as tools in molecular biology and as possible future therapeutic agents for the benefit of mankind.

52   **53**   **54**   **55**

a, R_1 = H, R_2 = OH; **b**, R_1 = OH, R_2 = H

One may justifiably pose the question "why 4'-spiroalkylated nucleosides?" First and foremost, we have developed synthetic methodology which we fully expect to deliver **52–55** concisely and at the 100 % ee level of optical purity. Beyond such practical considerations, extensive crystal structure data, most notably that for DNA and RNA fragments, reveal the existence of considerable void space in the region below C4' of each nucleoside building block. The empty space being referred to is sufficiently voluminous to accommodate more than the string of three methylene groups under consideration here. Accordingly, there are no pertinent concerns regarding nonbonded steric superimposition.

Damage to DNA can occur oxidatively by means of radical attack. [39] Whereas hydroxyl radicals are not at all chemoselective, [40] several agents whose active center is bound to the minor groove preferentially abstract hydrogen atoms from the 4'- and/or 5'-position of a deoxyribose. [39,41–43] In **54a** and **54b**, the 4'-hydrogen is no longer present, being replaced by a carbon center that forms part of the spirocyclopentane ring. In addition, the C5' site in **54** is neopentylic in nature, a feature that should certainly lessen the propensity for attack at that position for steric reasons. A significantly reduced hydrolysis rate at the phosphodiester linkages would be a bonus.

The torsion angles around the bonds of the sugar-phosphate DNA backbone are of decisive importance for the secondary structure of DNA as well as for base-base recognition. [44] For antisense agents to be effective inhibitors of protein expression *in vivo*, they have to resist the action of DNA-degrading enzymes and bind to their target *m*RNA sequences with high affinity and in a sequence-specific manner. [45,46] Therefore, identification of the conformational preference exhibited by any new nucleoside or nucleotide is warranted. For the present purposes, it is informative to compare thymidine with **54a** (B = Th) and **54b** (B = Th) at the computational level (Amber). The overlap of **54b** on natural thymidine is notably remarkable (RMS = 0.007); for **54a**, the departure from direct superimposition is greater (RMS = 0.058), but still very acceptable (Fig. 3–7). [47]

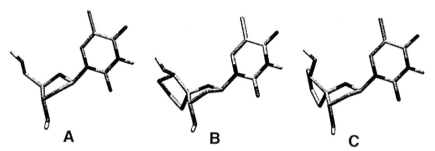

A   **B**   **C**

Fig. 3.7 The minimum energy conformations of thymidine (**A**), **54b** (B = Th) (**B**), and **54a** (B = Th) (**C**).

Although the preceding analysis provides no insight into reduced conformational flexibility, enhanced duplex stability is anticipated as a consequence of the structural preorganization provided by spirocyclization. This entropic advantage should be matched with reduced pairing enthalpy terms having compensatory effects on the free energy of duplex formation. Indeed, when recourse was again made to the Amber program for energy minimization of the three homodinucleotides based on thymidine, **54a** (B = Th), and **54b** (B = Th), strong base-pairing was uniformly seen in all three examples. Clearly, the correspondence is very close. In light of these facts, the incorporation of **54a** and **54b** into DNA is likely to have a significantly positive biological impact.

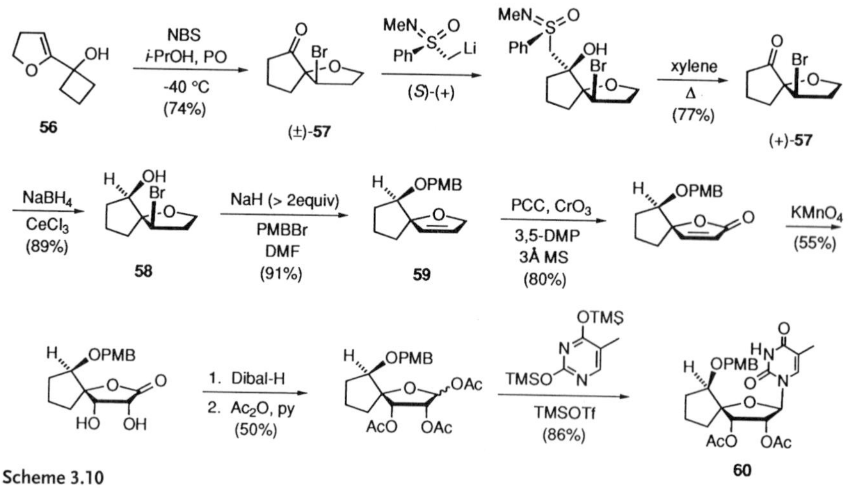

Scheme 3.10

Recognition that N-bromosuccinimide is capable of transforming carbinol **56** stereo-selectively into racemic bromo-ketone **57** under controlled conditions was instrumental in dictating the synthetic route to be pursued. [48] The resolution of **57** has been accomplished by proper adaptation of Johnson's sulfoximine technology [49] (Scheme 3–10). The sodium borohydride reduction of (+)-**57** is sterically controlled, leads exclusively to **58**, and makes possible direct conversion to **59**. [47] Through a sequence of steps culminating in a Vorbrüggen coupling, [50] one arrives uneventfully at **60**.

Related pursuits in the α series have been facilitated by the fact that the dehydro-bromination of (+)-**57** as a first maneuver gives rise to **61**, thereby channeling subsequent hydride reduction exclusively in the complementary direction as shown in Scheme 3–11. [51] The continued conversion of **61** into **62** proceeds in a modestly satisfactory manner. Both synthetic schemes are plagued by minor problems that shall be rectified following more critical scrutiny. The significant point is that the two protocols are capable of delivering their intended targets.

Gene products of increasing variety are being discovered that are regulated by antisense RNA molecules expressed intracellularly. [52,53] It is hoped that the spiro-cyclic restriction approach outlined above will ultimately prove feasible for therapeutic applications in the context of gene therapy. We recognize that increasing RNA binding affinity to synthetic oligonucleotides does not alone lead to biological efficacy. These compounds must also penetrate cells and reach cytoplasm where they must bind effectively to m-RNA and particularly the crucial codon region. [54] The conservative spirocyclic modifications proposed by us are apt to possess improved properties such as suitably enhanced lipophilicity, resistance to alkaline hydrolysis, as well as stability toward nuclease degradation. These and many related issues will be pursued well into the next century.

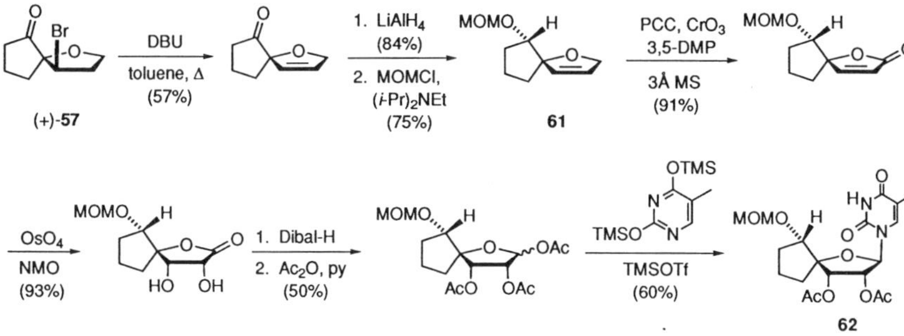

Scheme 3.11

3.5
Acknowledgement

The author trusts that he has transmitted the intellectual insight and tireless energy of those co-workers who have been associated with him in this undertaking. Their names can be found in the references. Financial support has been provided by the National Science Foundation, Hoechst Marion Roussel, and the Paquette Research Fund.

References

1 For example: (a) IIMORI, T.; STILL, W. C.; RHEINGOLD, A. L.; STALEY, D. L. *J. Am. Chem. Soc.* **1989**, *111*, 3439. (b) ERICKSON, S. D.; STILL, W. C. *Tetrahedron Lett.* **1990**, *31*, 4253.

2 Hydrogen bonding capability: Berthelot, M.; BESSEAU, F.; LAURENCE, C. *Eur. J. Org. Chem.* **1998**, 925 and relevant references therein.

3 Gas-phase proton affinity: BORDEJE, M. C.; MO, O.; YANEZ, M.; HERREROS, M; ABBOUD, J. L. M. *J. Am. Chem. Soc.* **1993**, *115*, 7389.

4 Rate of proton attachment in solution: PINES, E.; FLEMING, G. R. *Isr. J. Chem.* **1993**, *33*, 179 and relevant references therein.

5 WINSTEIN, S.; HOLNESS, N. J. *J. Am. Chem. Soc.* **1955**, *77*, 5562.

6 ELIEL, E. L.; WILEN, S. H.; MANDER, L. N. *Stereochemistry of Organic Compounds*, John Wiley and Sons, Inc.: New York, 1994.

7 PEDERSON, C. J.; FRENSDORFF, H. K. *Angew. Chem., Int. Ed. Engl.* **1972**, *11*, 16.

8 PAQUETTE, L. A.; NEGRI, J. T.; ROGERS, R. D. *J. Org. Chem.* **1992**, *57*, 3947.

9 (a) PHILLIPS, L.; WRAY, V. *J. Chem. Soc., Chem. Commun.* **1973**, 90. (b) WOLFE, S.; RAUK, A.; TEL, L. M.; CSIZMADIA, I. G. *J. Chem. Soc. B.* **1971**, 136. (c) WOLFE, S.; TEL, L. M.; CSIZMADIA, I. G. *Can. J. Chem.* **1973**, *51*, 2423. (d) WOLFE, S.; TEL, L. M.; HAINES, W. J.; ROBB, M. A.; CSIZMADIA, I. G. *J. Am. Chem. Soc.* **1973**, *95*, 4863. (e) WHANGBO, M.-H.; WOLFE, S. *Can. J. Chem.* **1976**, *54*, 949, 963. (f) VAN-CATLEDGE, F. A. *J. Am. Chem. Soc.* **1974**, *96*, 5693.

10 (a) ZEFIROV, N. S.; GURVICH, L. G.; SHASHKOV, A. S.; KRIMER, M. Z.; VOROB'EVA, E. A. *Tetrahedron* **1976**, *32*, 1211. (b) ZEFIROV, N. S.; SAMOSHIN, V. V.; SUBBOTIN, O. A.; BARANENKOV, V. I.; WOLFE, S. *Tetrahedron* **1978**, *34*, 2953. (c) ZEFIROV, N. S.; SAMOSHIN, V. V.; PALYULIN, V. A. *Zh. Org. Khim.* **1983**, *19*, 1888.

11 Reviews: (a) WOLFE, S. *Acc. Chem. Res.* **1972**, *5*, 102. (b) ZEFIROV, N. S. *Tetrahedron* **1977**, *33*, 3193.

12 (a) ALLEN, L. C. *Chem. Phys. Lett.* **1968**, *2*, 597. (b) JUARISTI, E. *Introduction to Stereochemistry and Conformational Analysis*; Wiley-Interscience: New York, 1991, Chapter 18.

13 (a) EPIOTIS, N. D.; CHERRY, W. R.; SHAIK, S.; YATES, R. L.; BERNARDI, F. *Top. Curr. Chem.* **1977**, *70*, 1. (b) BRUNCK, T. K.; WEINHOLD, F. *J. Am. Chem. Soc.* **1979**, *101*, 1700. (c) DIONNE, P.; ST.-JACQUES, M. *J. Am. Chem. Soc.* **1987**, *109*, 2616. (c) JUARISTI, E.; ANTUMEZ, S. *Tetrahedron* **1992**, *48*, 5941.

14 PAQUETTE, L. A.; BRANAN, B. M.; ROGERS, R. D.; BOND, A. H.; LANGE, H.; GLEITER, R. *J. Am. Chem. Soc.* **1995**, *117*, 5992.

15 PAQUETTE, L. A.; BRANAN, B. M.; FRIEDRICH, D.; EDMONDSON, S. D.; ROGERS, R. D. *J. Am. Chem. Soc.* **1994**, *116*, 506.

16 BRANAN, B. M.; PAQUETTE, L. A. *J. Am. Chem. Soc.* **1994**, *116*, 7658.

17 PAQUETTE, L. A.; BRANAN, B. M. *Heterocycles* **1995**, *40*, 101.

18 PAQUETTE, L. A.; BRANAN, B. M.; STEPANIAN, M. *Tetrahedron Lett.* **1996**, *37*, 1721.

19 NEGRI, J. T.; ROGERS, R. D.; PAQUETTE, L. A. *J. Am. Chem. Soc.* **1991**, *113*, 5073.

20 PAQUETTE, L. A.; TAE, J.; HICKEY, E. R.; TREGO, W. E.; ROGERS, R. D. manuscript in preparation.

21 (a) COREY, E. J.; CARPINO, P. *Tetrahedron Lett.* **1990**, *31*, 7555. (b) BURGESS, K.; CASSIDY, J.; HENDERSON, I. *J. Org. Chem.* **1991**, *56*, 2050.

22 PAQUETTE, L. A.; TAE, J.; HICKEY, E. R.; ROGERS, R. D. *Angew. Chem. Int. Ed.* **1999**, *38*, 1409.

23 CRAM, D. J.; LEIN, G. M. *J. Am. Chem. Soc.* **1985**, *107*, 3657.

24 LEHN, J.-M.; SAUVAGE, J. P. *J. Am. Chem. Soc.* **1975**, *97*, 6700.

25 PAQUETTE, L. A.; TAE, J.; BRANAN, B. M.; EISENBERG, S. W. E.; HOFFERBERTH, J. E. *Angew. Chem. Int. Ed.* **1999**, *38*, 1412.

26 PAQUETTE, L. A.; BRANAN, B. M.; BOLIN, D. G., manuscript in preparation.

27 PAQUETTE, L. A.; TAE, J.; BRANAN, B. M.; EISENBERG, S. W. E., manuscript in preparation.

28 PAQUETTE, L. A. (review) in: Recent Research Developments in Chemical Sciences, Transworld Research Network, Trivandrum, India, 1977, p. 1.

29 PAQUETTE, L. A.; STEPANIAN, M.; MALLAVADHANI, U. V.; CUTARELLI, T. D.; LOWINGER, T. B.; KLEMEYER, H. J. *J. Org. Chem.* **1996**, *61*, 7492.

30 CAHIEZ, G.; ALEXAKIS, A.; NORMANT, J. F. *Tetrahedron Lett.* **1978**, 3013.

31 PAQUETTE, L. A.; BOLIN, D. G.; STEPANIAN, M.; BRANAN, B. M.; MALLAVADHANI, U. V.; TAE, J.; EISENBERG, S. W. E.; ROGERS, R. D. *J. Am. Chem. Soc.* **1998**, *120*, 11603.

32 PAQUETTE, L. A.; STEPANIAN, M.; BRANAN, B. M.; EDMONDSON, S. E.; BAUER, C. B.; ROGERS, R. D. *J. Am. Chem. Soc.* **1996**, *118*, 4504.

33 FITJER, L.; STEENECK, C.; GAINI-RAHIMI, S.; SCHRÖDER, U.; JUSTUS, K.; PUDER, P.; DITTMER, M.; HASSLER, C.; WEISER, J.; NOLTEMEYER, M.; TEICHERT, M. *J. Am. Chem. Soc.* **1998**, *120*, 317 and relevant references therein.

34 LEE, H. W.; KISHI, Y. *J. Org. Chem.* **1985**, *50*, 4402.

35 TAE, J. unpublished results, The Ohio State University.

36 ALTMANN, K.-H.; IMWINKELREID, R.; KESSELRING, R.; RIKS, G. *Tetrahedron Lett.* **1994**, *35*, 7625.

37 (a) MARQUEZ, V. E.; SIDDIQUI, M. A.; EZZITOUNI, A.; RUSS, P.; WANG, J.; WAGNER, R. W.; MATTEUCCI, M. D. *J. Med. Chem* **1996**, *39*, 3739 and earlier papers cited therein. (b) MARQUEZ, V. E.; EZZITOUNI, A.; RUSS, P.; SIDDIQUI, M. A.; FORD, H., JR.; FELDMAN, R. J.; MITSUYA, H.; GEORGE, C.; BARCHI, J. J., JR. *J. Am. Chem. Soc.* **1998**, *120*, 2780.

38 (a) TARKOY, M.; BOLLI, M.; LEUMANN, C. J. *Helv. Chim. Acta* **1994**, *77*, 716. (b) STEFFENS, R.; LEUMANN, C. J. *J. Am. Chem. Soc.* **1997**, *119*, 11548.

39 PRATVIEL, G.; BERNADOU, J.; MEUNIER, B. *Angew. Chem. Int. Ed. Engl.* **1995**, *34*, 746.

40 VON SONNTAG in: The Chemical Basis of Radiation Biology, Taylor and Francis, London, 1987.

41 STUBBE, J.; KOZARICH, J. W. *Chem. Rev.* **1987**, *87*, 1107.

42 GOLDBERG, I. H. *Acc. Chem. Res.* **1991**, *24*, 191.

43 NICOLAOU, K. C.; DAI, W.-M. *Angew. Chem., Int. Ed. Engl.* **1991**, *30*, 1387.

44 ESCHENMOSER, A.; DOBLER, M. *Helv. Chim. Acta* **1992**, *75*, 218.

45 UHLMANN, E.; PEYMAN, A. *Chem. Rev.* **1990**, *90*, 543.

46 MILLIGAN, J. F.; MATTEUCCI, M. D.; MARTIN, J. C. *J. Med. Chem.* **1993**, *36*, 1923.

47 OWEN, D. unpublished observations, The Ohio State University.

48 LANTER, J. C. PH.D. Dissertation, The Ohio State University, 1997.

49 JOHNSON, C. R. *Aldrichimica Acta* **1985**, *18*, 3.

50 VORBRÜGGEN, H.; RÜH-POHLENZ, C. *Org. React.* **1999**, Vol. 55.

51 BIBART, R. T. unpublished results, The Ohio State University.

52 TAKAYAMA, K. M.; INOUE, M. *Crit. Rev. Biochem.* **1990**, *25*, 155.

53 DELIHAS, N. *Mol. Microbiol.* **1995**, *15*, 411.

54 MONIA, B. P.; LESNIK, E. A.; GONSALEZ, C.; LIMA, W. F.; MCGEE, D.; GUINOSSO, C. J.; KAWASAKI, A. M.; COOK, P. D.; FREIER, S. M. *J. Biol. Chem.* **1993**, *268*, 14514.

4
Heterogeneous Catalysis: from "Black Art" to Atomic Understanding

Gerhard Ertl

4.1
Introduction

A series of observations in which the presence of an apparently unaffected substance initiated the progress of a chemical reaction prompted the Swedish chemist Berzelius in 1835 to introduce the term "catalyst", but it was only towards the end of the last century that a clear definition could be given by Ostwald, namely "a catalyst is a substance which affects the rate of a chemical reaction without appearing in the final products". This is achieved through the formation of intermediate compounds between the catalyst and the molecules involved in the reaction whereby an alternative path is offered which may be passed through with higher probability, i.e.

Heterogeneous Catalysis

$$\frac{dn_i}{dt} \longrightarrow \boxed{} \longrightarrow \frac{dn_i'}{dt}\ \frac{dn_j}{dt}$$

Reactor

i : Reactants
j : Products

Steady-state reaction rate:

$$\frac{dn_j}{dt} = v = f\,(p_i,\, p_j,\, T,\, catalyst)$$

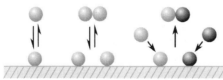

non-diss.   dissociative

Chemisorption   Surface reaction

Fig. 4.1 Elementary processes involved in heterogeneous catalysis (schematic).

enhanced rate. In heterogeneous catalysis this role is played by the interface between the (solid) catalyst and the adjacent (gaseous or liquid) phase containing the reacting species, and the elementary processes involved are sketched schematically in Fig. 4.1. Today heterogeneous catalysis underlies the cleaning of the exhausts from cars and industrial plants as well as the technologies for the production of chemicals and fuels. Despite this enormous economic significance our knowledge about the molecular processes is rather limited and practical catalysis still exhibits some aspects of "black art". [1,2] Progress in our understanding of the principles governing heterogeneous catalysis has, however, been made in recent years through the so-called "surfaces science approach" in which detailed investigations were made by applying the arsenal of modern surface physical methods to well-defined single crystal surfaces under high vacuum conditions acting as model systems [3] as will be illustrated by a few selected examples.

4.2
A Case Study: Ammonia Synthesis

One of the most important and also oldest industrial processes based on heterogeneous catalysis is the synthesis of ammonia from hydrogen and nitrogen, $N_2 + 3 H_2 \rightarrow 2 NH_3$. [4–6] The current worldwide annual capacity of about 150 million tons is largely (87 %) used for the production of fertilizers while the rest serves for the formation of a large variety of nitrogen-containing technical products. This process was first realized on a laboratory scale in 1909 by F. Haber, and only four years later the first industrial plant developed by C. Bosch started its operation. For this purpose a catalyst found by A. Mittasch by testing thousands of different samples was employed. Most remarkably, essentially the same kind of catalyst is still in use in almost all industrial plants. This catalyst is prepared from magnetite, Fe_3O_4, with small amounts of aluminum, potassium and calcium oxides as additives which improve the activity as "promoters". Only quite recently a catalyst based on ruthenium was found to be a possible alternative, but has so far been installed only in a single plant. [7] In the latter case [8] as well as for the iron-based "traditional" catalyst [9,10] the reaction mechanism could be elucidated in some detail through extended experimental as well theoretical efforts using the surface science approach. Even more, the kinetic parameters derived from these model studies served to successfully evaluate the ammonia yields for the conditions of "real" catalysis. [11–13] This process thus represents one of the rare examples of a catalytic reaction for which full mechanistic understanding could be achieved.

The sequence of reaction steps and the schematic energy diagram for the progress of the reaction on the iron-based catalyst are depicted in Fig. 4.2. [9] The decisive (rate-determining) step is the dissociative chemisorption of nitrogen, whereby splitting of the (strong) N-N bond is achieved through the formation of two (weaker) metal-N bonds. The resulting atomic configurations on an Ru(0001) single crystal surface for two different concentrations (coverages) are shown in Fig. 4.3 together with the change of the electronic density of states derived from density functional

$$N_2 + 2* \rightleftharpoons 2N-*$$

$$N-* + H-* \rightleftharpoons NH-* + *$$

$$NH-* + H-* \rightleftharpoons NH_2-* + *$$

$$NH_2-* + H-* \rightleftharpoons NH_3-* + *$$

$$NH_3-* \rightleftharpoons NH_3 + *$$

$$H_2 + 2* \rightleftharpoons 2H-*$$

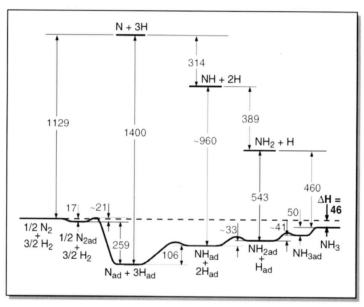

Fig. 4.2 Mechanism and energy diagram for ammonia synthesis on iron. (Energies in kJ/mole). [9]

theory (DFT) calculations which reveal that bonding is essentially achieved by formation of a bonding orbital derived from metallic d-states and N 2p-levels. [14] Also the variation of the bond strength with coverage as experimentally probed by thermal desorption spectroscopy [15] was reproduced by theory and found to be in agreement with the properties of a "real" supported Ru catalyst, [16] whereby the "materials" as well as the "pressure gap" could be bridged.

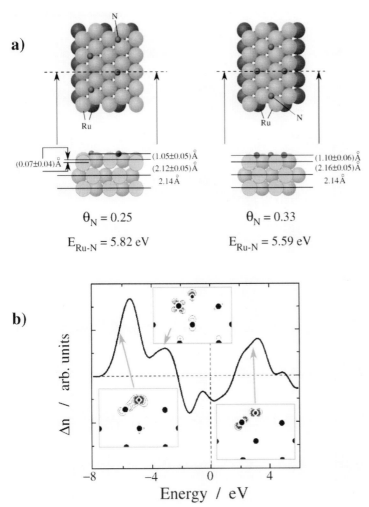

Fig. 4.3 (a) Structures of chemisorbed N on an Ru(0001) surface at coverages $\theta_N = 0.25$ and 0.33. (b) Calculated change of the electronic density of states upon N chemisorption and side views of the orbitals at the energies marked by arrows. [14]

4.3
The Surface Science Approach

Figure 4.4 shows a section of a perfect single crystal surface [such as Pt(111)] which is approached by a diatomic molecule (say O_2) undergoing dissociative chemisorption. The progress of this process is illustrated by a contour plot of the energetics as a function of the distance x of the molecule from the surface and of the separation y between the two atoms, together with the well-known one-dimensional Lennard-Jones potential diagram. (The molecular axis is assumed to be parallel to the surface

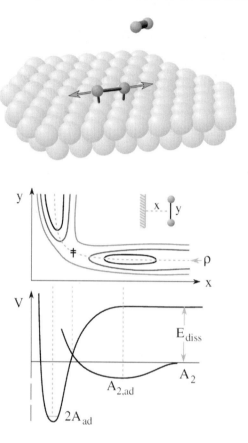

Fig. 4.4 Principle of dissociative chemisorption of a diatomic molecule.

and to exhibit the optimum configuration. A full description would require 6 coordinates, as can now be achieved by advanced theoretical methods.) [17] The molecule forms at first a relatively weak bond with the surface from where dissociation takes place by overcoming an activation barrier ($\neq$). Scanning tunneling microscopy (STM) enables direct imaging of the result on atomic scale. This is shown in Fig. 4.5 for the case of a small exposure of O_2 to a Pt(111) surface at 165 K. [18] The image exhibits pairs of additional features on the atomically resolved substrate surface which represent chemisorbed O atoms. Obviously the activation energy for dissociation is so small that this process takes place even at very low temperatures, which, on the other hand, suppress any mobility of the O atoms formed. Interestingly the latter are not located on neighboring sites but are separated from each other by about 0.5 to 0.8 nm, while the O-O separation in the molecule is only of the order of 0.1 nm. This effect is a consequence of the fact that the energy released by this process is preferentially channeled into the separation y parallel to the surface (cf. Fig. 4.4). A simple estimate reveals that the time necessary to damp this excess energy of the "hot" adatoms into the heat bath of the solid is of the order of 10^{-12} s, i. e. comparable with the period of a molecular vibration.

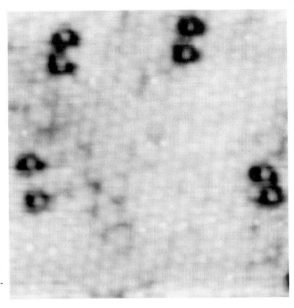

Fig. 4.5 Scanning tunneling microscopy (STM) image from a Pt(111) surface after dissociative chemisorption of O_2. [18]

Further inspection of Fig. 4.5 demonstrates that the O atoms are surrounded by dark zones. The STM technique probes not only the atomic topography but also the electronic structure, and the dark zones reflect the modification of the local electronic density in the vicinity of the adsorbates, this modification being responsible for the operation of "indirect" interactions (which may be either repulsive or attractive) between adsorbed particles mediated through the substrate.

In the present example obviously the whole surface area (exhibiting the perfect densely packed configuration of Pt atoms) is active in dissociating the molecules. In general, however, this activity will be influenced by the local coordination of the surface atoms giving rise to the "structure sensitivity" in catalysis. An impressive example is offered by NO interacting with an Ru(001) surface. Figure 4.6a shows an STM image recorded 6 min after exposure of this surface to a small amount of NO at room temperature. [19] The dark line represents a monoatomic step running across the otherwise perfect (0001) surface. (Such steps are the most abundant structural imperfections on "flat" surfaces.) Near the step, dark pointlike features are visible which can be identified with chemisorbed N atoms resulting from dissociation of NO. The O atoms, on the other hand, show up as faint streaks also further away from the step: obviously dissociation takes place at the step from where the products diffuse away, as becomes evident from Fig. 4.6b recorded 120 min after gas exposure where the N atoms have spread further across the flat terraces. The mobility of the O atoms is much higher, so that they pass below the STM tip during scanning and give rise to the short streaks. Obviously the step atoms with their lower coordination (and hence higher degree of unsaturated valency) act as "active centres" – a concept introduced by H. S. Taylor as early as 1925. [20] Closer inspection reveals, however, that the situation is somewhat more complex. It turns out that the surface consists

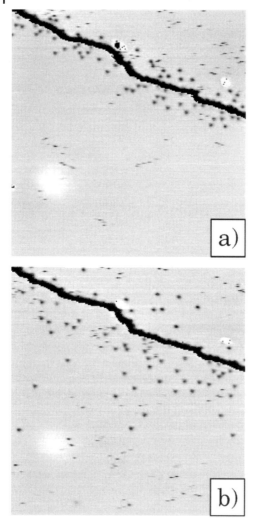

Fig. 4.6 STM images from an Ru(0001) surface with a monoatomic step after adsorption of NO at 300 K: (a) 6 min after exposure, (b) 120 min after exposure. [21]

in fact of an alternating sequence of less active (I) and more active (II) steps. For a given exposure to gaseous NO the number of N + O atoms per unit step length increases linearly with the widths of the adjacent terraces for steps II, while it is lower and independent of terrace width for steps I. It has to be concluded that the impinging NO molecules are (non-dissociatively) adsorbed on the whole surface from where they diffuse rapidly until they collide with a step where they may dissociate. Hence the increase of activity of steps II with terrace width. The O atoms formed occupy threefold coordinated sites with another Ru atom in the layer below. With steps II these sites are formed from one edge atom and two atoms from the second row while with steps I the situation is just opposite (Fig. 4.7). The latter thus exhibit a higher degree of unsaturation and bond the O atoms more strongly. Once

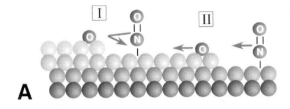

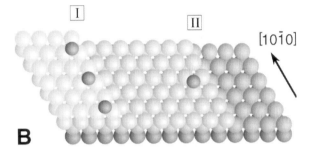

Fig. 4.7 Chemisorption of O atoms near the two types of monoatomic steps on an Ru(0001) surface

formed they do not diffuse away but block the "active center", while this inhibitory effect does not occur with steps II and enables a continuous activity.

Surface diffusion of adsorbates occurs through thermally activated hopping to one of the neighboring adsorption sites, and the residence time at a specific site may be expressed as $\tau = \tau_0 \exp(E^*_{diff}/k_BT)$, whereby the preexponential τ_0 is typically of the order of a vibrational period ($\sim 10^{-13}$ s) and E^*_{diff} is the activation energy for surface diffusion, which may be affected by interactions with other adsorbates as well as by structural defects. As a consequence, adsorbate diffusion is in general a quite complex phenomenon. For the present system the residence time for an isolated N adatom at 300 K is about 45 s and E^*_{diff} (N) = 0.95 eV. [21] The O atoms are much faster and their dynamics became experimentally accessible only by markedly increasing the scanning speed of the STM (to up to 20 images per second!): τ(O) = 0.07 s at 300 K, and E^*_{diff}(O) = 0.7 eV. [22]

The operation of interactions between adsorbed particles as a consequence of their influence on the local electronic structure of the substrate also affects their residence times. While an O atom on an Ru(0001) surface far (i.e. more than 3 lattice constants) away from another adatom stays there on the average for 0.07 s, this time increases to 0.22 s in the presence of another particle at distance of $2a_o$. [23] Such effects are quite general and usually cause the formation of ordered phases depending on coverage, such as in the present case a fluctuating 2 × 2-phase embedded into a quasi-gaseous diluted phase as depicted by the series of snapshots in Fig. 4.8. [22] The determination of the detailed structure can in such cases most conveniently be achieved by applying the technique of low energy electron diffraction (LEED). [24]

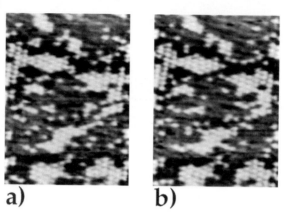

a) b)

Fig. 4.8 STM snapshots from an O-covered Ru(0001) surface ($\theta_O = 0.09$) recorded in an interval of 0.17 s. [22]

4.4
The Atomic Mechanism of a Catalytic Reaction: Oxidation of Carbon Monoxide

The oxidation of carbon monoxide, $2CO + O_2 \rightarrow CO_2$ is one of the reactions (apart from reduction of nitric oxides and oxidation of unburnt hydrocarbons) taking place in the car exhaust catalyst. The latter consists of small noble metal particles (Pt, Rh or Pd) on a ceramic support. The reaction proceeds through the following steps: [25]

$$CO + * = CO_{ad}$$
$$O_2 + 2* = O_{2,ad} \rightarrow 2\,O_{ad}$$
$$O_{ad} + CO_{ad} \rightarrow CO_2 + 2*$$

* denotes schematically a free adsorption site which has, however, a quite different meaning for the two surface species involved, CO_{ad} and O_{ad}, as becomes evident from inspection of the adsorbate structures.

Figure 4.9 shows the respective structures for a Rh(111) surface as determined by LEED. [26] (Similar geometries are found for other surfaces.) Figure 4.9a represents the configuration of adsorbed CO molecules for a coverage $\theta_{CO} = 1/3$. (The coverage denotes the ratio of the density of adsorbed particles to that of surface atoms.) The CO molecules are bound through the C atoms to single Rh atoms ("on top"). Although the coverage can become still higher, this arrangement represents an already fairly dense packing if the van der Waals radius of CO is taken into account. As a consequence, a surface at this stage is no longer able to chemisorb oxygen. O_{ad} is required for the reaction, but its formation is inhibited by the second reactant unless the temperature is high enough ($\geq 200\,°C$) to give continuous desorption of CO, enabling O_2 adsorption sites to become available. That is why the car exhaust catalyst is inefficient as long as it is cold.

The situation is quite different for oxygen adsorption. The O_{ad} atoms form the relatively open mesh of a 2×2 structure (Fig. 4.9b) where these are located in three-fold coordinated sites separated from each other by $2a_o = 0.56$ nm. At this stage further uptake of oxygen is pronouncedly slowed down, but the surface is still read-

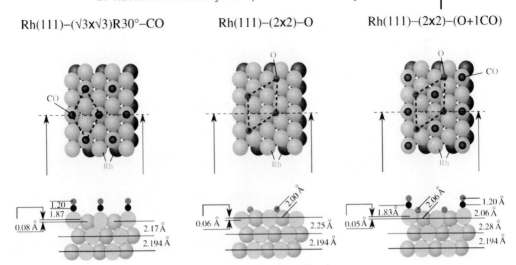

Rh(111)–(√3x√3)R30°–CO   **Rh(111)–(2x2)–O**   **Rh(111)–(2x2)–(O+1CO)**

Fig. 4.9 Structures of ordered phases on an Rh(111) surface formed by chemisorption of CO (a), O (b), and coadsorption of O + CO (c). [26]

ily able to adsorb CO as becomes evident from Fig. 4.9c. In this mixed phase a CO molecule occupies the empty "on top" position within the unit cell. Reaction may proceed from such a configuration and takes at 200 °C (on the average) about 10^{-4} s, whereby the CO_2 formed is spontaneously released into the gas phase and the empty adsorption sites are again available for reaction.

Figure 4.10 reproduces the result of a theoretical treatment illustrating the progress of the elementary process $O_{ad} + CO_{ad} \rightarrow CO_2$ on a Pt(111) surface. [27] The

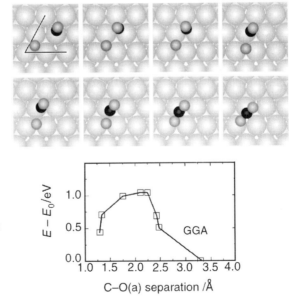

Fig. 4.10 Theoretical treatment of the progress of the reaction $O_{ad} + CO_{ad} \rightarrow CO_2$ on a Pt(111) surface. The inset represents the variation of the total energy with C-O separation for the various configurations shown. [27]

inset shows the variation of the total energy for the different configurations, rationalizing the occurrence of an activation barrier prior to product formation.

Experimentally, the kinetics of this reaction has been monitored by STM as reproduced in Fig. 4.11. [28] A Pt(111) surface precovered with a substantial amount of O_{ad} (a) was exposed to a very low partial pressure (5×10^{-8} mbar) of CO. The latter molecules were adsorbed within the 2×2 O unit cells as well as in the empty space between the O_{ad} patches. The progress of the reaction is reflected by the continuous shrinking of the O_{ad} islands. Since the two adsorbates form different ordered phases

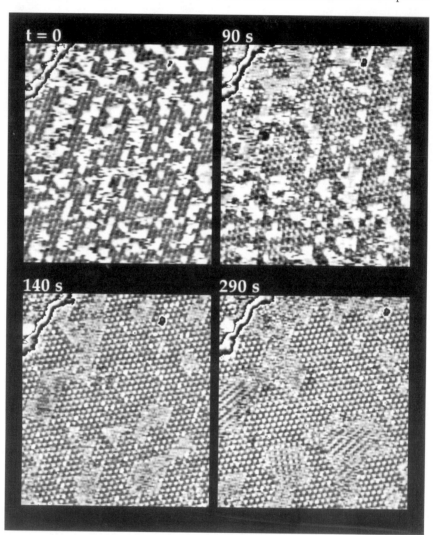

Fig. 4.11 A series of STM images recorded during progressing reaction $O_{ad} + CO_{ad} \rightarrow CO_2$ on a Pt(111) surface at 247 K. The O_{ad}-precovered surface ($t=0$) was continuously exposed to CO with a partial pressure of about $5 \cdot 10^{-8}$ mbar, corresponding to an impingement rate of about 1 monolayer per 100 s. Image size 180 Å. [28]

they can also be readily identified, and determination of the reaction rate on the atomic scale becomes feasible by simply counting the number of O_{ad} particles over a larger area (whose ratio yields the coverage θ_O) and its decrease with time.

The usual concept of chemical kinetics formulates the rate of the reaction $CO_{ad} + O_{ad} \rightarrow CO_2$ in terms of the concentrations (coverages) as $r = -\frac{d\theta_O}{dt} = k \cdot \theta_O \cdot \theta_{CO}$, and hence the ratio $r/\theta_O \times \theta_{CO}$ should be equal to the rate constant k (which depends only on temperature). The result of such an analysis of the data of Fig. 4.11 is plotted in Fig. 4.12 as a function of time (=progress of reaction) as a dashed line which obviously by no means yields a constant. The reason for the failure of this approach is quite obvious. It is that the underlying mean-field approximation is based on the assumption of a random distribution of reacting particles which might be fullfilled by gaseous or liquid systems but not by reactions on solid surfaces where the interactions between the adsorbed species cause the formation of ordered phases. With the present example the reaction occurs predominantly at the boundaries between CO_{ad} and O_{ad} patches, and the rate is rather expected to be proportional to the total length of domain boundaries L, i.e. $r = k' \times L$. If the experimental data are analyzed in this way the solid line is obtained for the ratio r/L, which is now indeed constant.

Under macroscopic conditions the information on the domain lengths is, of course, not available. Description of the kinetics on the basis of the mechanism of the reaction ("microkinetics") is instead based on correlations between the coverages of the various surface species and the external parameters of partial pressures and temperature. (This was also the approach underlying the successful modeling of ammonia synthesis mentioned above.) With the present system, the dashed line in Fig. 4.12 demonstrates that this approach fails for high (low t) and low O coverages (high t), but that, on the other hand, $r = k \cdot \theta_O \cdot \theta_{CO}$ is a good approximation over the intermediate coverage range. In previous molecular beam experiments an O-covered Pt(111) surface had been exposed to a modulated CO molecular beam and the formation of CO_2 was monitored by mass spectrometry. Analysis of the data in terms of the above mean-field approximation over the intermediate coverage range indeed yielded the same kinetic parameters (activation energy and preexponential) as the microscopic evaluation based on the STM data [29] and validates this approach.

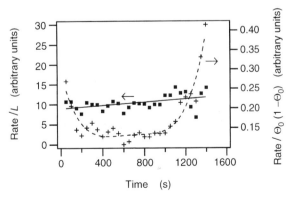

Fig. 4.12 Evaluation of the rate of CO_2 formation from STM images, normalized to the product of O and CO coverages which is equal to $\theta_O(1-\theta_O)$ (*crosses* and *dashed line*), as well as to the length of the boundaries between O and CO domains (*squares* and *solid line*). [28]

In "real" catalysis the actual situation will even be far more complex. Energetic heterogeneity due to the participation of various structural elements of the surface and interactions between adsorbed species are just a few of the complicating factors coming into play. Nevertheless it is concluded that adequate description of the kinetics may be achieved on the basis of the outlined strategy as long as the analysis is restricted to a limited range of parameters, which condition will frequently be fullfilled with practical reaction situations.

4.5
Further Aspects

The sketched examples represent just the tip of the iceberg called heterogeneous catalysis. An excellent account of the historical development as well as of the current state of the art can for example be found in a recent review by J. M. Thomas. [30] A few of the additional effects for which understanding can also be sought on the basis of the surface science approach will only be briefly listed:

- The occurrence of kinetic instabilities as well as oscillatory and even chaotic temporal behavior of a catalytic reaction under steady-state flow conditions can be traced back to the nonlinear character of the differential equations describing the kinetics coupled to transport processes (diffusion and heat conductance). Studies with single crystal surfaces revealed the formation of a large wealth of concentration patterns of the adsorbates on mesoscopic (say μm) length scales which can be studied experimentally by suitable tools and theoretically within the framework of nonlinear dynamics. [31]

- Since the bond between the surface and the adsorbates is of comparable strength to that between the surface atoms, the latter are usually displaced from their original positions (relaxation), or the geometry of the surface unit cell is altered (reconstructed) or even new crystal planes are formed (faceting). [32] The catalytic reaction digs its own bed. These effects will be more significant under the high pressure conditions of "real" catalysis and may represent a source for the "pressure gap". In situ investigations of the surface properties under such conditions become e.g. possible by novel optical techniques which recently came into use. [33, 34] The synthesis of chiral molecules is of enormous significance for the production of pharmaceuticals and may, for example, be achieved by nickel catalysts treated with tartrate. The mechanism of this enantioselective catalysis is still widely unclear and might well be due to a proper structural modification of the metal surface – an effect which still awaits detailed elucidation. [35]

- Catalysts consisting of more than one component are often superior to monometallic samples. Model studies with potassium on Fe surfaces revealed, for example, the role of the "electronic" promoter in ammonia synthesis. A particularly remarkable case was recently reported for a surface alloy formed by Au on a Ni(111) surface where the combination of STM experiments and theoretical studies led to the development of an improved practical catalyst for steam reforming. [36]

- Practical metal catalysts frequently consist of small metal particles on an oxide support. Suitable model systems can be prepared by growing small metal aggregates onto single crystal oxide films, a technique whereby the role of the particle size or of the support material may be studied. [37] A quite remarkable example of the variation of the catalytic activity with particle size has recently been found for finely dispersed Au on a TiO_2 support, which was revealed to be highly reactive for combustion reactions. [38] On the basis of STM experiments it was concluded that this phenomenon has to be attributed to a quantum size effect determined by the thickness of the gold layers.

Apart from metallic catalysts, there exist further important classes of catalytically active materials, including mixed oxides, shape-selective zeolites and other complex materials. [30] The development of new catalysts will also in the future be largely empirical, but will be guided by increasing insight into the underlying principles and elementary processes. The introduction of combinatorial techniques will possibly facilitate this search. [39] Heterogeneous catalysis will continue to be a challenge for future multidisciplinary research ranging from surface physics to chemistry and reaction engineering. Its economic importance is illustrated by Fig. 4.13, which shows the growing market for catalysts in the different fields of applications. [40]

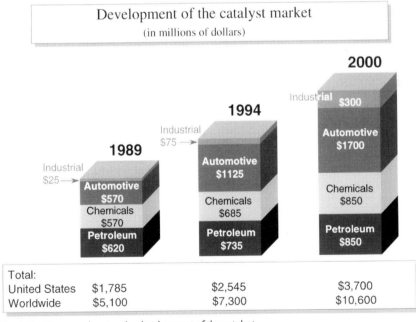

Fig. 4.13 Diagram showing the development of the catalyst market. [40]

References

1 J. M. THOMAS and W. J. THOMAS, "Principles and practice of heterogeneous catalysis", VCH, Weinheim (1996)

2 G. ERTL, H. KNÖZINGER and J. WEITKAMP (eds.), "Handbook of heterogeneous catalysis", Vols. 1–5, Wiley-VCH, Weinheim (1997)

3 G. A. SOMORJAI, "Introduction to surface chemistry and catalysis", Wiley, New York (1994)

4 J. R. JENNINGS (ed.), "Catalytic ammonia synthesis. Fundamentals and practice", Plenum Press, New York and London (1991)

5 M. APPL, "Ammonia, principles and industrial practice", Wiley-VCH, Weinheim (1998)

6 R. SCHLÖGL, in ref. [2], p. 1697

7 J. R. TENNISON, in ref. [4], p. 303

8 O. HINRICHSEN, F. ROSOWSKI, M. MUHLER and G. ERTL, *Chem. Eng. Sci.* 51 (1996), 1683

9 G. ERTL, in ref. [4], p. 109

10 J. J. MORTENSEN, L. B. HANSEN, B. HAMMER and J. K. NØRSKOV, *J. Catal.* (in press)

11 P. STOLTZE and J. K. NØRSKOV, *J. Catal.* 110 (1988), 1

12 M. BOWKER, I. PARKER and K. C. WAUGH, *Surf. Sci.* 197 (1988), L223

13 L. M. APARICIO and J. M. DUMESIC, *Top. Catal.* 1 (1994), 233

14 S. SCHWEGMANN, A. P. SEITSONEN, H. DIETRICH, H. BLUDAU, H. OVER, K. JACOBI and G. ERTL, *Chem. Phys. Lett.* 264 (1997), 680

15 H. DIETRICH, K. JACOBI and G. ERTL, *J. Chem, Phys.* 105 (1996), 8944

16 A. HORNUNG, M. MUHLER and G. ERTL, *Catal. Lett.* 53 (1998), 7

17 A. GROSS, *Surf. Sci. Rep.* 32 (1998), 291

18 J. WINTTERLIN, R. SCHUSTER and G. ERTL, *Phys. Rev. Lett.* 77 (1996), 123.

19 T. ZAMBELLI, J. WINTTERLIN, J. TROST and G. ERTL, *Science* 273 (1996), 1688

20 H. S. TAYLOR, *Proc. Roy. Soc.* A108 (1925), 105

21 T. ZAMBELLI, J. TROST, J. WINTTERLIN and G. Ertl, *Phys. Rev. Lett.* 76 (1996), 795

22 J. WINTTERLIN, J. TROST, S. RENISCH, R, SCHUSTER, T. ZAMBELLI and G. ERTL, *Surf. Sci.* 394 (1997), 159.

23 S. RENISCH, R. SCHUSTER, J. WINTTERLIN and G. ERTL, to be published

24 G. ERTL and J. KÜPPERS, "Low energy electrons and surface chemistry" 2nd ed., VCH, Weinheim (1985).

25 T. ENGEL and G. ERTL, *Adv. Catal.* 28 (1979), 1.

26 S. SCHWEGMANN, H. OVER, V. DE RENZI and G. ERTL, *Surf. Sci.* 375 (1997), 91.

27 A. ALAVI, P. HU, T. DEUTSCH, P. L. SILVESTRELLI and J. HUTTER, *Phys. Rev. Lett.* 80 (1998), 3650.

28 J. WINTTERLIN, S. VÖLKENING, T.V.W. JANSSENS, T. ZAMBELLI and G. ERTL, *Science* 278 (1997), 1931

29 C. T. CAMPBELL, G. ERTL, J. KUIPERS and J. SEGNER, *J. Chem. Phys.* 73 (1980), 5862.

30 J. M. THOMAS, *Angew. Chem. Int. Ed. Engl.* 33 (1994), 913

31 R. IMBIHL and G. ERTL, *Chem. Rev.* 95 (1995), 697

32 G. A. SOMORJAI, *J. Mol. Catal.* A107 (1996), 39

33 D. W. GOODMAN, *J. Phys. Chem.* 100 (1996), 13090

34 P. L. J. GUNTER, J. W. NIEMANTSVERDRIET and G. A. SOMORJAI, *Catal. Rev.* 39 (1997), 77

35 A. Baiker and H. U. Blaser, in ref. [2], p. 2422

36 D. W. Goodman, *J. Mol. Catal.* A119 (1997), 307

37 H. J. Freund, *Angew. Chem. Int. Ed. Engl.* 36 (1997), 452.

38 M. Valden, X. Lai and D. W. Goodman, *Science* 281 (1998), 1647

39 R. Schlögl, *Angew. Chem. Int. Ed. Engl.* 37 (1998), 2333

40 J. A. Cusumano, in "Perspectives in catalysis", J. M. Thomas and K. I. Zamaraev, eds., Blackwell Oxford (1992), p. 1.

5

Drugs for a New Millennium

Solomon H. Snyder

5.1
Introduction

A millennium, a century, even a decade is a long time frame for speculation about anything. Advances in biomedical research in the last few decades have been so extraordinary and escalating at an ever-accelerating pace that any prophecy is a risky proposition. However, it is possible to divine the big unanswered questions and envisage ways in which they might reasonably be approached in the next few decades, a task which I will try to essay.

So many drugs treat so many different medical conditions that a detailed and comprehensive coverage would likely be tiresome. Instead, I will address certain broad themes and diseases that offer both immense challenges and great potential for advances. Rather than review detailed experimental issues, I will confine myself to the "big picture" issues, providing examples of specific research only in a few instances, drawing largely from areas I know best.

5.2
Cell Death

A simple-minded approach to future drug development would be to assume that we will find the molecular etiology of all diseases, which in many instances will involve specific abnormalities in one gene or a group of genes. Treatment would then consist of replacing the missing or abnormal gene. While this approach may ultimately work in a good number of instances, I suspect that we may be able to make major inroads without such information. Moreover, knowing the causes of diseases may not do much for cures. Let me provide a few examples. Immunization is one of the most important advances in all medicine. We now approach immunization by isolating the offending organism, cloning the genes for key proteins and then developing vaccines based on these proteins or related fragments. However, the giant breakthroughs took place hundreds of years ago prior to the isolation of any organism. Vaccination commenced with tissue extracts of individuals infected with cowpox long before cowpox, smallpox or any viruses were understood. The major antibiotics were developed without any understanding of how they might kill bacteria. Similarly, the major drugs in psychiatry emerged from serendipitous events in the absence of molecular insight.

Cell death is fundamental to diseases that afflict the largest number of people and for which treatment is presently gravely inadequate. I am thinking primarily of cancer as well as stroke and neurodegenerative diseases. In the case of cancer, we wish to kill malignant but not normal cells. In the case of stroke and neurodegenerative diseases as well as myocardial infarcts, we wish to prevent cell death. Accordingly, understanding what is going on in cell death and developing agents that will augment or inhibit the death process may provide therapeutic agents even when we do not know the "causes" of the individual disease entities.

The decade of the 1990s has witnessed an explosion in cell death research. One of the first advances was the appreciation that there are at least two distinct ways in which cells die. One of these is called programmed cell death or apoptosis. The other is un-programmed cell death, generally referred to as necrosis. The best known examples of apoptotic cell death occur during normal developmental processes. For instance, in the brain about twice as many neurons are generated in embryonic life than can be accommodated in the adult brain. Neurons that successfully make contact with their target sites survive while the others die by apoptosis. Similar processes take place in other parts of the body. There is much debate as to the type of cell death that occurs in pathologic conditions such as stroke or myocardial infarct. One simple-minded formulation posits that cells will die by an apoptotic route if they have leisure, while necrosis takes place if the tissue insult is overwhelming and there is no time for the delicately coordinated apoptotic path.

In apoptosis a series of events takes place in an orderly sequence involving the activation of various proteases which are called caspases, for cysteine and aspartate proteases. Several distinct caspases act in a cascade vaguely reminiscent of the blood-clotting cascade of complement proteins. If one wishes to interfere with the apoptotic process, then one strategy would be to develop drugs that inhibit various caspases, a current effort underway in the pharmaceutical industry.

Besides the caspases, other proteins either augment or inhibit the death process. The first well-characterized one was an oncogene called bcl-2. Bcl-2 is anti-apoptotic. Other proteins in the bcl-2 and related families are pro-apoptotic. Accordingly, if one wishes to kill cancer cells, a productive approach might be to develop drugs that will block the effects of bcl-2, while if one wishes to preserve certain cells, antagonists of the "death genes" would be indicated. Like the caspase area, drug development based on the bcl-2 family is brisk.

5.3
Stroke and Myocardial Infarct

While stroke and myocardial infarct involve very different organs, they display notable similarities. Most prominently, both are usually caused by clots that occlude blood vessels. Moreover, in both of these conditions efforts to modulate cell death offer therapeutic promise.

Stroke is one of the three leading causes of death in Western societies, trailing only cancer and myocardial infarct. It is one of the few major diseases for which there remains virtually no treatment. In recent years the clot-dissolving protein tissue plasminogen activator (TPA) has been approved for the acute treatment of stroke. TPA dissolves clots in cerebral vessels just as in coronary vessels. However, TPA can prevent myocardial damage when administered up to half a day after the onset of coronary symptoms, but is effective in stroke victims only if given within three hours of the first symptoms. While patients with acute chest pain hasten to the hospital, initial symptoms of stroke are typically vague involving dizziness and minor confusion which in most patients has happened in the past and has gone away without any treatment. Hence, most stroke patients do not enter the hospital for a day or two. TPA augments bleeding so that damage caused by vascular occlusion is sometimes made worse by TPA-induced hemorrhage. Additionally, many strokes are caused initially by hemorrhage rather than clots in cerebral vessels.

Because of these considerations, stroke presents a major therapeutic lacuna. When I was a medical student, it was thought that blood vessel occlusion in both stroke and myocardial infarct is followed by rapid death of the oxygen-deprived, hypoxic tissue so that treatment is only palliative. Greater optimism has emerged in the last two decades as we have learned that the major tissue damage involves cell death that occurs gradually over a day or two. In stroke, various chemical changes associated with tissue hypoxia trigger a massive release of the amino acid glutamate (Fig. 5.1; Lipton and Rosenberg, 1994). Besides its role as a constituent of protein and a key player in intermediary metabolism, glutamate is a neurotransmitter. Glutamate is the major excitatory neurotransmitter and likely the most abundant of all transmitters. In animals, strokes lead to a 50–100 fold increase in the release of glutamate. It is felt that this glutamate almost literally excites partially hypoxic cells to death. Evidence for this notion came with the development of drugs that block receptors for glutamate and which in animals reduce neural damage in stroke by 50–60 %. Some of these drugs are in clinical trial so that in the early decades of the next millennium we may anticipate major benefit. Glutamate synapses are regulated by a complex array of proteins and small molecules, several of which are appropriate targets for drug development. Hence, even if initial glutamate antagonists are not effective drugs, I am confident that this synaptic constellation will lead to valuable treatments.

Since glutamate mediates excitation under physiologic as well as pathologic circumstances, glutamate antagonists may also be useful hypnotic anti-convulsants and anti-anxiety drugs, while substances mimicking glutamate may be effective cognition enhancers and general stimulants.

Exactly how does the "over-excitation" by glutamate kill cells? A variety of research indicates that a novel gaseous neurotransmitter, nitric oxide, is responsible for doing much of the dirty work of glutamate in stroke and other forms of neurotoxicity (Fig. 5–1; Lincoln et al., 1997). The nitric oxide neurons have receptors for glutamate, and glutamate stimulates the formation of nitric oxide, which can be released to adjacent cells. Drugs that block the formation of nitric oxide prevent stroke damage. Moreover, mice with targeted deletion of the gene for nitric oxide formation are

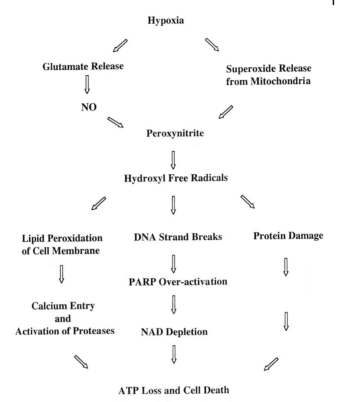

Fig. 5.1 Mechanisms of neuronal damage in stroke.

resistant to stroke damage. Hence, blocking nitric oxide formation may be as good or even better than blocking glutamate effects, especially as nitric oxide is closer to the death process than glutamate.

But substituting nitric oxide for glutamate as a proximal culprit in neuronal death merely begs the question as to "what kills?" Nitric oxide itself has properties that fit with an executioner's role. Nitric oxide is a free radical; hence it is highly reactive chemically and capable of attacking proteins and other macromolecules. However, nitric oxide is not a particularly lethal free radical. Its ability to kill derives from its interactions with other molecules, especially superoxide. Superoxide is an altered form of oxygen that leaks out of partially hypoxic mitochondria. Superoxide combines quite readily with nitric oxide to form a far more toxic molecule called peroxynitrite. Peroxynitrite in turn degenerates into the hydroxyl free radical, one of the most toxic of all such chemicals. There is an abundance of solid experimental evidence indicating that peroxynitrite and hydroxyl free radicals are the key agents that do the killing.

Even in the absence of nitric oxide, hypoxic mitochondria will generate lethal free radicals. Superoxide can evolve into hydroxyl free radical without the intervention of nitric oxide. Hence, the machinery for cell death via free radicals is present in non-neural tissues like heart, that are never exposed to the neurotransmitter actions of nitric oxide. Heart researchers have appreciated for some time that a major portion of the damage following coronary artery occlusion does not simply represent hypoxic tissue dying on the spot. Instead, following the occlusion, blood from collateral vessels pours into the damaged area so that the partially hypoxic tissue is overwhelmed with a superabundance of oxygenation. This leads to high densities of oxygen free radicals, which kill gradually over a period of many hours in a situation analogous to stroke. Superoxide has been extensively implicated in myocardial infarct as well as in stroke. Evidence for this includes the finding that infusion of the enzyme superoxide dismutase, which destroys superoxide, reduces myocardial infarct damage in both animals and humans.

Thus, both in myocardial infarct and in stroke, free radical damage is crucial. How do the free radicals kill cells? Here things become murkier. When a cell dies, everything falls apart. The cell membrane disintegrates as oxidation damages the lipids that are key components of the membranes. Cell membrane barriers, which normally keep ions such as calcium outside of cells, become weakened so that calcium rushes into the cell and can cause damage by activating protein-degrading enzymes, proteases, which damage crucial proteins in cells. Free radicals also damage DNA.

Is there any way of distinguishing between the relative importance of these cellular executioners? Much research has focused on calcium-activated proteases such as calpain. Drugs that inhibit these enzymes do offer protection against cell damage in a variety of experimental models. Notable success has also come from exploring the consequences of DNA damage. Coping with DNA damage presents some challenging choices to cells. Limited amounts of DNA damage occur constantly, so that cells have evolved extensive enzymatic systems to repair the damage. However, excess damage may not be worth repairing. By analogy, if an automobile has sustained excess damage, it is better to discard the car rather than fix it with the attendant risk of unrepaired defects. Similarly, efforts to repair large amounts of DNA damage in cells may be imperfect, leading to cancerous changes.

Nature seems to have evolved a mechanism which initiates the repair of modest amounts of DNA damage, while the same system destroys cells with excess damage. The mechanism involves an enzyme in the nucleus called poly(ADP-ribose)polymerase (PARP) (Fig. 5.1; Pieper et al., 1999). PARP has negligible activity under basal circumstances but is activated by DNA damage. PARP utilizes the oxidative co-factor NAD to donate groups called ADP-ribose to a variety of nuclear proteins, especially PARP itself. These activities expose the damaged DNA molecules to other more conventional repair enzymes. When excess damage to DNA takes place, PARP is massively activated. Its heightened activity consumes the great bulk of NAD in cells. In order to replace the missing NAD, ATP, the universal energy donor, is consumed, and the cell dies of energy depletion. While PARP over-activation may be useful in eliminating cells with excess DNA damage, its over-activation in stroke and myocardial infarction might contribute to cell death by energy depletion. Evidence in favor of this notion comes from findings that PARP inhibitors can reduce stroke and myocardial infarct damage in experimental animals substantially. Moreover, mice with targeted deletion of the gene for PARP display 60–80% reductions in stroke damage. PARP inhibitor drugs may be important therapeutic agents and are being explored by the pharmaceutical industry.

5.4
Schizophrenia

Emotional distress is so prevalent that, depending on one's definition of psychiatric illness, more than half of the population has been afflicted at one time or another in their lives. The psychiatric diagnostic and statistical manual contains a broad array of diagnoses. However, one can subsume most of these under three major headings, schizophrenia, manic-depressive illness, and anxiety-related disorders. Neurotic and anxiety-associated conditions are most common, especially if one is fairly broad in diagnostic definition. Depression is also prevalent, with serious forms afflicting 10% or more of the population. Rigorous definitions of schizophrenia lead to an incidence of about 1–2% of the population. Thus, schizophrenia is one of the less common psychiatric disorders, with an incidence similar to diabetes. However, it is the most devastating psychiatric disease. Schizophrenia typically appears in late adolescence or early adult life. Victims are so disabled that genuinely productive employment is often not possible for most of their lives. The toll in terms of anguish to the patients and their families is incalculable.

The symptoms of schizophrenia are so foreign to most of us that it is difficult to appreciate their horror. Patients feel that they have lost control of their minds with thoughts being inserted into their brains. Their delusions are usually terrifying. The common auditory hallucinations are horrifically accusatory. Many psychiatrists characterize the entire adult lives of schizophrenics as a living death. The behavior of schizophrenics is so disruptive that lives of their parents and siblings are often devastated. Thus, whether calculated in terms of dollars lost to a country's economy or human suffering, schizophrenia may well rank as the number one mental illness.

What is to be done? A plethora of genetic studies over the last fifty years has established a powerful genetic predisposition for the disease. Some of the best evidence comes from twin studies. Monozygotic or identical twins have identical genes whereas dizygotic or fraternal twins are no more similar genetically than siblings. Twin studies provide a useful approach to separating environmental from genetic factors. If one of a pair of monozygotic twins is schizophrenic, the likelihood of the other twin also developing schizophrenia is 60–80%. For dizygotic twins the comparable "concordance" rate is only about 15–20%. Studies of monozygotic twins separated at birth have also been informative. If one co-twin becomes schizophrenic, the odds are greater than 50% that the other will also succumb to the disease.

Many research groups have been searching for the presumed abnormal gene(s). Strategies employed are similar to those that have successfully identified aberrant genes in diseases such as cystic fibrosis and muscular dystrophy. Briefly, one obtains blood cells from most members of large families that have a high incidence of the disease. Then, one seeks commonalities in the genes of individuals inflicted with the illness. There have been hints of specific abnormalities on particular chromosomes of schizophrenics but nothing definitive. In coming decades I would expect researchers to identify one or more genes whose aberrant structure leads to a predisposition to develop schizophrenia. As is thought to be the case for cancer, developing schizophrenia may require alterations in more than one gene. Also as in cancer, schizophrenia may be a collection of different diseases each with its own abnormal genes. Such a situation would make the search for the "cause" an excruciating challenge.

Even if one finds the abnormal gene, this does not necessarily translate into curative therapy. The abnormal gene might code for an enzyme or a receptor so that one could develop drugs to activate or inhibit the target protein. However, it is equally possible that the disease gene codes for a protein whose function is not readily appreciated. An analogous case involves Huntington's disease. After a fifteen year long collaborative investigation by a number of excellent groups, the pathologic gene in Huntington's disease was identified as one coding for a single large protein called huntingtin. Unfortunately, links between huntingtin, disease symptoms and therapy have been elusive.

The disease is characterized by deterioration of selected brain regions, most prominently the caudate-nucleus, which regulates the coordination of movement. Hence, patients with Huntington's disease display grossly abnormal movements, which led to the earlier name for the condition, "Huntington's Chorea." One would expect huntingtin to be highly concentrated in the caudate nucleus. But huntingtin occurs in similar concentrations throughout the brain, in glial cells as well as in neurons and in tissues all over the body. This is perplexing, considering that the symptoms are brain-specific. The abnormality in the huntingtin protein in patients involves large repeated sequences of the amino acid glutamine. Researchers speculate that destruction of brain tissue in patients might arise from a "gain of function" in which the glutamine repeats bind to other proteins which are brain specific. No direct evidence that such huntingtin-associated proteins cause disease symptoms has yet been obtained. The example of Huntington's disease teaches us that finding the abnormal gene is often only the first step in a search for pathophysiology and novel therapy.

The history of drug development for schizophrenia reflects the value of serendipity and may point toward future advances.

5.4.1
Neuroleptic Drug Development

The introduction of the phenothiazine neuroleptic drug chlorpromazine in the treatment of schizophrenia is regarded by many as the most important event in 20th century psychiatry (Table 5.1; Swazey, 1974). Prior to chlorpromazine most schizophrenics could look forward to a lifetime in a state mental hospital. Though chlorpromazine and its successor neuroleptic drugs do not "cure" the disease, they favorably influence the fundamental symptoms so much that most patients can function reasonably well. Together with the advent of the community mental health move-

Tab. 5.1 History of chlorpromazine development.

1883 Phenothiazine synthesized as a dye.
1946 Promethazine introduced as a phenothiazine antihistamine.
1950 Promethazine employed as a preanesthetic calming drug by Laborit.
1950 Synthesis of chlorpromazine, a very sedating phenothiazine antihistamine .
1951 Laborit replaces promethazine with chlorpromazine for pre-surgical anesthesia.
1952 Delay and Deniker observe antipsychotic actions of chlorpromazine.

ment, the neuroleptics have almost literally emptied the wards of large state mental hospitals and restored to loved ones patients who in earlier generations would have been lost forever.

The term phenothiazine refers to a three-ringed chemical structure with a sulfur (thia) and a nitrogen (zine) in the center ring flanked by two phenyl (pheno) groups. Phenothiazine was first introduced as a dye, a product of the 19th century revolution in the dye industry brought about by advances in organic chemistry. The early 20th century witnessed the beginnings of the modern pharmaceutical industry as organic chemists transformed their expertise in modifying chemical ring structures from dyes to drugs. Among the first drugs developed in this effort were agents designed to block actions of the neurotransmitter acetylcholine and epinephrine (the true neurotransmitters of sympathetic neurons, norepinephrine was not known until the late 1940s). The prototypic anti-cholinergic drug, atropine, is the active product of the plant extract Belladonna, used for hundreds of years in treating various gastro-intestinal and other complaints. The discovery by Sir Henry Dale in 1911 of hista-mine and his prescient conviction that it played a role in allergies led to an interest in drugs that might block the effects of histamine. The first structures evaluated as antihistamines were somewhat related to atropine. One of them, promethazine, incorporated the phenothiazine ring.

In the late 1940s Henri Laborit, a French neurosurgeon, speculated that the auto-nomic, involuntary nervous system employs histamine and is disordered during sur-gical anesthesia. Accordingly, he decided to utilize antihistamines as a key constitu-ent in a pre-anesthetic cocktail of drugs. He was struck with the effectiveness of promethazine in calming patients before surgery – not surprising because of the sedating properties of most antihistamines, especially promethazine. He asked the Rhone-Poulenc Drug Company to provide him with another antihistamine which might be even more sedating. Chlorpromazine had been synthesized as an antihis-tamine but was discarded because it was much too sedating for daily use. Laborit who was so struck by the "beatific quietude" the drug elicited in his patients that he urged his psychiatrist colleagues Jean Delay and Pierre Deniker to try it in psychia-tric patients. Delay and Deniker administered chlorpromazine to manic-depressives and found its calming effect remarkable. They published their first paper in Decem-ber 1952 and, within a year, the drug was being extensively employed. Psychiatrists tried it in numerous conditions and soon appreciated that it exerted a unique anti-schizophrenic effect (Ayd and Blackwell, 1981).

Tab. 5.2 How neuroleptics act – sequence of insights.

Neuroleptics at therapeutic doses elicit Parkinsonian effects.
Parkinson's disease involves dopamine deficiency.
Neuroleptics block dopamine (D2) receptors.
Neuroleptics relieve positive not negative symptoms.
Clozapine relieves negative and positive symptoms.
Clozapine successors block serotonin 5-HT2 subtype receptors.
Future: What transmitter systems best explain drug effects? Are they aberrant in schizophrenia?

Initially, clinicians thought that chlorpromazine and related drugs were acting primarily as sedatives. However, when minimally sedating neuroleptics were developed, they were equally therapeutic, while non-neuroleptic sedatives were not effective in schizophrenia. Moreover, analysis of symptom responses indicated selective improvement in uniquely schizophrenic symptoms such as delusions and hallucinations. Accordingly, understanding the molecular mechanism of drug activity might shed light on the disease's pathophysiology.

Clues to the mechanism of action of chlorpromazine came in the first clinical studies of Delay and Deniker (Table 5.2). They had no idea as to what would be an appropriate dose and merely titrated patients till they saw benefit. At therapeutically effective doses, which could vary fivefold in different patients, they invariably noticed neurologic side effects mimicking the symptoms of Parkinson's disease. They dubbed the drugs "neuroleptics" from the Greek meaning "to clasp the neuron", suggesting that something involving specific neuronal groups in the brain was crucial for therapeutic effects.

Subsequently, work in numerous laboratories established that neuroleptics act by blocking dopamine receptors, especially a subtype designated D-2 (Carlsson, 1988; Snyder, 1974). The most prominent group of dopamine neurons in the brain terminates in the caudate nucleus and degenerates in Parkinson's disease, so that dopamine deficiency is widely regarded as the proximal cause of symptoms. By blocking dopamine receptors, neuroleptics create a functional deficiency of dopamine, explaining the Parkinsonian side effects. By blocking dopamine receptors in areas of the brain controlling emotional behavior, such as the limbic system, the drugs presumably exert their anti-schizophrenic actions.

Knowledge that the drugs act by blocking dopamine receptors led to a "dopamine hypothesis" of schizophrenia suggesting that excess dopamine activity in the brain is involved in the genesis of symptoms. This notion was supported by observations that amphetamines, which act by releasing dopamine, markedly worsen schizophrenic symptoms. Numerous studies of post-mortem schizophrenic brains and imaging studies of dopamine receptors in patients have failed to find evidence for specific biochemical abnormalities of dopamine systems in schizophrenia. Nonetheless, even if disruptions in dopamine neurotransmission are not etiologic, it seems likely that dopamine plays some role, at least indirectly, in modulation of symptoms. Certainly, designing drugs to interfere with dopamine neurotransmission provides the surest way of obtaining agents that will be therapeutic in schizophrenia.

Despite their importance in psychiatry, the neuroleptics are by no means cure-alls. Even patients that respond extremely well to neuroleptics remain disturbed. Specifically, though their florid hallucinations and delusions, the "positive" symptoms of schizophrenia, are alleviated, patients remain emotionally detached from the environment. This "wallflower" syndrome and related symptoms are commonly designated the "negative symptoms of schizophrenia" and often are the most disabling ones. The first glimmer of effective treatment of such negative symptoms came with the drug clozapine.

Clozapine was designed as a "me too" neuroleptic (Healy, 1996). Because of major side effects it was withdrawn from the market in most countries after having been available for a relatively short period of time. Then a peculiar thing happened. Psychiatrists all over Europe (the drug had never been introduced in the United States) began complaining that they were losing their most valued agent. Patients that failed to respond to conventional neuroleptics would do well with clozapine. More strikingly, clozapine was relieving the negative symptoms, transforming patients who had been emotionally dead for years, into relatively warm, interactive human beings. After controlled trials confirmed these unique properties, the drug was reintroduced to the market with appropriate precautions for monitoring the possibility of agranulocytosis, loss of white blood cells, a potentially life-threatening side effect.

A search began to ascertain molecular actions that might account for the extraordinary effect of the drug on negative symptoms. By simply looking for drugs that would mimic the biochemical and behavioral profile of clozapine in animals, researchers developed a number of clozapine-like agents. In the United States, three are presently marketed: risperidone, olanzapine, and quetiapine. These drugs exert beneficial effects on negative as well as positive symptoms and do not cause agranulocytosis, though most psychiatrists feel that they are not as effective as clozapine in dealing with the negative symptoms.

Determining the molecular mechanism whereby clozapine affects negative symptoms should afford major insight into their genesis. Thus far, there have been many theories but no direct proof. One of the best guesses is that clozapine relieves negative symptoms by blocking subtypes of receptors for the neurotransmitter serotonin. Hence, drugs that block this form of serotonin receptor as well as dopamine D-2 receptors in the right proportions might be ideal.

5.4.2
Drug Psychoses

Drug-induced psychoses provide another approach to understanding schizophrenia without finding the causal genetic abnormality (Hollister, 1968). Numerous drugs are psychotomimetic (Table 5.3). For drugs that produce psychoses which closely mimic abnormalities in schizophrenic mentation, one might hope that drug-induced molecular aberrations will closely resemble those of schizophrenic brain. As already mentioned, amphetamines can exacerbate schizophrenic symptoms. Non-schizophrenics ingesting very large doses of amphetamines invariably become psychotic, with disturbances closely resembling an acute paranoid form of schizo-

Tab. 5.3   Psychotomimetic drugs as clues to mental illness.

LSD, mescaline and other psychedelic drugs cause psychosis, but their perceptual and cognitive changes do not resemble schizophrenia.

Amphetamine and cocaine psychoses mimic subtypes of schizophrenia, but there is no classic thought disorder and there are no negative symptoms.

Phencyclidine (PCP) psychosis faithfully masquerades as schizophrenia, though some say it resembles mania. PCP is discovered to block NMDA subtypes of glutamate receptors. Glycine and cycloserine, which stimulate NMDA receptors, are antipsychotic.

phrenia. Some researchers have utilized this information to buttress the dopamine theory of schizophrenia. Others argue that the psychosis caused by amphetamine is not close enough to the symptomotology of schizophrenia to be regarded as a valid model of the illness. Psychedelic drugs, such as LSD and mescaline, cause reproducible psychotic behavior that once was thought to model schizophrenia but has subsequently fallen into disrepute as a valid mimicker of the illness (Grinspoon and Bakalar, 1979).

A potentially more fruitful drug psychosis approach to schizophrenia has emerged in recent years from studies of the drug phencyclidine (PCP) (Carlsson and Carlsson, 1990). Phencyclidine was first developed as an anesthetic but was found to be psychotomimetic in human subjects. For many years no one knew how it acted. In the 1980s research into the influences of glutamate, the major excitatory neurotransmitter in the brain, provided clues. Glutamate acts via a number of receptor subtypes of which one of the most studied is the *N*-methyl-D-aspartate (NMDA) receptor. The NMDA receptor is an ion channel that admits sodium as well as calcium ions following the binding of glutamate. Phencyclidine influences the NMDA receptor at its ion channel site to block transmission. This anti-excitatory effect accounts for the anesthetic qualities of the drug. For reasons that are not at altogether clear, blocking NMDA receptors at this site also leads to psychotic manifestations. Psychiatrists debate whether phencyclidine psychosis resembles schizophrenia, manic-depressive illness, or neither. The present consensus is that phencyclidine psychosis is the best drug model of schizophrenia yet developed.

Evidence favoring this model has emerged from therapeutic trials in schizophrenia. The model implies that drugs stimulating NMDA receptors should be therapeutic. Since glutamate is an excitatory neurotransmitter, glutamate derivatives that penetrate into the brain and directly activate the receptor would augment activity at all types of glutamate receptors and might cause seizures. Researchers have taken advantage of sites on the NMDA receptor that are unique to this subtype of glutamate receptor. NMDA receptors cannot function unless a unique site on the receptor that interacts with glycine or the D isomer of serine is occupied. A number of groups have administered large doses of glycine or a derivative, cycloserine, to schizophrenics and obtained notably therapeutic responses.

Where might the NMDA research take us? Genes for the various subtypes of NMDA receptors have been cloned. One might ascertain whether schizophrenics display abnormalities in any of these genes. Even if they do not, the possibility of developing therapeutic agents through this receptor is tantalizing. Glycine and cycloserine, the agents already tested, are unsatisfactory drugs. They are electrically charged and do not penetrate the blood brain barrier very well. Derivatives that have better drug-like qualities may be more effective therapeutic agents and provide tools to further probe the NMDA system in patients. PET scanning and other imaging strategies permit us now to readily quantify and localize neurotransmitter receptors in humans. Such imaging studies comparing schizophrenics and control subjects may be useful.

5.5
Drugs of Abuse

One major change in societal attitudes towards drugs in the last decade of the 20th century involves judgements about the importance of various classes of drugs. Traditional thinking holds that drugs are important if they relieve symptoms of major illnesses. With the advent of "managed care" and other pressures on medical costs, a new discipline, "pharmacoeconomics", has emerged. Pharmacoeconomics attempts to quantify the exact financial impact of each drug. For instance, a drug that relieves congestive heart failure will justify its cost if it reduces the duration of hospitalization.

Viewed from the perspective of pharmacoeconomics, a case can be made, especially in the United States, that the most important pharmaceutical agents are not those that treat disease but those that are employed recreationally and usually illicitly – *drugs of abuse*. Life for most drug addicts is so dominated by their chosen agents that they do not work and require support by the welfare system. More importantly, addicts often devote their lives to criminal activities that pay for the drug habit. In 1995 it was estimated that 80 % of the inmates of federal and state prisons in the United States were there because of crimes related to drug abuse. The cost to the United States from the loss of property associated with this criminal activity is estimated to be $50×10^9$ per year. For heroin users, crime is focused on securing money to pay for drugs. By contrast, cocaine renders its users paranoid and aggressive, leading to crimes of violence including senseless assaults and murders. Since drug-related crime occurs most in inner cities, the resulting devastation of civil society has led to desertion by many citizens of the central cities of the United States. The financial loss from this exodus is so great that its financial impact is probably incalculable.

5.5.1
Definitions and Varieties

Drug abuse is a term generally used to designate the self-administration of a drug for recreational rather than therapeutic ends. The term "abuse" implies that there are adverse effects of such drug use. Agents employed may be legal, such as alcohol and nicotine, or illegal, such as marijuana, heroin and cocaine.

The term "addiction" implies the repeated use of drugs leading to tolerance, physical dependence and craving. Tolerance simply means that a chronic user will require more of a drug to obtain the desired effect than when the agent was first employed. Because drugs are metabolized more rapidly with chronic use, such "metabolic tolerance" can occur with non-abusable agents such as antibiotics. With psychoactive drugs, one gets tolerance even when brain and blood levels of the drug are the same in the addicted and the naïve subject, a phenomenon called "cellular tolerance."

Dependence refers to the existence of withdrawal symptoms when drug administration is abruptly terminated. Withdrawal symptoms can be physical, such as the shivering and trembling following withdrawal of opiates, or psychological, such as the depression displayed by individuals stopping the chronic use of cocaine. Thus, there has been differential designation of "physical dependence" and "psychological dependence." Older definitions of addiction required the presence of physical dependence. Since there are not many obvious physical symptoms of withdrawal from cocaine, some writers maintained that cocaine was not technically an side effect drug. Amphetamines, whose actions are quite similar to those of cocaine, were also designated as "non-side effect" by their manufacturers. However, most authorities now agree that psychological and physical dependence are of comparable importance.

If tolerance and dependence were the totality of addiction, then we would not have a pervasive problem. One can readily withdraw an addict from his or her drug and after a period of weeks, withdrawal symptoms abate. Heroin addicts incarcerated at federal addiction centers such as those in Lexington, Kentucky, are maintained in the drug-free state for one or two years. Unfortunately, when they are returned to their home environment, they often return just as rapidly to the use of the offending drug. Some of this recidivism is due to socioeconomic forces. However, a number of studies have substantiated that individuals return to drug abuse because of compulsive drug seeking behavior which appears to have a biological basis, and which can be demonstrated in various animal species including rodents. Thus, compulsive drug-seeking behavior is a crucial component in the definition of addiction. It accounts for well-known truisms such as the inability of abstaining alcoholics to take even a single drink for fear of triggering a return to alcoholism.

What are the major classes of side effect drugs? Alcohol and nicotine are probably the most widely abused agents and, in many ways, are more destructive of human life than more tainted substances such as heroin and cocaine. Alcohol is thought to act by facilitating the actions of the major inhibitory neurotransmitter in the brain, γ-aminobutyric acid (GABA), and by diminishing the effects of glutamate, the major excitatory neurotransmitter in the brain. The most extensive research favors a primary action by facilitating the effects of GABA on its receptor sites. In this way, ethanol appears to act in the same or closely similar fashion to other "downer" drugs such as barbiturates and benzodiazepines (Valium® and related substances).

Nicotine is the psychoactive ingredient of cigarettes and definitely accounts for the side effect quality of cigarette smoking. Nicotine mimics the neurotransmitter acetylcholine at one of the subclasses of the nicotinic acetylcholine receptors. Nicotinic stimulation causes enhanced alertness, an apparently beneficial effect which may account in part for the attraction of cigarettes. When first administered to an individual, nicotine also causes unpleasant dizziness and nausea, but individuals who become regular cigarette smokers quickly become tolerant to these effects. Nicotine probably causes greater compulsive drug seeking behavior than any other addicting substance, as it is more difficult to withdraw cigarette smokers than users of any other drug of abuse.

Opiate drugs encompass morphine and its many synthetic derivatives including heroin. Opiates act by mimicking endogenously occurring morphine-like neuro-transmitter peptides termed the enkephalins or endorphins.

Many other mind-altering drugs are abused, but most of them are not truly side effect. Examples include marijuana and the psychotomimetic drugs such as LSD and phencyclidine, also known as PCP or angel dust. Though many people use these agents on a regular basis, withdrawal symptoms are minimal and there is little compulsive drug seeking.

5.5.2
Approaches to Treatment: Focus on Cocaine

The major classes of abused drugs described above cause different subjective effects and act via different receptor systems. However, those which are addicting all display the same formal properties of tolerance, dependence and compulsive drug seeking. All of them are, by definition, "rewarding" in that they make people feel "good" and lead human and animals to go to considerable lengths to obtain drugs for repeated self-administration. Researchers have looked hard to identify commonalities amongst these drugs at a molecular level. There is a body of evidence suggesting that a single brain region, the nucleus accumbens in the "emotional" limbic area of the brain, is a major reward center for these drugs and utilizes the neurotransmitter dopamine to secure these effects. In animal studies lesions of the nucleus accumbens and/or destruction of its dopamine neurons greatly diminish drug-seeking behavior for numerous reinforcing agents including nicotine, alcohol, cocaine, amphetamines and barbiturates.

A molecular basis for addiction has been notably elusive. As of this writing, there is no accepted mechanism to account for the major features of addiction. With the rapid escalation of new findings in the field, I suspect that the coming one or two decades may well witness major breakthroughs into an understanding of addiction. Such insights may then lead to new therapeutic agents that could influence the addicting process for all drugs of abuse.

In terms of agents to treat drug abuse, the one success story involves methadone for heroin addiction. In the 1960s the pharmacologist Vincent Dole and his wife Marie Nyswander administered large doses of the opiate methadone as replacement therapy for heroin. Methadone is orally active and long acting so that a single oral dose will provide high blood levels and greatly diminish or abolish craving for a period of 24 h. By contrast, heroin must be injected intravenously and is so short acting that, 2–4 h following an injection, the addict is already experiencing withdrawal symptoms and seeking frantically for a new dose. Because of this, a heroin user's life is dominated by his/her drug. The methadone maintenance program developed by Dole and Nyswander, now universally employed in the United States and Europe, has greatly diminished the adverse sequelae of heroin addiction. Once stabilized on methadone, an addict can lead a fulfilling personal and professional life. Being addicted to an opiate does not *per se* interfere with one's functioning. Indeed, many physician addicts have led effective lives for decades. Though some individuals are concerned that methadone treatment merely replaces addiction to one opiate with addiction to another one, the consensus is that the treatment is of immense value both to the addicts and society. The fact that the patients must remain on methadone for many years is regarded as no worse than the requirement of a diabetic for daily doses of insulin.

The situation with cocaine is quite different. Presently there is no adequate treatment for cocaine addiction. However, recent advances in understanding the drug's molecular actions offer hope. Cocaine inhibits the reuptake inactivation of the neurotransmitter dopamine. Most neurotransmitters, including dopamine, are inactivated after release by being pumped back into the nerve ending that had released them. By blocking this "dopamine transporter" cocaine facilitates the synaptic actions of dopamine especially in the limbic system. Studies employing mice with targeted deletion of genes for neurotransmitter transporters indicate that inhibition by cocaine of the serotonin transporter may also contribute to its actions.

Uhl and associates have provided insight into exactly how cocaine influences the dopamine transporter, suggesting potential treatments for cocaine addiction (Kitayama et al., 1996). Cocaine does not compete directly at the dopamine recognition site of the transporter (Fig. 5.2). Instead, it binds to a separate site, which indirectly or "allosterically" influences the conformation of the dopamine recognition site to inhibit dopamine transport. Thus, in principle, one should be able to develop a drug that would compete with cocaine for its recognition site but not influence that site in the way that cocaine affects it. Instead, such a drug would merely passively occupy the cocaine recognition site and block the access of cocaine. Such a drug would be regarded as a cocaine antagonist. It could be employed in treating cocaine overdoses. In theory, one might be able to utilize a "pure" cocaine antagonist to treat addicts such that whenever they administer the drug they would not feel its effects and would soon lose interest. Of course, the addicts might simply not take the antagonist drug so that it would be of use only if a formulation were developed that could be implanted and act for one or two months. Alternatively, one might develop an agent that partially mimics the effects of cocaine to provide a mild level of behavioral stimulation but also acts as an antagonist to prevent the effects of self-administered cocaine. Such mixed agonist/antagonist drugs exist already in the opiate system and are being evaluated as alternatives to methadone treatment. Limited studies in rodents indicate that cocaine antagonists can be developed which by themselves have no behavioral effects but which interrupt the influences of cocaine.

Since cocaine is presently regarded as the most dangerous of addicting drugs, approaches to its control by new pharmacologic agents would be a great boon.

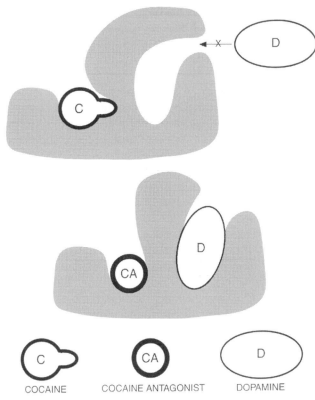

Fig. 5.2 Model of dopamine (DA) transporter. The transporter possesses separate but overlapping recognition sites for DA and cocaine. Cocaine inhibits DA transport indirectly as its binding alters the conformation of the DA site to block DA access. A cocaine antagonist (ant.) binds to the cocaine recognition site but does not elicit any conformational change in the DA site so that cocaine effects are blocked while DA transport remains normal.

Rapidly accelerating research into such drug development suggests that within one or two decades the scourge of cocaine will be abated.

5.6
Conclusions and New Directions

I have deliberately restricted my coverage to a few topics that illustrate important themes that are likely to lead to major drug advances in the coming century and to diseases that I know best. It seems to me that understanding cell death will provide molecular tools to impact the most common and debilitating diseases, such as cancer, stroke and myocardial infarction. Advances in pinpointing drug-sensitive enzymatic steps in the cell death process have been breathtaking. A decade ago next to nothing was known about this area. Now, an extensive cascade of programmed cell death pathways has been elegantly mapped, and what we know today appears to be the tip of an iceberg. Mental illness poses more complex challenges, which I have addressed through a historical perspective.

For want of space I have not addressed one other broad area that impacts many diseases, angiogenesis. Angiogenesis has been extensively studied in cancer research. Tumors cannot develop unless supported by the growth of new blood vessels. Anti-angiogenic agents do reduce tumor growth in various animal models. Conversely, lack of blood vessel perfusion of tissue is pathogenic in coronary artery disease and in peripheral vascular disease. A number of protein growth factors stimulate blood vessel growth. Examples include basic-fibroblast growth factor and vascular endothelial growth factor. In animals these agents cause sprouting of new blood vessels in the coronary artery system and in peripheral vascular beds and augment blood flow.

Stroke and myocardial infarction are specific disease entities in which pathogenic substances that cause cell death have been identified such as glutamate, nitric oxide, and various oxygen free radicals. Drugs affecting these processes offer dramatic relief in animal models. The next decades will likely witness the development of even better agents and their application in humans. Except for modest therapeutic effects of tissue plasminogen activator, there is no treatment at all for vascular stroke, the third leading cause of death in the United States. I expect that alleviation of stroke damage will be one of the most important therapeutic advances in the next few decades.

Neurodegeneration occurs in numerous conditions besides stroke. Such entities include Huntington's and Alzheimer's diseases. Alzheimer's may well yield to molecular interventions in the coming decades, though the path to new drugs is not as clear cut as for stroke. There is growing acceptance that the deposition of amyloid plaques is key for the pathophysiology of the disease. Amyloid plaques result from the deposition of a 40–42 amino acid peptide designated amyloid-beta. The amyloid-beta peptide derives from improper processing of a large precursor protein designated APP.

Though the majority of cases are sporadic and non-familial, the most important advances have come from the relatively rare genetic forms of Alzheimer's disease. Several forms of familial Alzheimer's disease have been ascribed to specific mutations in APP processing. Even more common forms of familial Alzheimer's disease have been ascribed to mutations in a different family of proteins, the presenilins. Recent evidence indicates that presenilins play a role in processing APP, as pathogenic abnormalities in the presenilins lead to increased formation of the amyloid-beta peptide. A third predisposition to Alzheimer's disease involves increased formation of one of the subtypes of the common blood protein apolipoprotein-E (ApoE). Individuals with high levels of ApoE4 have a high incidence and early onset of Alzheimer's disease whereas individuals with high levels of ApoE2 are protected. Interestingly, the different subtypes of ApoE also affect APP processing.

Thus, Alzheimer's disease reflects a common phenotype which can be brought about by different biochemical pathways. One strategy for developing therapeutic agents would be to inhibit the enzymes that give rise to the abnormal plaque-forming processing of APP. Despite vast efforts in numerous laboratories, the exact enzyme proteins have yet to be identified, though the processing steps have been delineated. A step designated alpha-secretase provides physiologic processing of APP. Pathologic processing to the amyloid beta-peptide involves the actions of two enzymes referred to as beta- and gamma-secretase respectively. Most researchers feel that inhibitors of beta- and/or gamma-secretase will block the formation of the amyloid-beta peptide and retard or prevent the onset of the disease. Though the enzymes have yet to be cloned, several pharmaceutical companies have already identified inhibitors of the beta- and gamma-secretase which can lower amyloid-beta peptide levels in rodent brain.

In contrast to the well-defined therapeutic strategies in cancer, myocardial infarction, stroke, and neurodegenerative diseases, things are far murkier in psychiatry. Numerous groups seem to be coming close to the aberrant genes in schizophrenia and manic-depressive illness. However, there have been many false starts. Even if such genes are identified, they may not lead to meaningful therapeutic strategies. Though fundamental understanding of psychiatric diseases lags behind the neurologic disabilities, we already have valuable therapeutic agents. I have not addressed the area of drugs and depression. Effective antidepressants have been available since the mid 1950s, and the recent generation of selective serotonin reuptake inhibitors has provided therapeutic benefit with minimal side effects. Pharmacologic strategies continue to point toward novel modes of obtaining more effective and safer drugs so that suffering from these diseases may be considerably alleviated before we have any fundamental understanding of their causation.

In summary, the second half of the 20th century has witnessed a revolution in the development of important therapeutic agents for many different diseases, especially those afflicting the nervous system. The escalating rate of scientific advance portends far more effective drugs, many of which will likely revolutionize treatment and perhaps lead to cures of the major scourges of the human species.

5.7
Acknowledgements

Supported by USPHS grants DA-000266, MH-18501, Research Scientist Award DA-00074, and a grant of the Theodore and Vada Stanley Foundation.

General References

1 LIPTON S.A., ROSENBERG P.A. Excitatory amino acids as a final common pathway for neurologic disorders. *N. Eng. J. Med.* **330**, 613–622 (1994).

2 LINCOLN J., HOYLE C.H.V., BURNSTOCK G. *Nitric Oxide in Health and Disease.* Cambridge University Press (1997).

3 PIEPER A.A., VERMA A., ZHANG J., SNYDER S.H. Poly (ADP-ribose) polymerase, nitric oxide and cell death. *Trends Pharmacol. Sci.* **20**, 171–179 (1999).

4 SWAZEY J.P. *Chlorpromazine in Psychiatry.* MIT Press, Cambridge, MA (1974).

5 AYD F.J., BLACKWELL B. (eds). *Discoveries in Biological Psychiatry.* Lippincott, Philadelphia, PA (1981).

6 CARLSSON A. The current status of the dopamine hypothesis of schizophrenia. *Neuropsychopharmacology* **1**, 179–180 (1988).

7 SNYDER S.H. *Madness and the Brain.* McGraw Hill, NY (1974).

8 HEALY D. *The Psychopharmacologists.* Chapman and Hall, London (1996).

9 HOLLISTER L.E. *Chemical Psychoses.* C.C. Thomas, Springfield, IL (1968).

10 GRINSPOON L., BAKALAR J. *Psychedelic Drugs Reconsidered.* Basic Books, NY (1979).

11 CARLSSON M., CARLSSON A. Interactions between glutaminergic and monaminergic systems within the basal ganglia – implications for schizophrenia and Parkinson's disease. *Trends Neurosci.* **13**, 272–276 (1990).

12 KITAYAMA S., MORITA K., DEHI T., WANG J.B., DAVIS S., UHL G.R. Dissection of dopamine and cocaine binding sites on the rat dopamine transporter expressed in COS cells. *Ann. NY Acad. Sci.* **801**, 388–393 (1996).

6
Protein Folding and Beyond

Yong Duan, Stephen C. Harvey and Peter A. Kollman

6.1
Introduction

How do proteins fold into their native structures? To study the mechanism of protein folding is to seek an answer to this question. The astronomically large number of possible conformations suggests that proteins use some sort of "directed" mechanisms to fold. Under evolutionary pressure, these mechanisms have been "refined" such that proteins can fold reasonably quickly to well-defined functionally useful structures. We will review briefly the history of protein folding studies and discuss the theoretical studies in more detail. We will concentrate on recent developments and give a brief perspective of the future.

The ultimate goal of studying protein folding is to elucidate the mechanisms governing the protein folding process, i.e., how proteins fold into their native states from random coil states. Related but independent goals of computational research on proteins include structure prediction and protein design. Structure prediction involves predicting the protein structures from their sequences without necessarily following the folding mechanisms, often using the knowledge of existing protein structures. CASP (Critical Assessment on Structure Prediction) is a forum that serves to assess critically the accuracy of the structure prediction methods and has been reviewed recently. [1] The fascination of designing new proteins to perform biological function has propelled the advancement of protein design. Exciting progress has been made in this area that has enabled redesign of small proteins of known structures. [2–4]

Proteins can fold into and subsequently maintain well-defined structures under physiological conditions, and their structures are determined by their sequences. [5] The mechanisms governing the folding process have been subject of extensive experimental and theoretical studies for almost half a century. Because of the close association between protein structures and their functions, this problem has often been referred to as the second half of genetics. Since understanding the protein-folding mechanism requires a detailed understanding of the entire protein-folding process, including the formation of native structures, such an understanding should improve the accuracy of structural prediction. This would have a direct impact in structural biology, protein design, biochemistry, and drug design.

Proteins can change their conformations when they bind to substrates and ligands. These conformational changes are often associated with protein function and are important for a detailed understanding of biochemical processes. Because these induced conformational changes are governed by the same physical principles as protein-folding processes, understanding protein folding would inevitably lead to the understanding to these (induced) structural changes. The close association between protein structures and their functions also argues for the importance of reliable structural prediction methods in protein design. Furthermore, knowledge gained from the study of protein-folding mechanism can be the foundation of such design. Given the potential application of protein design in medicine and bioengineering, a significant impact on the biological sciences in general would be expected if a full and thorough understanding of protein-folding mechanisms is achieved. This, together with the growing large quantity of genomic data, strongly supports the need for reliable structure prediction methods that would complement the existing experimental approaches such as X-ray crystallography and NMR spectroscopy. The discovery of folding-related diseases reinforces such a need. [6–8] Despite great progress made using a variety of approaches, it is still difficult to establish detailed descriptions of the protein-folding processes, and such descriptions are the necessary steps toward the comprehensive understanding of the mechanism of folding. Recent progress has included a microsecond molecular dynamics simulation using atomic level representation of both solvent and protein, [9] and the simulation was only an order of magnitude shorter than the estimated folding time of the protein.

Experimental studies have been conducted to study various aspects of protein-folding processes. By monitoring the change of heat capacities during folding and unfolding using highly accurate calorimetric measurements, the energetics of folding can be studied and the change of enthalpy and entropy can be obtained. [10] Spectroscopic studies, on the other hand, can monitor low-resolution structural changes in the folding processes, such as overall helicity, burial of certain side chains, and protection of amide hydrogens. Mutagenesis studies can characterize the contributions of individual residues to the kinetics and thermodynamics. [11] Structural studies, particularly multi-dimensional NMR spectroscopy, can provide atomic-level structures of presumed intermediates in protein folding under mild denaturation conditions. [12] NMR spectroscopy can also give rich information on the unfolded and partially folded states. [13] A recent direction, partly due to the discovery of the molten-globule states, [14,15] has been the study of early stage events and the formation of molten globules and early intermediate states in protein folding. [16] Exciting developments include the ultra-fast laser-jump technique, which can monitor the folding process to as fast as the sub-microsecond time scale with a dead time of tens of nanoseconds. [17–19] We would expect, with the further development of both simulation and experimental methods, that the simulation and experimental time scales would have more overlap and that more direct comparisons could be made, particularly since both methods can be applied to study fast-folding small proteins. Such small fast-folding proteins are ideal prototypical systems for our understanding of the folding mechanisms. A common feature of these experimental methods is that the results are macroscopic and represent the behavior of an ensemble of molecules. Recent advances have made it possible to monitor the kinetic processes of individual molecules. [20]

6.1.1
Computational Protein Folding

Theoretical studies of protein folding have come of age. The first wave of theoretical efforts included the attempts, based on the understanding of polymer physics, to build simple models such as the framework model, the diffusion-collision model, the nucleation-condensation model, and the hydrophobic collapse model. [21–25] These models have since been revised and adapted to explain new data, and some have been extended to include the ability to make quantitative predictions, particularly to predict kinetic behavior. A general feature of these highly simplified models is the emphasis on a particular feature of the protein-folding process and the generalization from the observations of low-resolution experimental results based on fundamental physical principles. These models are sometimes referred to as the "theoretical models" and have helped to build abstract concepts from the complex folding processes. Recent development included the introduction of folding funnels, [26–28] which have gained popularity among theorists and experimentalists alike. Rather than describing the folding processes, the folding funnel describes the overall shape of the free energy landscape, and its ruggedness intimately relates to protein-folding kinetics. A smooth funnel would rule out the possibility of intermediate states, and folding would consist of simple monotonic downhill processes; a rough funnel would imply multiple intermediate states, and folding would be close to multidimensional diffusive processes among these states. Since then, existing models have started to merge under the framework of the folding funnel. [29,30]

The second wave of efforts involved the application of low-level representation computer models [25,31] to the study of the folding process in which each protein residue is represented by one or two particles. These simulations were developed when it became clear that simple models such as those described above were not sufficient to yield necessary information for a detailed understanding of the process, including structures. Since the ability to predict the native structures is a critical test of any protein-folding theory, the utilization of computer models in this field, even at a simplified level of representation, is a significant step forward. These models have been proven to be useful, and simulations using them have provided great insights into both energetics and kinetics of the protein-folding process and have served as effective tools for the understanding of the general features of protein-folding free energy landscapes. [25,29,32] Furthermore, these methods have been applied to the prediction of protein structures and have produced encouraging results. [31,33,34] Because of their simple representations, these methods can sample conformational space efficiently and are much less expensive than other methods (e.g., explicit atomic-level representations). With the improvement of simulation parameters, we can expect that these types of methods will be important in the field of structural prediction and study of protein-folding mechanisms, particularly when combined with high-level representations.

6.1.2
All-atom Simulations of Protein Unfolding and Short Peptide Folding

Finally, perhaps, the third wave of protein-folding "theories" has just begun, marked by the detailed atomic-level representation of both solvent and protein. The initial attempts at using such an approach were the study of small peptide fragments. [35] Because the detailed models require both a large number of particles, which is typically more than 10 000, and a small time step of one to two femtoseconds (10^{-15} second), direct simulation of the folding processes, which take place on a microsecond or longer time scale, has been difficult. Therefore, such models have been applied to study the unfolding processes of small proteins [36–39] that can be accelerated substantially by raising the simulation temperature. The detailed representation has allowed direct comparisons with experiments, and encouraging results have been obtained. [40] Limited refolding simulations were also attempted starting from partially unfolded structures generated from the unfolding simulations, and considerable fluctuations were observed. [39] These short-time refolding simulations have also identified the transition states in the vicinity of the native state. [40,41] Care should be taken, though, because the short-time refolding simulations can only sample the conformational space in the vicinity of the unfolding trajectories. The significant entropy-enthalpy imbalance [42] of the typical high-temperature unfolding simulations implies some degree of distortion of the free energy landscape. Multiple simulations starting from the same structure can reduce the ambiguity. [41] Equilibration of water in this type of short-time refolding simulation is needed to avoid simulating a trivial collapse process driven by the equilibration of water that takes place when the system is quenched to room temperature and to restore faithfully the room-temperature free energy profile around the starting structure. The entropy-enthalpy imbalance inherent in the typical unfolding simulations may also be reduced by conducting the unfolding simulations at mild unfolding temperatures [43] such that both temperature and pressure can be maintained at experimentally relevant conditions.

An extension of unfolding simulations is the attempt to reconstruct the free energy landscape from restrained simulations using the weighted histogram method, [44] which demonstrated funnel-shaped free energy landscapes and showed that the shapes of such funnels are also dependent on the type of proteins (i.e., α-helical or α/β). [45] It is somewhat misleading to call these types of calculations "unfolding simulations" since these are free energy calculations and can, in principle, be conducted in both directions by varying the restraining functions. In fact, one must be cautious in direct application of the WHAM method in analysis of unfolding trajectories. In the WHAM method, the number of events is used to calculate probabilities under the assumption that sufficient number of transitions takes place among the sampled states in each simulation. Because unfolding simulations often produce monotonic trajectories, where the proteins move monotonically away from the native state, [30] the restraining function becomes instrumental in maintaining reversibility, particularly in the case of protein folding, where extensive reversible transitions among all states are still difficult to model. Restraining functions ensure a sufficient number of transitions between neighboring states, which is a requirement for the estimation of the free energies of these states and to ensure that the average is obtained under equilibrium conditions, and which is the foundation of histogram-based free energy calculations. Indeed, this approach is significantly different from other unfolding approaches and can, in principle, yield a highly accurate description of the free energy surface. Because of their high sampling efficiency and quantitative nature, approaches such as the weighted histogram methods are expected to play important roles in computational studies of protein folding and conformational changes. Nevertheless, the weighted histogram method has also been applied in the analysis of unfolding trajectories. [30]

Encouraging developments have been made in the area of direct simulations of the folding processes of small peptide fragments with explicit representations of both solvent and peptides [35] in attempts to study the mechanisms of secondary structure formation. Tobias and Brooks studied the formation of a β-turn in aqueous solution. [35] Case and co-workers studied the transitions between two conformations of a β-turn motif and most of the simulated distances agree with NMR data. [46] Daura et al. [47] studied the folding and unfolding processes of a short β-peptide in methanol at temperatures below, around, and above T_m of the peptide. They observed reversible formation of secondary structure in a simple two-state manner within 50 ns. The estimated folding free energies, based on the population of the states, are in qualitative agreement with experimental observations. Recently, Chipot et al. studied the formation of undecamer peptides at the water-hexane interface. [48,49] These studies provided detailed atomic-level descriptions on the formation of isolated secondary structure motifs in their respective solvents.

6.2
All-Atom Simulations of Folding of Small Proteins

An exciting development was the application of such models in the direct simulation of the early stages of the folding process of a small protein, 36-residue villin headpiece subdomain, for 1 μs, [9] since such "top-down" simulations were perceived as impossible for the study of protein folding using present computational resources. [50] The villin headpiece subdomain is a helical protein with well-defined tertiary and secondary structures. [51] Its three helices form a unique type of fold with the helix-1 aligned perpendicular to the plane formed by helices 2 and 3. The three helices are held together by a tightly packed hydrophobic core. Its melting temperature T_m is about 70 °C, [52] and the estimated folding time is about 10 μs, [53,54] making it one of the fastest-folding small stable proteins.

6.2.1
Concomitant Hydrophobic Collapse and Partial Helix Formation

In this simulation, the initial collapse phase was accompanied by partial formation of the native helices and reduction of hydrophobic surface in an apparent simple downhill process. The observed time scale of helix formation, 60 ns, corresponds well to experimental observations. [18,19] The importance of the initial collapse phase has been somewhat understated in the past and has often been termed as the dead time "burst phase", perhaps due to the fact that experimental studies of these ultra-fast processes have been difficult until recently. [17–19] Because the early stage species can often form in the "burst phase" and may lead to the subsequent formation of other species and intermediates, the characteristics of these early stage species may affect the folding kinetics, whether or not they themselves are productive intermediates. The observed concomitant formations of both helical domains and hydrophobic clusters in the simulation suggest a way to lower the entropy cost in the subsequent folding processes by reducing the protein internal entropy in the early stages. The reduced protein entropy can be partly compensated by the entropy gained due to releasing water from the surface, as shown in the simulation by the strong correlation between the R_g and the solvation free energy (SFE) and by the large decrease of the SFE during the collapse process. These early stage nascent "domains" may (1) dissipate, (2) aggregate, or (3) grow. Some of these "domains" may well be the "nuclei" for the later stage intermediate structures. Formation of native-like "domains" in the early stage helps the formation of the later-stage intermediate structures and perhaps the formation of the native structure. Conversely, the non-native "domains" will eventually dissipate, and those that are retained in the later stage intermediate species will be difficult to dissipate and will contribute to the free energy barriers. Correlation calculations indicated that the collapse was driven by both burial of hydrophobic surface and a lowering of the internal energy of the protein. [55]

Considerable fluctuations between compact and extended states were also observed in the simulation, suggesting a shallow free energy landscape in the vicinity of extended conformations. The residence times of the compact conformations are much longer than that of the extended conformations, suggesting that the compact conformations are energetically more favorable than the extended conformations. These clearly demonstrated that the free energy landscape is very rugged and that the existence of intermediate states is highly likely.

6.2.2
A Marginally Stable Intermediate State

Proteins can fold from fully unfolded states to the native state by going through many intermediate states of varying degrees of stability. In fact, since these intermediate states can be referred to as the "landmarks" of the protein-folding free energy landscape, studying the mechanism of protein folding, in a sense, consists of studying these intermediate states, the relationship between them, and the relationship between them and the native state. One microsecond, even though it is more than two orders of magnitude longer than the longest simulation on proteins in water conducted up to 1998, is still an order of magnitude shorter than the shortest estimated folding time for a protein; thus, it is unrealistic to expect the protein to reach the native state during such a simulation. However, such a time scale appears to be sufficient to observe the marginally stable intermediates, if they exist. This was indeed the case in the simulation. A marginally stable intermediate was observed in the simulation and lasted for about 150 ns. The main-chain structure of the intermediate showed remarkable similarity to the native structure, including partial formation of helices 2 and 3 and a closely packed hydrophobic cluster (see Fig. 6.1). The solvation free energy reached a level comparable to that of the native structure. Two pathways approaching the intermediate were observed in the simulation, which may imply a general feature of the folding free energy landscape in which the native state, being thermodynamically favorable, may require such features as limited pathways to achieve kinetic stability for carrying out its designated biological functions.

The existence of intermediate states on the folding free energy landscape, in addition to its overall funnel shape, is perhaps one of the few things that has gained acceptance among both experimentalists and theorists. An interesting paradox is the observation that the folding processes of some proteins are monophasic. One way to resolve such a paradox is the recognition of fast interconversion among these states. [56] Such a fast conversion would be possible if unfolded states and these intermediate states are on the same broad basin of the free energy landscape and are separated from the native state by a high free energy barrier. Another way to

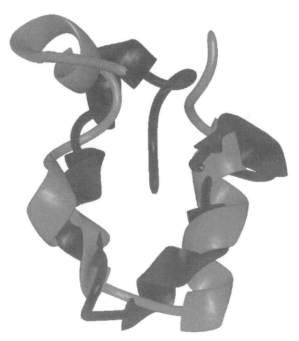

Fig. 6.1 Main chain structures of the native (grey) and the intermediate states (black). The intermediate state shown here was the most stable state observed in the simulation whose residence time was about 0.15 μsec. Only the average NMR structure of the native conformation is shown here.

resolve such a paradox is that only a single (highly stable) dominant intermediate state can form within the "burst phase" (i.e., within the dead time of experiments). Such a scenario is quite counterintuitive given the complexity of protein structures. Somewhat surprisingly, this latter view has dominated until recently. The third possibility is somewhat similar to the one proposed by Zwanzig. [56] In this scenario, multiple intermediate states, such as the one observed in the simulation, may exist and may be able to interconvert rapidly among themselves. However, only one (or a few) is (or are) close to the transition state (or states) that allow(s) the formation of all native contacts. These intermediate states are compact and have significant native-like features (such as the one observed in the simulation). The multiple intermediate states imply that the free energy barrier to the folded state will be entropic and packing, and side chain contacts play an important role. [54] The reduction of multiple states to a single state is certainly entropically unfavorable and, unlike the early stage collapse process, such an entropy loss can no longer be compensated by the release of water, at least not as much as in the collapse phase. Because denaturant would reduce the stability of the intermediate states, enhance the rate of the interconversion processes, and reduce the free energy of activation of the transition state for folding, the folding rate may be enhanced in such conditions. [57,58] This scenario would explain the monophasic behavior for folding observed in many proteins. Multiphasic folding processes would correspond to multiple types of intermediate states with high free energy barriers between the adjacent types of the intermediate states and possible fast conversion among the intermediate states of the same class. We believe that the simulation is most consistent with the third scenario. Furthermore, the interconversion processes between intermediate states may involve breaking of both native and non-native contacts in a way that is similar to unfolding the intermediate structures. This was evident in the intermediate state observed in the simulation. The mispacked side chains of this highly compact structure indicated that direct conversion from this intermediate state to the native state would be energetically prohibitive and that partially unfolding this structure was the only way to continue the search. They also suggested that reaching specific side-chain packing, rather than main-chain topology, may be the bottleneck of the folding process, and further implied that the transition state structure (or structures) proceeding to the native state may be less compact than that of the native structure and perhaps than the structures of many later-stage intermediate states, these being similar to those observed in the unfolding simulations of CI2, [40] such that partial unfolding (i.e., expansion) from structurally similar intermediate states would be necessary to enable the transition from the transition state to the native state to take place.

6.3
A Perspective View

Encouraging progress has been made recently, and we are now in an era of active application of molecular dynamics simulation to study the folding process. Because of the vital importance of water in protein folding and in the cell, the explicit representation of both solvent and proteins, with the promise of even more accurate models, is expected to play an increasingly important role in our understanding of protein-folding mechanisms.

The subject of protein folding is perhaps one of the most challenging problems in biophysical chemistry. Given the complexity of protein structures, diverse folding processes should be expected, including the possible role of certain folding-assisting domains within large proteins. [59] Furthermore, the complexity of *in vivo* folding processes is as yet one more challenge. [60] Therefore, full understanding of protein-folding mechanisms is indeed a daunting task. However, despite this, recent developments have made us believe that an eventual solution to such a problem may lie ahead in the next few decades. From a theoretical perspective, an immediate objective is to accurately replicate the complete folding process of small fast-folding proteins on computers including atomic detail. Such simulation results would inevitably provide the data for developing abstract models at a conceptual level that describe general and unambiguous features of the protein-folding mechanisms. The success of such simulations would itself be a strong testimony to the accuracy of the method and parameters. The diversity of protein structures and the complexity of the *in vivo* folding process can, in principle, be dealt with by a combination of experiments and further simulations. We may also be able to answer questions such as whether or not a particular part of protein is designed to assist the folding of the rest of the protein, such as found in so-called intra-molecular chaperons, and how this assistance occurs. The mechanism of chaperon-"assisted" folding processes can also be better understood. But without the basic understanding of the folding process of single-domain small fast-folding proteins, understanding of more complex folding processes is more difficult. With the development of both experimental and simulation methods, such a goal should be achievable in the future.

It is interesting to consider the rate of progress in macromolecular simulations and to ask what might be accomplished if the historical rate of increase in computational power is continued. Just for illustrative purposes, let us ask when someone might hope to carry out a molecular dynamics simulation treating every atom in an organism over the organism's entire life cycle. To pose this as a concrete question, we ask when one could expect to see the results of a simulation on an intact *E. coli* covering 20 min, which is the approximate time for replication under optimal conditions.

We extrapolate from two simulations, the 10 ps simulation on bovine pancreatic trypsin inhibitor (BPTI) reported over twenty years ago [61] and the recent 1 μs simulation on the villin headpiece subdomain. [9] Each of these was a state-of-the-art simulation, using the best algorithms and the most powerful hardware available at the time.

The computational power, P, of a molecular dynamics simulation is proportional to the length of the time period being simulated (t, measured in seconds) and to the rth power of the number of heavy (nonhydrogen) atoms, n. The value of r depends on how one calculates the interatomic interactions. There are about $n^2/2$ pairwise interactions, so $r=2$ if all interactions are included in the calculation. At the other extreme, a common approach is to use a finite cutoff distance, in which case there are mn interactions, m being the average number of atoms within the cutoff distance of any given atom; since $m \ll n$, r approaches one, at least in the limit of very large systems. In short, $1 \le r \le 2$.

An exact determination of the relative values of P for the BPTI and villin simulations is not possible, because some algorithmic developments reduce computational costs (particularly methods that allow one to increase the size of the time step and to efficiently treat long-range interactions), while others increase the costs (e.g., more detailed force fields and appropriate boundary conditions). But we can place reasonable bounds on the historical growth rate of P by using $r=1$ and $r=2$ as lower and upper limits on the costs of calculating interatomic interactions.

As in all computer applications for the past few decades, P has been growing exponentially with time. It is a straightforward calculation to show that in the 21 years between the BPTI simulation ($t \approx 10^{-11}$s; $n \approx 600$) and the villin headpiece simulation ($t \approx 10^{-6}$ s; $n \approx 12\ 000$), P has been increasing by a factor of 10 every 3–4 years. A simple extrapolation would thus predict that a simulation covering the replication cycle of an intact *E. coli* with a volume of about 1 µ^3 ($t \approx 1000$ s; $n \approx 3 \times 10^{10}$) could be expected some time in the second half of the next century if computational power continued to grow exponentially at historical rates.

Clearly, such a projection is fanciful. Although it is widely believed that computer power will continue to grow at current rates for about the next decade or so, it is not possible to predict how long such a growth rate can continue. On the other hand, it is not necessary to simulate the full life cycle of an intact bacterium. Such a simulation would be extreme overkill, since it would describe the synthesis and folding of every molecule in the cell, with atomic detail.

It is, however, important to note what might be accomplished if the growth rate calculated above can be continued for another 10–20 years. If this happens, simulations on proteins with 100–200 residues can be expected to reach into the millisecond domain, and simulations covering a full second would be attainable within about twenty years. Recent demonstrations of the accuracy of modern molecular force fields [62,63] hold the promise that unbiased molecular dynamics simulations could follow the folding process all the way from the completely unfolded state to the native state, a truly exciting prospect.

References

1 KOEHL P. and LEVITT M. "A brighter future for protein structure prediction." Nat. Struct. Biol. (1999) 6(2): 108–111.

2 STRUTHERS M. D., CHENG R. P. and IMPERIALI B. "Design of a monomeric 23-residue polypeptide with defined tertiary structure." Science (1996) 271(5247): 342–345.

3 DAHIYAT B. I. and MAYO S. L. "De novo protein design: Fully automated sequence selection." Science (1997) 278(5335 (Oct. 3)): 82–87.

4 REGAN L. and WELLS J. "Recent adventures in molecular design." Curr. Op. Str. Biol. (1998) 8(4): 441–442.

5 ANFINSEN C. B. "Principles that govern the folding of protein chains." Science (1973) 181: 223–230.

6 PRUSINER S. B. "Molecular biology of prion deseases." Science (1991) 252: 1515–1522.

7 PRUSINER S. B. "Prion diseases and the BSE crisis." Science (1997) 278(5336): 245–251.

8 BAUM J. and BRODSKY B. "Folding of peptide models of collagen and misfolding in disease." Curr. Op. Str. Biol. (1999) 9: 122–128.

9 DUAN Y. and KOLLMAN P. A. "Pathways to a protein folding intermediate observed in a 1 μs simulation in aqueous solution." Science (1998) 282: 740–744.

10 MAKHATADZE G. I. and PRIVALOV P. L. "Energetics of protein structure." Adv. Prot. Chem. (1995) 47: 307–425.

11 FERSHT A. R. "Characterizing transition states in protein folding: an essential step in the puzzle." Curr. Op. Str. Biol. (1995) 5: 79–84.

12 ELIEZER D., YAO J., DYSON H. J. and WRIGHT P. E. "Structural and dynamic characterization of partially folded states of apomyoglobin and implications for protein folding." Nat. Struct. Biol. (1998) 5(2): 148–155.

13 ZHANG O., KAY L. E., SHORTLE D. and FORMAN-KAY J. D. "Comprehensive NOE characterization of a partially folded large fragment of staphylococcal nuclease Delta131Delta, using NMR methods with improved resolution." J. Mol. Biol. (1997) 272(1): 9–20.

14 KUWAJIMA K., YAMAYA H., MIWA S., SUGAI S. and NAGAMURA T. "Rapid formation of secondary structure framework in protein folding studied by stopped-flow circular dichroism." FEBS Lett. (1987) 221(1): 115–118.

15 KUWAJIMA K. "The molten globule state as a clue for understanding the folding and cooperativity of globular-protein structure." Proteins (1989) 6: 87–103.

16 PTITSYN O. B. "Structure of folding intermediates." Curr. Op. Str. Biol. (1995) 5: 74–78.

17 BALLEW R. M., SABELKO J. and GRUEBELE M. "Direct observation of fast protein folding: the initial collapse of apomyoglobin." Proc. Natl. Acad. Sci., USA (1996) 93: 5759–5764.

18 WILLIAMS S., CAUSGROVE T. P., GILMANSHIN R., FANG K. S., CALLENDER R. H., WOODRUFF W. H. and DYER R. B. "Fast events in protein folding: helix melting and formation in a small peptide." Biochemistry (1996) 35(3): 691–697.

19 GILMANSHIN R., WILLIAMS S., CALLENDER R. H., WOODRUFF W. H. and DYER R. B. "Fast events in protein folding: relaxation dynamics of secondary and tertiary structure in native apomyoglobin." Proc. Natl. Acad. Sci., USA (1997) 94(8): 3709–3713.

20 LU H. P., XUN L. Y. and XIE X. S. "Single-molecule enzymatic dynamics." Science (1998) 282(5395): 1877–1882.

21 Go N. "Theoretical studies of protein folding." Annu. Rev. Biophys. Bioeng. (1983) **12**: 183–210.

22 Bashford D., Weaver D. L. and Karplus M. "Diffusion-collision model for the folding kinetics of the lambda-repressor operator-binding domain." J. Biomol. Str. & Dyn. (1984) **1**: 1243–1255.

23 Dill K. A. "Dominant forces in protein folding." Biochemistry (1990) **29**: 7133–7155.

24 Karplus M. and Weaver D. L. "Protein-folding dynamics – the diffusion-collision model and experimental-data." Prot. Sci. (1994) **3**(4): 650–668.

25 Dill K. A., Bromberg S., Yue K. Z., Fiebig K. M., Yee D. P., Thomas P. D. and Chan H. S. "Principles of protein-folding – a perspective from simple exact models." Prot. Sci (1995) **4**(4): 561–602.

26 Onuchic J. N. and Wolynes P. G. "Energy landscapes, glass transitions, and chemical-reaction dynamics in biomolecular or solvent environment." J. Chem. Phys. (1993) **98**(3): 2218–2224.

27 Bryngelson J. D., Onuchic J. N., Socci N. D. and Wolynes P. G. "Funnels, pathways, and the energy landscape of protein-folding – a synthesis." Proteins (1995) **21**(3): 167–195.

28 Wolynes P. G., Onuchic J. N. and Thirumalai D. "Navigating the folding routes." Science (1995) **267**(March 17): 1619.

29 Dill K. A. and Chan H. S. "From Levinthal to pathways to funnels." Nat. Struct. Biol. (1997) **4**(1): 10–19.

30 Lazaridis T. and Karplus M. "'New view'of protein folding reconciled with the old through multiple unfolding simulations." Science (1997) **278**(N5345): 1928–1931.

31 Skolnick J. and Kolinski A. "Simulations of the folding of a globular protein." Science (1990) **250**: 1121–1125.

32 Sali A., Shakhnovich E. and Karplus M. "How does a protein fold." Nature (1994) **369**(6477): 248–251.

33 Ortiz A. R., Kolinski A. and Skolnick J. "Fold assembly of small proteins using Monte Carlo simulations driven by restraints derived from multiple sequence alignments." J. Mol. Biol. (1998) **277**(2): 419–448.

34 Ishikawa K., Yue K. and Dill K. A. "Predicting the structures of 18 peptides using Geocore." Prot. Sci. (1999) **8**(4): 716–721.

35 Tobias D. J., Mertz J. E. and Brooks C. L., III "Nanosecond time scale folding dynamics of a pentapeptide in water." Biochemistry (1991) **30**(24): 6054–8.

36 Tirado-Rives J. and Jorgensen W. L. "Molecular dynamics simulations of the unfolding of an alpha-helical analogue of ribonuclease A S-peptide in water." Biochemistry (1991) **30**(16): 3864–71.

37 Daggett V. and Levitt M. "A model of the molten globule state from molecular dynamics simulations." Proc. Natl. Acad. Sci., USA (1992) **89**(11): 5142–5146.

38 Daggett V. and Levitt M. "Protein folding unfolding dynamics." Curr. Op. Str. Biol. (1994) **4**(2): 291–295.

39 Alonso D. O. V. and Daggett V. "Molecular dynamics simulations of protein unfolding and limited refolding: characterization of partially unfolded states of ubiquitin in 60% methanol and in water." J. Mol. Biol. (1995) **247**: 501–520.

40 Li A. and Daggett V. "Identification and characterization of the unfolding transition state of chymotrypsin inhibitor 2 by molecular dynamics simulations." J. Mol. Biol. (1996) **257**(2): 412–429.

41 Pande V. S. and Rokhsar D. S. "Molecular dynamics simulations of unfolding and refolding of a β hairpin fragment of protein G." (1999) (preprint).

42 Finkelstein A. V. "Can protein unfolding simulate protein folding?" Prot. Eng. (1997) **10**(8): 843–845.

43 Wang L., Duan Y., Shortle R., Imperiali B. and Kollman P. A. "Study of the stability and unfolding mechanism of BBA1 by molecular dynamics simulations at different temperatures." Proteins Science (1999) **8**: 1292–1304.

44 Boczko E. M. and Brooks C. L., III "Constant-temperature free-energy surfaces for physical and chemical processes." J. Phys. Chem. (1993) **97**(17): 4509–4513.

45 Sheinerman F. B. and Brooks C. L., III "Calculations on folding of segment B1 of streptococcal protein G." J. Mol. Biol. (1998) **278**(2): 439–456.

46 DEMCHUK E., BASHFORD D. and CASE D. A. "Dynamics of a type VI reverse turn in a linear peptide in aqueous solution." Fold. & Des. (1997) **2**(1): 35–46.

47 DAURA X., JAUN B., SEEBACH D., VAN GUNSTEREN W. F. and MARK A. E. "Reversible peptide folding in solution by molecular dynamics simulation." J. Mol. Biol. (1998) **280**: 925–932.

48 CHIPOT C. and POHORILLE A. "Structure and dynamics of small peptides at aqueous interfaces – a multi-nanosecond molecular dynamics study." Theochem.-J. Mol. Str. (1997) **398**: 529–535.

49 CHIPOT C. and POHORILLE A. "Folding and translocation of the undecamer of poly-L-leucine across the water-hexane interface. A molecular dynamics study." J. Am. Chem. Soc. (1998) **120**(46): 11912–11924.

50 SHAKHNOVICH E. I. "Theoretical studies of protein-folding thermodynamics and kinetics." Curr. Op. Str. Biol. (1997) **7**(1): 29–40.

51 McKNIGHT C. J., DOERING D. S., MATSUDAIRA P. T. and KIM P. S. "A thermostable 35-residue subdomain within villin headpiece." J. Mol. Biol. (1996) **260**: 126–134.

52 McKNIGHT C. J., MATSUDAIRA P. T. and KIM P. S. "NMR structure of the 35-residue villin headpiece subdomain." Nat. Struct. Biol. (1997) **4**(3): 180–184.

53 PLAXCO K. W. (1998) (personal communication).

54 PLAXCO K. W., SIMONS K. T. and BAKER D. "Contact order, transition state placement and the refolding rates of single domain proteins." J. Mol. Biol. (1998) **277**(4): 985–994.

55 DUAN Y., WANG L. and KOLLMAN P. A. "The early stage of folding of villin headpiece subdomain observed in a 200-nanosecond fully solvated molecular dynamics simulation." Proc. Natl. Acad. Sci., USA (1998) **95**(17): 9897–9902.

56 ZWANZIG R. "Two-state models of protein folding kinetics." Proc. Natl. Acad. Sci., USA (1997) **94**(1): 148–150.

57 JACOB M., SCHINDLER T., BALBACH J. and SCHMID F. X. "Diffusion Control in an Elementary Protein Folding Reaction" Proc. Natl. Acad. Sci., USA (1997) **94**: 5622–5627.

58 TAO P. and SOSNICK T. R. "Intermediates and kinetic traps in the folding of a large ribozyme revealed by circular dichroism and UV absorbance spectroscopies and catalytic activity." Nat. Struct. Biol. (1997) **4**(11): 931–938.

59 SOHL J. L., JASWAL S. S. and AGARD D. A. "Unfolded conformations of alpha-lytic protease are more stable than its native state." Nature (1998) **395**(6704): 817–819.

60 SIGLER P. B., XU Z. H., RYE H. S., BURSTON S. G., FENTON W. A. and HORWICH A. L. "Structure and function in GroEL-mediated protein folding." Annu. Rev. Biochem. (1998) **67**: 581–608.

61 McCAMMON J. A., GELIN B. R. and KARPLUS M. "Dynamics of folded proteins." Nature (1977) **267**: 585–590.

63 CHEATHAM T. E., III and KOLLMAN P. A. "Observation of the A-DNA and B-DNA transition during unrestrained molecular dynamics in aqueous solution." J. Mol. Biol. (1996) **259**: 434–444.

64 CHEATHAM T. E., III, MILLER J. L., FOX T., DARDEN T. A. and KOLLMAN P. A. "Molecular dynamics simulations on solvated biomolecular systems: the particle mesh Ewald method leads to stable trajectories of DNA, RNA, and proteins." J. Am. Chem. Soc. (1995) **117**(14): 4193–4194.

7
The Enzymology of Biological Nitrogen Fixation

Robert H. Burris

7.1
Early History

The history of biological nitrogen fixation goes back to ancient times, as leguminous crops have been cultivated for their inherent values, and because it was recognized that other crops grown in association with legumes benefited from the association. The first rational explanation for the basis of this benefit was given by Boussingault in 1838. His greenhouse experiments indicated an increase in the nitrogen content of peas and clover with no appreciable increase in wheat or oats. His later experiments in the field verified that the legume lucerne (alfalfa) increased in nitrogen content, whereas potatoes, wheat, turnips, mangel, oats, and rye did not increase appreciably. His conclusions were challenged, as they did not conform to the beliefs of his contemporaries. Liebig was particularly critical. As Liebig was the leading chemist of his day, he apparently did not feel it necessary to do any experiments, as his word that Boussingault was wrong would be adequate to debunk the concept that N_2 could be fixed from the air by plants. He advanced the idea that ammonia from the air supplied the necessary nitrogen to the plants, but he supplied no experimental evidence in support of his position.

The French and Germans seldom saw eye to eye on matters, so Ville in France performed a number of carefully controlled experiments and verified that his plants exhibited gains in nitrogen content. To resolve the controversy, the French Academy of Science appointed a commission to check the matter. The group wanted to perform the experiments properly, so they cleaned the greenhouse and placed the plants on a newly painted lead sheet. The plants succumbed to the paint fumes. Experiment by committee obviously has its drawbacks. A repeat experiment was attended by the mishap of ammonia being evaporated near the water supply. Eventually tests satisfactory to the commission were completed, and not surprisingly they supported the position of Boussingault rather than the position of Liebig.

Boussingault reran some of his experiments but omitted some of his better nitrogen-fixing plants. His results were less convincing than those of his earlier work, and he became less supportive of their validity. In 1857 the Rothamsted group of Lawes, Gilbert, and Pugh decided to do some experiments. In their field plots they had observed apparent benefits to yields in the presence of leguminous plants. In their controlled greenhouse tests they grew the plants in closed chambers in soil that had been heated to a high temperature. They scrubbed the air and added back carbon dioxide. They supplied distilled water that was free of ammonia. At the end of the experiment they analyzed the plants, the soil, and even broke up the pots for analysis. They found no evidence of nitrogen fixation. Apparently their drastic measures had killed the nitrogen-fixing bacteria. Their experiments were so carefully executed that they were accepted as definitive for years.

Some questions were raised, and probably the best experimental evidence for fixation was offered by Atwater working at New Haven, Connecticut. In rather extensive experiments in the greenhouse and field, he found that peas exhibited a marked increase in total nitrogen during their period of growth. The most convincing experiments, however, were presented by Hellriegel and Wilfarth in 1886 at a meeting in Berlin. They had observed spotty results with their tests of the N_2-fixing ability of pea plants. They observed that the pea plants that grew well always had nodules on their roots. These nodules carried many bacteria. When plants were grown on sterilized soil they grew poorly, and the occasional thrifty plants were nodulated by chance contamination. Plants to which untreated soil was added grew well and were uniformly nodulated. It is interesting that Gilbert, of Lawes, Gilbert, and Pugh, chaired the 1886 meeting at which Hellriegel and Wilfarth presented their data. Their results were published in 1887 and 1888. [1] It was 1888 before the Rothamsted group repeated the experiments of Hellriegel and Wilfarth. Lawes and Gilbert [2] published verification of the Hellriegel and Wilfarth observations in 1891, so this opened the field for general investigation. Pugh in the meantime had gone back to Penn State, and he soon joined the administration there. He later became president of the institution. Although we hear little subsequently of Pugh in science, there is a street named after him in State College, Pennsylvania.

7.2
Practical Applications

The acceptance that leguminous plants in association with root-nodule bacteria can fix N_2 was followed by investigation of practical applications of the findings to agricultural practice. Certain bacteria can fix N_2 on their own without being associated with a plant. The first clear-cut case of such free-living fixation was reported in 1893 by Winogradsky [3] who found that the anaerobic organism *Clostridium pasteurianum* could fix N_2 as a free-living organism. In 1901, Beijerinck [4] reported that the aerobic organism *Azotobacter chroococcum* also was a vigorous N_2 fixer. However, it could not be demonstrated that adding these free-living fixers to the soil conferred any benefit to plant growth.

For some years, work in the area of biological nitrogen fixation centered on practical agronomic aspects and the microbiology of the symbiotic association. It was recognized that exploitation of the process could be of substantial benefit to agriculture, so the rhizobia were isolated, cultured, and added to the leguminous seeds at the time of planting. The rhizobia were variously supplied as liquid cultures, agar cultures, or cultures that had been transferred to solids such as powdered peat.

There was varied success with these inocula, as a number of unscrupulous operators supplied ineffective inoculants. Some states in the United States established laws that required inoculant suppliers to have their inoculants certified annually by a state agency. This generally was successful in discouraging distribution of ineffective inocula.

Media were devised that supported consistent growth of the rhizobia, and mannitol often was favored as the energy source in the media. Yeast extract commonly was added to supply micronutrients. Selection of effective strains of the rhizobia was an empirical process based on greenhouse and field testing. Usually, several effective strains were grown on liquid media and then mixed on a solid support for distribution.

7.3
Biochemistry of N_2 Fixation

Elucidation of the biochemistry of nitrogen fixation certainly did not interest early workers in the field. Perhaps the first serious concern in this area was evidenced when Dean Burk initiated studies with Meyerhof. [5] They made measurements of the response in nitrogen fixation by *Azotobacter* to changes in the partial pressure of N_2. Although the data were somewhat inaccurate, they set the stage for subsequent studies by Wilson and collaborators. These early studies have been reviewed thoroughly by Wilson in his book on *The Biochemistry of Symbiotic Nitrogen Fixation*. [6] The book also gives an excellent review of earlier investigations in the field. For his studies, Wilson chose the symbiotic system of red clover and *Rhizobium trifolii*, although this posed many complications not inherent in studies with a free-living organism such as *Azotobacter chroococcum*. Typically, Wilson's group grew red clover in sterilized sand in 9-liter serum bottles. The seeds were surface sterilized, germinated aseptically, inoculated with *Rhizobium trifolii*, and planted in the serum bottles. These bottles were stoppered, supplied with a suitable gas atmosphere, and then supplied with CO_2 as needed. Depletion of CO_2 was indicated by a change in color of cresol red indicator in 0.001 M sodium bicarbonate solution contained in a vial hanging in the bottle. Gas atmospheres were renewed weekly. With this type of experimental set-up, Wilson and his collaborators established that the Michaelis constant for nitrogen fixation was about 0.05 atm. N_2. It also turned out that oxygen was inhibitory to N_2 fixation. Most surprising, however, was the observation that H_2 was inhibitory. Normally He was used as an inert gas to bring pressure in the serum bottles up to one atmosphere. However, on an occasion when the He supply was exhausted, H_2 was substituted with the assumption that it was inert. It appeared that it reduced the growth of N_2-fixing plants but did not have any effect on controls supplied with combined nitrogen. New tanks of H_2 gave the same response. Also, H_2 generated with aluminum plus sodium hydroxide, with zinc plus hydrochloric acid or electrolytically in the laboratory inhibited N_2 fixation. It had to be concluded that H_2 *per se* was inhibitory.

The first observations on H_2 inhibition were made in the 1930s. It was established that the inhibition was specific (H_2 had no effect on growth of the red clover plants on a fixed nitrogen compound such as an ammonium salt), and H$_2$ was a competitive inhibitor, i.e., H$_2$ competed with N$_2$ so that when the pN$_2$ was increased the effect of H$_2$ was decreased. It was a good many years, however, before the real nature of the inhibition was established in more detail. [7]

7.4
First Product of N$_2$ Fixation

In the 1930s, a controversy that attracted a good bit of attention was initiated when Virtanen and coworkers reported that leguminous plants excreted nitrogen from their root systems, and the fixed nitrogen was primarily in the form of aspartic acid. From this, they developed the hypothesis that nitrogen was fixed as hydroxylamine, which then was combined with oxalacetic acid to form an oxime. The oxime in turn was reduced to aspartic acid and was excreted. Virtanen assembled most of his data in a book entitled *Cattle Fodder and Human Nutrition*. [8] The problems with his hypothesis were that others could not obtain excretion when they repeated the experiments, and they could find no evidence for the postulated oxime. P.W. Wilson went to Virtanen's laboratory and over a period of months was unable to observe excretion in the Helsinki laboratories. He returned to his laboratory in Madison, Wisconsin, with Virtanen's seeds, his bacterial inoculum, and even a large container of Virtanen's soil. As Virtanen attributed the failures to observe excretion to seasonal effects, Wilson's group in Wisconsin set up experiments monthly to cover this suggested difficulty. If there was any excretion of nitrogenous compounds in their many experiments, it was minimal. Other laboratories did not report verification of Virtanen's observations.

The other hypothesis suggested earlier was that ammonia rather than hydroxylamine was the product of biological nitrogen fixation. Winogradsky [9] had suggested this, and although the idea was perfectly logical, the evidence was indirect and minimal.

7.5
Studies with ^{15}N as a Tracer

Wayne Umbreit and I were graduate students in P. W. Wilson's laboratory, and we sometimes discussed the possibility of applying the stable isotope of nitrogen, ^{15}N, to studies of nitrogen fixation. Schoenheimer and Rittenberg [10] had been applying ^{15}N very successfully to the investigation of protein metabolism in animals. So I applied for and received a postdoctoral National Research Council fellowship for studies with ^{15}N in the laboratory of Harold Urey at Columbia University. He had concentrated ^{15}N in long exchange columns and had supplied the material to Schoenheimer and Rittenberg (a former student of Urey's) for their experiments. Dr. Urey was very helpful and pointed out that it would be advisable to first establish that nitrogenase catalyzed no exchange reaction that would confuse our results with

^{15}N. So Charles Miller and I [11] set up an experiment with N$_2$-fixing *Azotobacter vinelandii* under an atmosphere containing ^{15}N -enriched N$_2$ diluted with normal N$_2$ to give a non-equilibrium mixture of N$_2$ of masses 28, 29, and 30. If an exchange occurred that resulted in the dissociation and reassociation of the N$_2$ molecules, the non-equilibrium mixture would approach equilibrium and this equilibration could be measured with a mass spectrometer. There was no suggestion of equilibration in our tests, so we concluded that ^{15}N would be a valid tracer for our experiments. Ruben, Hassid and Kamen [12] had reported evidence from a 20-min exposure of barley plants to the radioactive tracer ^{13}N (half-life 10.5 min) that the barley plants fixed N$_2$. Being skeptical that a non-legume such as barley was capable of fixation, we repeated the experiment with ^{15}N, but stretched the exposure time from 20 min to 54 days and exposed inoculated clover and barley under the same atmosphere. The clover had an abundant enrichment in ^{15}N, whereas there was no detectable increase in ^{15}N concentration in the barley. [13]

To try to determine in what compounds the ^{15}N was concentrating after short ex-posures to ^{15}N$_2$, I joined Schoenheimer and Rittenberg in their laboratory, as they had experience in amino acid separations. A culture of *Azotobacter vinelandii* was exposed for 75 min to ^{15}N$_2$, and then the cells were collected and hydrolyzed, and their amino acids were recovered as various solid derivatives. These derivatives were digested and analyzed for their ^{15}N concentration. The highest concentration of ^{15}N was found in glutamic acid. [14] This suggested that ^{15}N$_2$ was fixed as ammonia and that the ammonia was assimilated by reductive amination of alpha-keto glutaric acid. Virtanen's hypothesis would have placed the highest ^{15}N concentration in aspartic acid rather than glutamic acid. It turned out later that assimilation of ammonia is primarily through the GOGAT system rather than by glutamic dehydro-genase, but the GOGAT system was unknown at the time the experiments were per-formed. In subsequent years, experiments comparable to those done with *Azotobac-ter vinelandii* were repeated with *Clostridium pasteurianum*, *Rhodospirillum rubrum*, *Chromatium* sp., *Chlorobium* sp., blue-green algae, and nodulated soybean plants. With every nitrogen-fixing agent the highest ^{15}N concentration was found in the glutamic acid derivative recovered. [15] This constituted strong support for the ammonia hypothesis of nitrogen fixation and for the uniformity of the process among nitrogen-fixing agents.

7.6
N$_2$ Fixation with Cell-Free Preparations

It obviously was difficult to establish details of the N$_2$ fixation process with intact organ-isms, as the ammonia fixed was rapidly assimilated into other compounds. So a search was initiated in several laboratories for a cell-free preparation that would fix N$_2$.

We tried preparations from a variety of N$_2$-fixing agents, but responses were vari-able. Sometimes we achieved rather good fixation with cell-free preparations, but the results were not consistent. For example, at a meeting in early 1958, we reported that cell-free preparations from *Clostridium pasteurianum* exposed under an atmosphere of $^{15}N_2$ assimilated the N_2 to give ^{15}N levels of 0.5 atom% ^{15}N excess. Earlier, Magee and Burris [16] had reported fixation by cell-free preparations from *Azotobacter vinelandii*. As we could detect 0.003 atom% ^{15}N excess with our mass spectrometer, we adopted 0.015 atom% excess as a reasonable level to indicate fixation. On this basis, the Magee paper reported five specific instances of fixation. However, the levels of fixation were much lower than those reported by Hoch and Westlake in 1958. [17] They found that cell-free preparations from *Clostridium pasteurianum* exposed to ^{15}N-enriched N_2 gave atom% ^{15}N excess, levels far above the arbitrary level of 0.015 accepted as evidence for fixation. They reported atom% ^{15}N excess values of 0.644, 0.812, 0.823, 0.301, 0.929, and 1.251 after exposure for 1 h to 0.05 atm. of 95 atom% $^{15}N_2$. The problem was not that we never obtained cell-free N_2 fixation, but that fixation was inconsistent. [18]

Carnahan et al. [19] in 1960 also reported N_2 fixation with cell-free preparations from *C. pasteurianum*. They dried their cells in a rotary evaporator at about 40 °C and then extracted them to make their preparations. Their other innovation was to supply high levels of pyruvate to support fixation. The method was reproducible, and we promptly verified their results. [20] Fixation for 3 h in our tests with added pyruvate gave 1.334 atom% ^{15}N excess. Fixation with other substrates was less vigorous; lactate supported fixation to give 0.028 and 0.033 atom% ^{15}N excess.

7.7
Nitrogenase Consists of Two Proteins

So by 1960 it was possible to achieve consistent fixation with the Carnahan et al. cell-free preparations of *C. pasteurianum* supplied with generous levels of pyruvate and incubated anaerobically. Mortensen [21] reported that the nitrogen-fixing enzyme system consisted of two or perhaps more proteins. A number of laboratories got into the business of purifying the enzyme system and verified that in fact two enzyme components were involved. Nomenclature became a problem, as different laboratories had different ideas on the logical names to attach to the two proteins. The two proteins could be separated on a DEAE cellulose chromatographic column, and the first component eluted often was referred to as component 1 and the second as component 2, or they were called protein 1 and protein 2. As they contained molybdenum and iron or iron only, 1 and 2 were sometimes called the MoFe and Fe proteins. Another variation was molybdofer and azofer. We have preferred the designation nitrogenase for the complete complex, dinitrogenase for the MoFe protein, and dinitrogenase reductase for the Fe protein. Although it has other functions, the main function of the Fe protein is the reduction of the MoFe protein.

As indicated, a number of laboratories got into the process of purifying the components of nitrogenase, so it is difficult to credit any particular laboratory with first success in the endeavor. In 1972, Tso *et al.* [22] reported purification of dinitrogenase and dinitrogenase reductase to the highest specific activities reported up to that time. Preparations from different organisms require modification of the techniques. Winter and Burris [23] reported a table of activities of components prepared up to 1976.

7.8
ATP Furnishes Energy for Fixation

In their 1960 paper, Carnahan et al. reported that ATP was inhibitory to nitrogenase activity in their cell-free preparations. Hence, when McNary and Burris [24] reported that ATP was needed to support fixation, the report was met with a good deal of skepticism. But experiments in a number of other laboratories verified the absolute need for ATP. Not only is ATP needed, it is needed in substantial amounts. Under ideal conditions 16 ATP are required per N_2 reduced to 2 NH_3. Under normal conditions in nature the requirement probably is in the 20 to 30 ATP per N_2 range. N_2 reduction is energy demanding whether it is accomplished chemically in the Haber process or enzymatically by the nitrogenase system.

7.9
H₂ an Obligatory Product of the Nitrogenase Reaction

It is interesting that not only is H_2 a specific inhibitor of N_2 fixation, but it also is an obligatory product of N_2 fixation. The N_2 fixation reaction can be formulated as:

$$N_2 + 8\,H^+ + 8\,e^- + 16\,ATP \rightarrow 2\,NH_3 + H_2 + 16\,ADP + 16\,Pi$$

This formulation indicates the stoichiometry under ideal conditions, and under most conditions the requirement for ATP will be considerably greater. As one increases the pN_2 the production of H_2 decreases. There were many discussions about whether one could completely suppress H_2 production at a very high pN_2, but there was a reluctance to perform the necessary experiments. Finally Simpson rigged up apparatus that allowed him to pressurize a reaction bomb unit up to 51 atm. of N_2. When his reaction was run with cell-free nitrogenase in the bomb, still about 27 % of the electrons were used in the production of H_2. [25] Extrapolation of the data indicate that 25 % of the electrons still will go to H_2 production at infinite pN_2. So one must accept that H_2 production is obligatory. Apparently the only recourse for improving the efficiency of N_2 fixation beyond the efficiency formulated above is to recycle the H_2 to recover some of its energy. As a number of N_2-fixing organisms possess hydrogenases, they have the potential to couple H_2 oxidation to the production of ATP and to utilize this ATP in the reduction of N_2. Evans et al. [26] have demonstrated higher fixation of N_2 in soybean plants infected with rhizobia possessing hydrogenase than in control plants infected with rhizobia devoid of hydrogenase.

Another interesting response in hydrogen metabolism was observed by Hoch et al. [27] They found that N_2-fixing organisms exposed to D_2 in the presence of N_2 formed HD. They referred to this as an exchange reaction. Later Bulen [28] indicated that it was improper to refer to this as an exchange reaction as it involved electron transfer. He also pointed to the factors in common between H_2 inhibition and HD formation.

Guth and Burris [7] used the HD reaction in reexamining the nature of H_2 inhibition. Their experiments indicated that H_2 inhibition results from an ordered, sequential reaction with H_2 binding before N_2. When D_2 is present, as an analog of H_2, it binds before N_2 forming ED_2. Then N_2 binds to produce the abortive complex ED_2N_2. It decomposes with the addition of 2 H^+ to yield $E + N_2 + 2$ HD. Thus D_2 (or H_2) competes with N_2 for the enzyme site, and if D_2 (or H_2) gets there before N_2, when the N_2 arrives it forms a complex that aborts before the N_2 can be reduced. If the enzyme binds N_2 first, the N_2 is reduced to 2 NH_3. So there is a competition for binding between N_2 and H_2 leading to the observed competitive inhibition.

7.10
N_2 and HD Formation

There has been some controversy about the need for N_2 in the formation of HD. Burgess et al. [29] reported that N_2 was not required. They used argon as their diluent gas and took the word of the supplier that it was free of N_2. Not only is commercial argon seldom free of N_2, but it is difficult to remove the last traces of N_2, and very little N_2 is required to support HD formation. To settle this difference in experimental observations, Li and Burris [30] made it a point to rid their diluent gas of contaminating N_2. One can absorb N_2 on molecular sieve at liquid N_2 temperature; the problem is that argon liquefies and freezes before you get down to the temperature of liquid N_2. So Li used neon as his inert gas and captured any contaminating N_2 on molecular sieve in a liquid N_2 bath. When the atmosphere above the nitrogenase system was carefully freed of N_2 there was no formation of HD.

7.11
Electron Transfer Sequence

With the availability of consistently active cell-free preparations of nitrogenase, it became feasible to investigate the enzymology of the system in greater detail. Cell-free preparations can be prepared from several N_2-fixing organisms, but *C. pasteurianum* and *A. vinelandii* have been used most commonly. It became apparent that the electron transfer sequence in the system was from dinitrogenase reductase to dinitrogenase to the substrate. Electrons were passed to dinitrogenase reductase (the Fe protein) and its potential was lowered to about −400 mv by the hydrolysis of ATP. At this potential it could pass electrons to dinitrogenase (the MoFe protein), and when dinitrogenase was reduced, it in turn could reduce the substrate N_2 to 2 NH_3. A convenient electron donor, commonly used in reconstructed systems, is sodium dithionite ($Na_2S_2O_4$). Physiological electron donors in intact systems can be ferredoxins, rubredoxins, and flavodoxins.

7.12
Alternative Substrates

Nitrogenase is a versatile enzyme system that can effect the reduction of a variety of substrates in addition to N_2. It was 1954 before the first substrate other than N_2 was recognized; Mozen and Burris [31] reported that nitrous oxide was a substrate for nitrogenase. Since then a number of other substrates have been reported including: H^+, acetylene, azide, cyanide, methyl isocyanide, cyclopropene, cyanamid and diazirine. Most of these have been only laboratory curiosities, but acetylene [32] has been a very useful substrate for measuring N_2 fixation both in the laboratory and in the field. Its reduction yields ethylene, and the ethylene can be measured easily by gas chromatography. Separation of ethylene on a gas chromatographic column is rapid, and a flame ionization detector for measurement has high sensitivity and reproducibility. One can inject acetylene into a container holding legume root nodules, walk across the room, inject gas from the container into a separating column, and after that short time, observe the ethylene peak. The sensitivity and simplicity of the method make it very useful in the field, so it has largely displaced other methods for detecting and quantitating N_2 fixation. For the greatest specificity, however, the reduction of ^{15}N-enriched N_2 still is the method of choice.

7.13
N_2 Fixation in Non-Leguminous Plants

Attention in symbiotic N_2 fixation has centered on the thousands of leguminous plants capable of N_2 fixation. This is hardly surprising when one considers the economic importance of soybeans, alfalfa, clover, beans, and many other legumes. However, the actinorhizal systems have substantial importance, not so much in crop plants as in maintaining the fixed nitrogen level in special areas and in fixation by shrubs and trees. Not only can a number of trees that are symbionts with actinorhizal organisms fix substantial amounts of nitrogen and grow rapidly in poor soils, but they often do well under arid conditions. They are being used increasingly in arid lands, along roadsides, and in plots for the production of wood. Alders, for example, have very wide distribution in a variety of habitats. *Comptonia peregrina* is being used extensively in plantings. *Leucaena leucocephala* exhibits phenomenally rapid growth and is being used in tropical areas.

The rhizobia are relatively easy to grow, and have been used as inocula for leguminous plants for well over a century. In contrast, the actinomycetes associated with non-leguminous plants resisted consistent culture until Callaham et al. [33] succeeded in culturing them. Since then, research on the actinorhizal systems has expanded rapidly, and non-leguminous N_2 fixers have been adopted widely as practical systems for cultivation. [34]

7.14
Control of Nitrogenase

Nitrogenase operates under an interesting control system. As indicated, N_2 fixation is an energy-demanding process, and if the organisms are supplied with an external source of fixed N, they often turn off their biological N_2 fixation systems. For example, if *Azotobacter vinelandii* is being grown on N_2, it turns off its nitrogenase immediately when a fixed nitrogen source, such as ammonium ion, is added to the medium. When the supply of ammonium ion is exhausted, the nitrogenase is turned back on promptly. [35] These transitions are effected by two enzymes termed dinitrogenase reductase ADP-ribosyl transferase (DRAT) and dinitrogenase reductase activating glycohydrolase (DRAG). [36] DRAT transfers an ADP-ribose moiety of NAD to arginine 101 of one of the two equivalent subunits of dinitrogenase reductase, and this inactivates the dinitrogenase reductase. Reactivation of the enzyme occurs when DRAG removes the ADP-ribose from the dinitrogenase reductase in a MgATP-dependent reaction.

7.15
Magnitude of Chemical and Biological N₂ Fixation

Chemical nitrogen fixation, primarily by the Haber-Bosch process, is a major chemical industry. Fixation worldwide may yield about 80×10^6 tons of fixed nitrogen yearly at the expense of large quantities of natural gas and with the generation of much CO_2. Biological fixation may yield about 120×10^6 tons of fixed nitrogen annually. This figure is not highly accurate, as biological fixation in the field is difficult to quantitate, and the data for the oceans are particularly subject to error. But it seems clear that the biological process, which depends primarily on photosynthesis for its energy supply, provides much more fixed nitrogen to our agricultural economy than does the fossil fuel-depleting, carbon dioxide-producing chemical fixation process.

The symbiotic N_2-fixing systems of legume associations and the actinorhizal associations account for most of the biological N_2 fixation on earth. There are free-living N_2 fixers that are important in certain niches. The cyanobacteria (or blue-green algae) are of substantial importance in fresh water and marine environments. Their contributions, particularly in the oceans, are difficult to quantitate. However, when one sees a wide expanse of *Trichodesmium* spp. surfacing on the ocean, one is impressed with their potential contribution of fixed nitrogen to the environment. Beach rock along the shoreline of the ocean in tropical areas often has a heavy cover of cyanobacteria that may fix N_2 and then be dislodged into the ocean by wave action.

Although free-living N_2 fixing bacteria such as the *Azotobacter* spp. and other species are readily recovered from the soil, there is little evidence that they contribute substantially to soil nitrogen. In the laboratory, they grow vigorously on N-free media, but in the field their growth is limited by the dearth of readily available energy-supplying substrates.

7.16
Associative Biological N_2 Fixation

There has been considerable interest recently in the importance of associative N_2 fixation (Becking [37] ; Dobereiner and Day [38]). In associative systems the microorganisms do not form specific N_2-fixing structures, such as nodules, on the plants. Rather, they grow externally, or sometimes internally, on plant roots, and utilize substrates excreted by the plants. These systems generally fix far less N_2 than the symbiotic systems, but when the plants supply an abundance of substrate, the fixation can be impressive. The most productive system that has been studied in some detail is associative N_2 fixation by *Acetobacter diazotrophicus* and sugar cane. Extensive studies in Brazil have demonstrated that sugar cane can be grown successfully without added fixed nitrogen, based on fixation by associated *A. diazotrophicus*. [39]

A. diazotrophicus is a remarkable organism, as it can grow in very concentrated sugar solutions and can fix N_2 at an unusually low pH. The optimal pH for most N_2 fixers is near neutrality, [40] and few fix below pH 5.0. *A. diazotrophicus* tolerates 10–20 % sugar solutions and grows on an N-free medium down to pH 3.0. Brazil has dramatically lowered its use of oil-based fuels by substituting ethanol-based fuels. Production of ethanol from sugar derived from a system that does not demand fixed nitrogen fertilizer contributes to an ecologically benign system.

7.17
Genetics of Biological N_2 Fixation

In recent years, much of the research on nitrogenase has centered on studies of the genetics of the system. Some 20 genes have been implicated in synthesis and control of nitrogenase. The structural genes for nitrogenase are HDK, with H and D being involved with dinitrogenase and K with dinitrogenase reductase. It was recognized rather early that Mo had some role in N_2 fixation. When the dinitrogenase component of the system was isolated and purified it proved to be a FeMo protein, and the implication was that these metals probably were at the active site of the enzyme. It proved possible to extract an FeMo unit from dinitrogenase (Shah and Brill [41]) and this was christened FeMoco, or iron-molybdenum cofactor. The genes responsible for the synthesis and insertion of FeMoco into the apoenzyme have been investigated in detail. [42] As indicated earlier, the FeMo unit, dinitrogenase, is reduced specifically by the Fe enzyme, dinitrogenase reductase, whereupon the reduced dinitrogenase can reduce N_2. At least 16 ATP are involved in this reduction of one N_2.

7.18
Composition and Structure of Nitrogenases

It is interesting that nature has developed dinitrogenases that utilize other metals as alternatives to Mo. Bishop and Premakumar [43] have established that vanadium-iron or iron-only dinitrogenases can function in place of Mo dinitrogenase. In the absence of Mo, the organisms can insert V into dinitrogenase instead of Mo; the active site of the enzyme then is at FeVco rather than at FeMoco. Likewise, FeFeco can serve as the all-iron active site.

Dinitrogenase has been crystallized and its tertiary structure determined by Kim and Rees. [44, 45, 46] As indicated, an Fe-Mo unit serves at the active site. Electrons are furnished to this active unit by the Fe enzyme dinitrogenase reductase. The two units together constitute nitrogenase.

7.19
Selection of N₂ Fixers

Although a great deal of information has accumulated regarding the structure, genetics and enzymology of nitrogenase in recent years, it has had only a minor influence on modifying or selecting rhizobia that function more efficiently or effectively in symbiotic N_2 fixation. We still depend upon empirical observations in the greenhouse or field in selecting bacterial strains. As indicated, rhizobia that can utilize H_2 that otherwise would be dissipated as a product of the nitrogenase reaction appear to be more effective symbionts than those devoid of hydrogenase, but to date this seems to be the only described property clearly associated with efficiency and effectiveness of symbiotic fixation. Apparently, even less is known about relative effectiveness among the actinomycetes that infect nonleguminous plants in the symbiotic actinorhizal systems.

References

1 HELLRIEGEL, H. and H. WILFARTH. Untersuchungen über die Stickstoffnahrung der Gramineen und Leguminosen. *Beilageheft zu der Ztschr. Ver. Rübenzucker-Industrie Deutschen Reichs*, 234 pp. (1888).

2 LAWES, J.B. and J.H. GILBERT. The sources of the nitrogen of our leguminous crops. *J. Roy. Agr. Soc., England, 3rd Ser.* 2, 657–702 (1891).

3 WINOGRADSKY, S. Sur l'assimilation de l'azote gaseux de l'atmosphere par les microbes. *C. R. Acad. Sci. Paris* 116, 1385–1388 (1893).

4 BEIJERINCK, M.W. Über oligonitrophile Mikroben. *Zentralbl. Bakteriol. Parasitenkd. Abt. II* 7, 561–582 (1901).

5 MEYERHOF, O. AND D. BURK. Über die Fixation des Luftstickstoffs durch *Azotobacter*. *Ztschr. Phys. Chem.* 139, 117–142 (1928).

6 WILSON, P.W. The Biochemistry of Symbiotic Nitrogen Fixation. Univ. of Wisconsin Press, pp. 302 (1940).

7 GUTH, J.H. and R.H. BURRIS. Inhibition of nitrogenase-catalyzed NH_3 formation by H_2. *Biochemistry* 22, 5111–5122 (1983).

8 VIRTANEN, A.I. Cattle Fodder and Human Nutrition. Cambridge University Press, London, 108 pp. (1938).

9 WINOGRADSKY, S. Sur l'origine de l'ammoniac dégagée par les fixateurs d'azote. *Zentbl. Bakt., 2Abt.* 97, 399–413 (1938).

10 SCHOENHEIMER, R. The Dynamic State of Body Constituents. Harvard University Press (1942).

11 BURRIS R.H. and C.E. MILLER. Application of N^{15} to the study of biological nitrogen fixation. *Science* 93, 114–115 (1941).

12 RUBEN, S., W.Z. HASSID and M.D. KAMEN. Radioactive nitrogen in the study of N_2 fixation by non-leguminous plants. *Science* 91, 578–579 (1940).

13 BURRIS, R.H. Failure of barley to fix molecular N^{15}. *Science* 94, 238–239 (1941).

14 BURRIS, R.H. Distribution of isotopic nitrogen in *Azotobacter vinelandii*. *J. Biol. Chem.* 143, 509–517 (1942).

15 WILSON, P.W. and R.H. BURRIS. Biological nitrogen fixation – A reappraisal. *Ann. Rev. Microbiol.* 7, 415–432 (1953).

16 MAGEE, W.E. and R.H. BURRIS. Oxidative activity and nitrogen fixation in cell-free preparations from *Azotobacter vinelandii*. *J. Bacteriol.* 71, 635–643 (1956).

17 HOCH, G.E. and D.W.S. WESTLAKE. Fixation of N_2 by extracts from *Clostridium pasteurianum*. *Fed Proc.* 17, 243 (1958).

18 BURRIS, R.H. Biological nitrogen fixation. *Ann. Rev. Plant Physiol.* 17, 155–184 (1966).

19 CARNAHAN, J.E., L.E. MORTENSON, H.F. MOWER and J.E. CASTLE. Nitrogen fixation in cell-free extracts of *Clostridium pasteurianum*. *Biochim. Biophys. Acta* 38, 188–189 (1960).

20 SCHNEIDER, K.C., C. BRADBEER, R.N. SINGH, L.C. WANG, P.W. WILSON and R.H. BURRIS. Nitrogen fixation by cell-free preparations from microorganisms. *Proc. Natl. Acad. Sci. USA* 46, 726–733 (1960).

21 MORTENSON, L.E. Nitrogen fixation in extracts of *Clostridium pasteurianum*. In: Non-heme Iron Proteins: Role in Energy Conversion, A. SAN PIETRO, Ed., Antioch Press, Yellow Springs, OH, pp. 243–269 (1965).

22 Tso, M.-Y., T. Ljones and R.H. Burris. Purification of the nitrogenase protein from *Clostridium pasteurianum*. *Biochim. Biophys. Acta* 267, 600–604 (1972).

23 Winter, H.C. and R.H. Burris. Nitrogenase. *Annu. Rev. Biochem.* 45, 409–426 (1976).

24 McNary, J.E. and R.H. Burris. Energy requirements for nitrogen fixation by cell-free preparations from *Clostridium pasteurianum*. *J. Bacteriol.* 84, 598–599 (1962).

25 Simpson, F.B. and R.H. Burris. A nitrogen pressure of 50 atmospheres does not prevent evolution of hydrogen by nitrogenase. *Science* 224, 1095–1097 (1984).

26 Evans, H.J., F.J. Hanus, R.A. Haugland, M.A. Cantrell, L.S. Xu, S.A. Russell, G.R. Lambert and A.R. Harker. Hydrogen recycling in nodules affects nitrogen fixation and growth of soybeans. Proceedings of the World Soybean Conference III, R. Shibles, Ed., Westview Press, Boulder, CO, pp. 935–942 (1985).

27 Hoch, G.E., K.C. Schneider and R.H. Burris. Hydrogen evolution and exchange and conversion of N_2O to N_2 by soybean root nodules. *Biochim. Biophys. Acta* 37, 273–279 (1960).

28 Bulen, W.A. Nitrogenase from *Azotobacter vinelandii* and reactions affecting mechanistic interpretations. *Proc. Int. Symp. Nitrogen Fixation*, 1st., pp. 177–186 (1976).

29 Burgess, B.K., S. Wherland, W.E. Newton and E.I. Stiefel. Nitrogenase reactivity: Insight into the nitrogen-fixing process through hydrogen-inhibition and HD-forming reactions. *Biochemistry* 20, 5140–5146 (1981).

30 Li, J.-L. and R.H. Burris. Influence of pN_2 and pD_2 on HD formation by various nitrogenases. *Biochemistry* 22, 4472–4480 (1983).

31 Mozen, M.M. and R.H. Burris. The incorporation of ^{15}N-labeled nitrous oxide by nitrogen fixing agents. *Biochim. Biophys. Acta* 14, 577–578 (1954).

32 Burris, R.H. The acetylene reduction technique. In: Nitrogen Fixation by Free-living Micro-organisms, Vol. 6, W.D.P. Stewart, Ed., Cambridge University Press, Cambridge, pp. 249–257 (1975).

33 Callaham, D., P. Del Tredici and J.G. Torrey. Isolation and cultivation *in vitro* of the actinomycete causing root nodulation in *Comptonia*. *Science* 199, 899–902 (1978).

34 Schwintzer, C.R. and J.D. Tjepkema. The Biology of Frankia and Actinorhizal Plants. Academic Press, San Diego, CA, 408 pp. (1990).

35 Ludden, P.W. and R.H. Burris. Removal of an adenine-like molecule during activation of dinitrogenase reductase from *Rhodospirillum rubrum*. *Proc. Natl. Acad. Sci. USA* 76, 6201–6205 (1979).

36 Ludden, P.W. and G.P. Roberts. The biochemistry and genetics of nitrogen fixation by photosynthetic bacteria. In: Anoxygenic Photosynthetic Bacteria, Kluwer Academic Publishers, Dordrecht, The Netherlands, pp. 929–947 (1995).

37 Becking, J.H. Fixation of molecular nitrogen by an aerobic *Vibrio* or *Spirillum*. *Antonie van Leeuwenhoek* 29, 326 (1963).

38 Döbereiner, J. and J.M. Day. Associative symbioses in tropical grasses: Characterization of micro-organisms and nitrogen-fixing sites. In: Proceedings of the 1st International Symposium on Nitrogen Fixation, W.E. Newton and C.J. Nyman, eds., Washington State University Press, Pullman, WA, Vol. 2, pp. 518–538 (1976).

39 Baldani, J.I., L. Caruso, V.L.D. Baldani, S.R. Goi and J. Döbereiner. Recent advances in BNF with non-legume plants. *Soil Biol. Biochem.* 29, 911–922 (1997).

40 Burris, R.H. Comparative study of the response of *Azotobacter vinelandii* and *Acetobacter diazotrophicus* to changes in pH. *Protoplasma* 183, 62–66 (1994).

41 Shah, V.K. and W.J. Brill. Isolation of an iron-molybdenum cofactor from nitrogenase. *Proc. Natl. Acad. Sci. USA* 74, 3249–3253 (1977).

42 Allen, R.M., R. Chatterjee, M. Madden, P.W. Ludden and V.K. Shah. Biosynthesis of the iron-molybdenum cofactor of nitrogenase. *Crit. Rev. Biotech.* 14, 225–249 (1994).

43 BISHOP, P.E. and R. PREMAKUMAR. Alternative nitrogen fixation systems. In: Biological Nitrogen Fixation, G. STACEY, R.H. BURRIS and H.J. EVANS, eds., Chapman and Hall, New York, pp. 736–762 (1992).

44 KIM, J. and D.C. REES. Structural models for the metal centers in the nitrogenase molybdenum–iron protein. *Science* **257**, 1677–1682 (1992).

45 KIM, J. and D.C. REES. Crystallographic structure and functional implications of the nitrogenase molybdenum–iron protein from *Azotobacter vinelandii*. *Nature* **360**, 553–560 (1992).

46 KIM, J. and D.C. REES. Nitrogenase and biological nitrogen fixation. *Biochemistry* **33**, 389–397 (1994).

8
The Chemistry of Nitrogen in Soils

Morris Schnitzer

8.1
Introduction

Except for small amounts of mineralogically fixed nitrogen (N), N is the only essential plant nutrient that is not formed from the weathering of minerals. Its source is the atmosphere, of which N_2 is the predominant constituent. N enters the soil through fixation of molecular N_2 by microorganisms and by the return of ammonia and nitrate in rainwater. From the agricultural point of view, N is required by most agricultural crops in relatively large amounts. The significance of N is indicated by the fact that it is an important constituent of proteins, nucleic acids, porphyrins, and alkaloids. In the soil, soil organic matter (SOM) acts as a storehouse and supplier of N to plant roots and microorganisms; 95 % and more of the total soil N is closely associated with SOM. [1]

While a considerable amount of research has been done on soil N over the years, most of this work has been limited to qualitative and quantitative determinations of proteinaceous materials, amino acids, and ammonia. This leaves about one-third to one-half of the soil N unidentified and poorly understood, so that there is a need for more research and information in this area.

The objective of this paper is to present an up-to-date account of what we know about the chemistry of soil N. The first part of this paper will deal with the distribution in different soils of proteinaceous materials, amino sugars, and ammonia, while the second part will focus on recent data on the identities and roles of heterocyclic N compounds, which also appear to play a significant role in supplying N to plant roots and microbes.

8.2
Nitrogen Fixation and Ensuing Reactions

Only a few soil microorganisms have the ability to use molecular N_2; all remaining living organisms require combined N for carrying out their life activities. Therefore, in the context of this discussion, the role of nitrogen-fixing bacteria, which can fix N_2 and reduce it to ammonia, becomes very important. Of special interest at this point are the following four reactions:

1. Nitrogen (N_2) in the atmosphere $\rightarrow$ organic N (nitrogen fixation)
2. Organic N $\rightarrow$ ammonia (mineralization or ammonification)
3. Ammonia $\rightarrow$ organic N (immobilization or assimilation)
4. Nitrate $\rightarrow$ organic N (nitrate assimilation or immobilization)

Nitrogen fixation (reaction 1) involves the reduction of elemental N_2 to the -3 oxidation state in NH_3. This biological process is catalyzed by nitrogenase, a large anaerobic metalloenzyme. The NH_3 produced is retained by the nitrogen-fixing cells and reacts with glutamate to form glutamine. Newly fixed NH_3 is only rarely released by healthy nitrogen-fixing cells but must pass through an organic form before entering the nitrogen cycle. [2]

The mineralization of organic N (reaction 2) is carried out mainly by microorganisms. Through this process, organically bound N is liberated as NH_3. Whether N is mineralized or immobilized by microorganisms depends on the magnitude of the C/N ratio of the substrate compared to that of the decomposer organisms. If the substrate has a low C/N ratio, N will be in excess and NH_3 will be liberated. If the C/N ratio of the substrate is higher than that of the decomposer organisms, N will be immobilized.

Immobilization (reaction 3) of nitrogen can occur through both biotic and abiotic processes. For example, NH_4^+ is efficiently immobilized by clay minerals in exchangeable and fixed forms. Exchangeable NH_4^+ is available for biological immobilization. While it has been assumed in the past that non-exchangeable NH_4^+, which is fixed in clay lattices, had a low biological availability, more recent data [3] show that in most soils 30–60 % of the fixed NH_4^+ is available for biological uptake.

Nitrification is a process through which nitrogen in NH_3 is oxidized to NO_2^- and/or NO_3^- by autotrophic nitrifying bacteria of the genera *Nitrosomonas* and *Nitrobacter*. This is a key reaction for determining the fate of nitrogen in soils. Nitrite and nitrate are more mobile than NH_3 and are therefore more readily lost through leaching. Nitrite can also be reduced to nitrous oxide (N_2O) and dinitrogen (N_2) by denitrifying bacteria. The main factor influencing the nitrification rate is the concentration of available NH_3.

The reduction of nitrate via nitrite to ammonia occurs in soils at low rates[3]. If it were possible to enhance the reduction of NO_3^- to NH_3 and its subsequent incorporation into SOM (reaction 4), large N losses resulting from denitrification and leaching could be presented so that the economy of N use would be improved.

8.3
Amino Acids, Amino Sugars, and Ammonia in Soils

The distribution of the total N in amino acids, amino sugars and ammonia in soils formed under widely differing climatic conditions was determined by Sowden et al. [4] The soil samples originated from the arctic, cool temperate, subtropical, and tropical regions of the earth's surface. All samples were analyzed by the same methods, which provided a degree of uniformity which had not been attained before and made it possible to gain new insights into the distribution of N in soils. Table 8.1

Tab. 8.1   Definitions of different types of N fractions.

Hydrolyzable N = % of total N hydrolyzed by hot 6 M HCl in 24 h

Nonhydrolyzable N (NH-N) = 100 – hydrolyzable N

Unidentified hydrolyzable N (UH-N) = % total hydrolyzable N – (% amino acid-N + NH$_3$-N + amino sugar-N)

Protein N = % amino acid N + 10 % amino acid N (to include amide N of asparagine and glutamine lost during 6 M HCl hydrolysis).

% Non-protein N = 100 – protein N

lists definitions of the different types of N fractions which the authors [4] characterized. While the total N contents of the soil samples analyzed were in the range 0.01–1.61 % (Table 8.2), the proportion of total N that could be hydrolyzed by hot 6M HCl was fairly consistent (84.2 to 88.9 %). Amino acid N ranged from 33.1 to 41.7 %, amino sugar N from 4.5 to 7.4 %, and ammonia N from 18.0 to 32.0 %. Some of the ammonia was probably formed during hydrolysis of amides, the decomposition of hydroxy- and other amino acids and amino sugars, the deamination of purines and pyrimidines, and the release of fixed NH$_4^+$ from clay minerals. It is noteworthy that amino acid N and amino sugar N constituted a greater proportion of total N in soils from the warmer regions, while the reverse was true for NH$_3$ N. The data in Table 8.2 also show that proportions of unidentified hydrolyzable N (UH-N) (16.5–17.8 %) and those of non-hydrolyzable N (NH-N) (11.1–15.8 %) were similar in all soils examined. Of special interest are the UH-N and NH-N fractions which constitute between 28 and 34 % of the total soil N. Little is known about the chemistry of these fractions except that the N in these materials is not protein N, peptide N amino acid N, nor amino sugar N nor NH$_3$ N. In the second part of this paper, further attention will be given to these fractions.

From the amino acid data in Table 8.2, protein N was calculated to make up 36, 40, 46 and 45 % of the total N in the arctic, cool temperate, subtropical and tropical soils, respectively, averaging 42 %. Thus, about 58 % of the total N in soils was non-protein. Amino sugars in the soils constituted between 4.5 and 7.4 % of the total N, with soils under warmer climates containing more amino sugars than those under cooler climates.

Since hot acid hydrolysis was required to release practically all of the amino acids and amino sugars from the soils, it is likely that the amino acids occur in soils in the form of peptides, polypeptides, and proteins closely associated with and protected by SOM and inorganic soil constituents such as clay minerals and hydrous oxides of iron and aluminum. Similarly, amino sugars do not appear to exist in soils as free compounds.

Sowden et al. [4] also did detailed amino acid and amino sugar analyses of the soils from the different climatic regions. The following amino acids were determined: *acidic amino* acids: aspartic and glutamic acids; *basic amino acids*: arginine, histidine, lysine and ornithine; *neutral amino acids*; phenylalanine, tyrosine, glycine, alanine, valine, leucine, isoleucine, serine, threonine, proline and hydroxyproline;

Tab. 8.2 Nitrogen distribution in soils from widely differing climatic zones (means and standard deviations). Modified from Sowden et al. [4]

Climatic zone	Soil N (range)	Total soil N	Percentage of total soil N					
			Hydrolyzable					
			Amino acid N	Ammonia N	Unidentified N (UH-N)	Amino Sugar N	Nonhydrolyzable N (NH-N)	
Arctic	0.02–0.16	86.1±6.6	33.1±9.3	32.0±8.6	16.5	4.5±1.7	13.9±6.6	
Cool temperate	0.02–1.06	86.5±6.4	35.9±11.5	27.5±12.9	17.8	5.3±2.1	13.5±6.4	
Subtropical	0.03–0.30	84.2±4.9	41.7±6.8	18.0±4.0	17.1	7.4±2.1	15.8±4.9	
Tropical	0.24–1.61	88.9±3.8	40.7±8.0	24.0±4.5	17.6	6.7±1.2	11.1±3.8	

sulfur-containing amino acids: methionine, cystine, cysteic acid and methionine sulf-oxide; *miscellaneous amino acids*: allo-isoleucine, γ-NH$_2$-butyric acid, 2,4-diaminobu-tyric acid, diaminopimelic acid, β-alanine and ethanolamine. In addition, the amino sugars glucosamine and galactosamine were also determined. Somewhat surprisingly, the distribution of N in each amino acid as a percentage of N in total amino acids was similar. For example, acidic amino acids constituted between 16.6 (arctic soils) and 26.2 % (tropical soils) of the total amino acid N, basic amino acids between 25.7 and 19.9 %, neutral amino acids between 53.8 and 50.8 %, sulfur-containing amino acids between 2.7 and 2.1 %, and miscellaneous amino acids between 1.2 and 1.9 %. As to amino sugars, glucosamine N constituted between 2.8 and 3.9 % of the total soil N, whereas N in galactosamine made up between 1.7 and 2.8 % of the total soil N.

A comparison of the mean amino acid composition of the soils with those of algae, bacteria, fungi, and yeasts showed the greatest similarity to that of bacteria. [4] This suggests, perhaps not too surprisingly, a major role for microorganisms in the synthesis in the soil of amino acids, peptides and proteins from plant and animal residues, and also explains the relatively uniform amino acid composition in different soils.

To establish whether refluxing with hot 6M HCl for 24 h hydrolyzed all proteinaceous materials in soils, Griffith et al. [5] hydrolyzed a number of soils first with hot 6M HCl, and subsequently hydrolyzed separate samples of acid-treated residues with either 0.2 M Ba(OH)$_2$ or 2.5 M NaOH under reflux. The results obtained showed that hot 6 M HCl released almost all of the amino acids in the soils in 24 h.

8.4
Nucleic Acid Bases in Soils

Guanine, adenine, cytosine, thymine, and uracil were identified in acid hydrolyzates of Scottish soils by Anderson. [6] Later, Cortez and Schnitzer [7] determined the distribution of purines (guanine and adenine) and pyrimidines (uracil, thymine, and cytosine) in 13 soils. Concentrations of purines plus pyrimidines ranged from 20.9 to 137.7 μg/g of air-dry soil. Quantitatively, the distribution in soils was: guanine > cytosine > adenine > thymine > uracil. The ratio of guanine + cytosine to adenine + thymine was > 2, and methyl cytosine was not detected, which suggested a microbial DNA origin for the nucleic acid bases. An average 3.3 % of the total N in inorganic soils but only 0.3 % of the total N in organic soils was found to occur in nucleic acid bases.

8.5
Bioavailability of the NH-N Fraction

To find out whether the N in NH-N fractions of soils was bioavailable, NH-N fractions isolated from several soils were incubated with a soil rich in clay minerals, a sandy soil, and a pure sand. [8] Most of the N in the NH-N fractions was found to be reduced to ammonia by biological activity. At pH 7.0, the order of biodegradation in the three substrates was sand > sandy soil > clay soil. Other workers [9–12] have also reported that NH-N fractions of soils were biodegradable. Mild chemical oxidation with peracetic acid converted up to 59% of the N in NH-N fractions from several soils to NH_3 and other N gases. [13] The available data show that the N in NH-N fractions of soils is not inert but can be converted biologically and chemically to NH_3 and other N gases, and so play a role in the soil N-cycle.

8.6
Chemistry of the UH-N Fraction

Schnitzer et al. [14] developed a chromatographic procedure that involved separation of the unidentified from the known N, so that they could take an unhindered look at the unidentified N compounds without interference from the known N compounds. Their procedure was as follows:

1. A number of soils were hydrolyzed with hot 6M HCl for 24h;
2. The soluble hydrolyzates was neutralized and the soluble materials separated on Sephadex G-25 gels;
3. The highest molecular weight fractions were further separated on Sephadex G-50 gels, and the second highest molecular weight fractions were separated on Sephadex G-15 gels.

In this manner, several fractions were prepared which contained between 97.5 and 98.6% unidentified N, but only 0.84% amino acid N, 0% amino sugar N, and 0.53% ammonia N.

In a subsequent study, Schnitzer and Spiteller [15] hydrolyzed each fraction with 2 M H_2SO_4. After neutralization of the soluble materials, the latter were reduced with $NaBH_4$ and then acetylated. The resulting acetates were analyzed by capillary gas chromatography/mass spectrometry, and identified by comparing their mass spectra with those of reference compounds of known structures and with literature data. Eighteen N-heterocyclics were identified. These compounds included hydroxy- and oxy-indoles, quinolines, isoquinolines, aminobenzofurans, piperidines, pyrrolines, and pyrrolidines. In addition, a number of benzylamines and nitriles were also identified. It is noteworthy that the N heterocyclics were isolated and identified without the use of pyrolysis.

8.7
Chemistry of the NH-N Fraction

Schulten et al. [16] identified the following N-containing compounds in NH-N fractions separated from several soils by pyrolysis-gas chromatography/mass spectrometry (Py-GC/MS): pyrrole (Ia), methyl pyrrole (Ib), pyridine (IVa), methylpyridine (IVb), indole (VIa), and benzothiazole (XI). The Roman numerals refer to the chem-

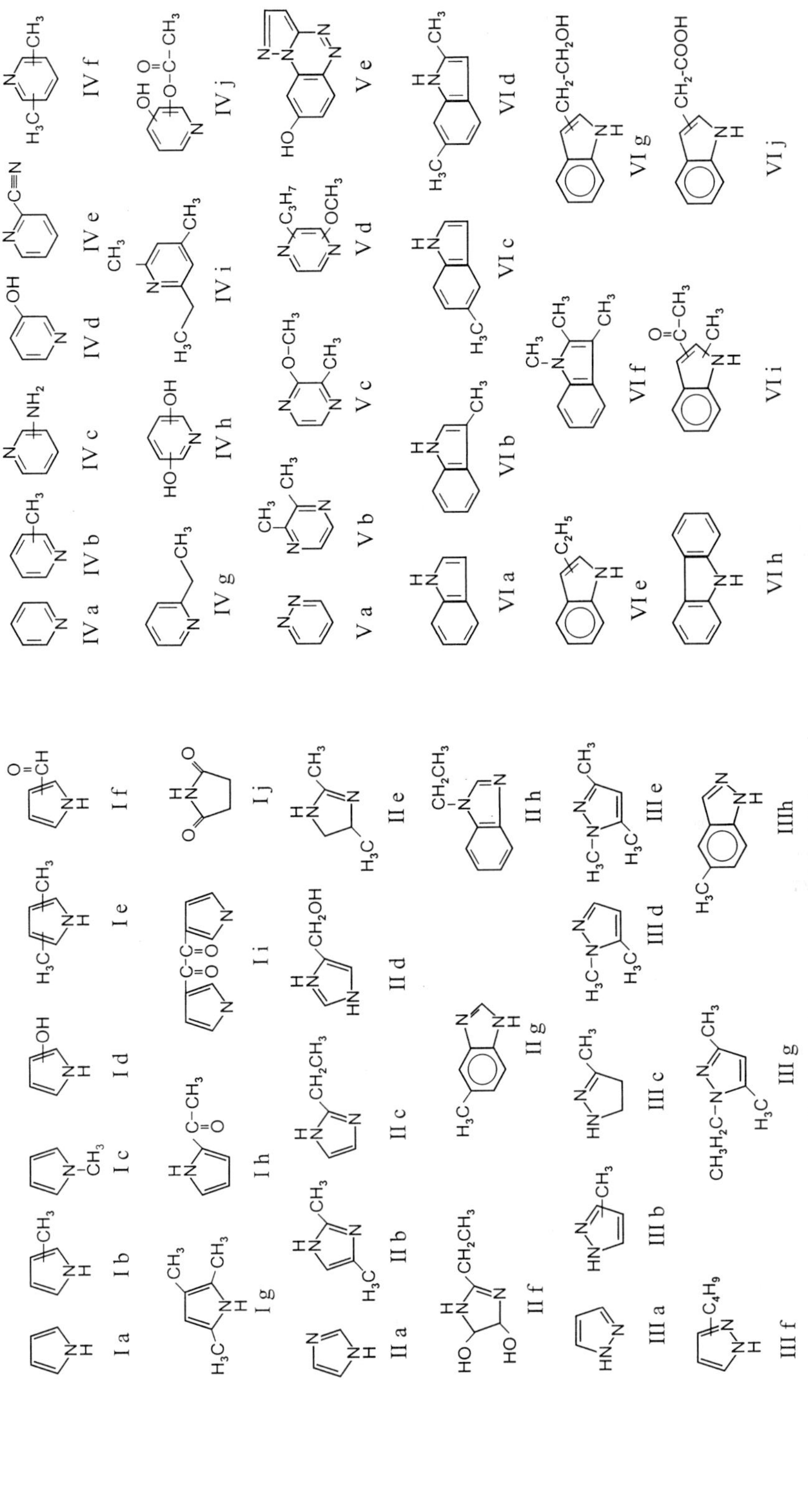

Fig. 8.1 Chemical structures of heterocyclic N compounds identified in soils. Modified from Schulten and Schnitzer. [17]

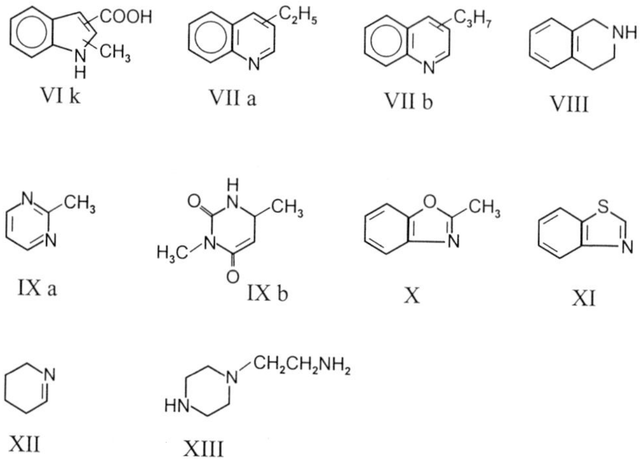

VI k VII a VII b VIII

IX a IX b X XI

XII XIII

Fig. 8.1 continued

ical structures of these compounds shown in Fig. 8.1. Other compounds identified were the dodecanenitrile, tetradecanenitrile, pentadecanenitrile and hexadecanenitrile, as well as a number of N derivatives of benzene such as benzeneamine, benzonitrile, and isocyanomethylbenzene.

8.8
Pyrolysis-field ionization mass spectrometry (Py-FIMS) and Curie-point pyrolysis-gas chromatography/mass spectrometry (CpPy-GC/MS) of soils

Schulten and Schnitzer [17] analyzed a number of soils by Py-FIMS and by CpPy-GC/MS with N-selective detection. Among the N-containing compounds identified (see Fig. 8.1 and Table 8.3) were pyrroles (Ia to Ij), imidazoles (IIa to IIh), pyrazoles (IIIa to IIIh), pyridines (IVa to IVg), pyrazines (Va to Ve), indoles (VIa to VIk), quinolines (VIIa, VIIb), isoquinoline (VIII), pyrimidines (IXa, IXb), methylbenzoxazole (X), benzothiazole (XI), tetrahydropyridine (XII), and piperazine ethanamine (XIII). Other compounds detected (not shown in Fig. 8.1) were *N-derivatives of benzene*: aminobenzene, benzonitrile, 1-amino-3-methyl-benzene, methylbenzonitrile, benezeneacetonitrile, isocyanomethylbenzene, 4-hydroxybenzonitrile, 2,5-dimethylaminobenzene, benzenepropanenitrile, methylaminobenzaldehyde, benzoacetonitrile, hydroxymethoxybenzonitrile, dimethylaminobenzaldehyde, dimethoxybenzonitrile, and 3-nitro-1,2-phthalic acid, *n-alkyl nitriles*: 2-propenenitrile, propanenitrile, 2-methylpropanenitrile, 3-methylbutanenitrile, 4-methylpentanenitrile, dimethylbutylnitrile, heptanenitrile, n-C$_{13}$ to n-C$_{19}$ alkylnitriles, and n-C$_{14}$ to n-C$_{19}$ dialkylnitriles.

Aside from the N-compounds listed above, a number of low-molecular-weight N compounds were also identified. [17] These included ammonia, hydrogen cyanide, dinitrogen (N$_2$), nitric oxide (NO), methylamine, nitrous oxide (N$_2$O), formamide, acetamide, and *N,N-dimethylmethanamine*.

Tab. 8.3 Identities of heterocyclic N compounds in soils shown in Fig. 8.1.

Compound	Identity	Compound	Identity
Ia	pyrrole	IVf	dimethylpyridine
Ib	methylpyrrole	IVg	2-ethylpyridine
Ic	N-methylpyrrole	IVh	dihydroxypyridine
Id	hydroxypyrrole	IVi	2-ethyl-4,5-dimethylpyridine
Ie	dimethylpyrrole	IVj	hydroxy-acetoxypyridine
If	2-formylpyrrole	Va	pyrazine
Ig	trimethylpyrrole	Vb	2,3-dimethylpyrazine
Ih	2-acetylpyrrole	Vc	2-methoxy-3-methylpyrazine
Ii	diketodipyrrole	Vd	methoxy-propylpyrazine
Ij	2,5-pyrrolidinedione	Ve	pyrazolo-benzotriazine-8-ol
IIa	imidazole	VIa	indole
IIb	2,4-dimethylimidazole	VIb	3-methylindole
IIc	2-ethyl-1H-imidazole	VIc	5-methylindole
IId	4-hydroxymethylimidazole	VId	2,6-dimethylindole
IIe	4,5-dihydro-2,4-dimethyl-1H-imidazole	VIe	ethylindole
		VIf	1,2,3-trimethylindole
IIf	2-ethyl-4,5-dihydroxyimidazole	VIg	indoleethanol
IIg	methylbenzoimidazole	VIh	2,3,4,5-tetrahydroindole
IIh	1-ethylbenzoimidazole	VIi	methyl-acetylindole
IIIa	pyrazole	VIj	indoleacetic acid
IIIb	methylpyrazole	VIk	methylindolecarboxylicacid
IIIc	4,5-dihydro-3-methylpyrazole	VIIa	ethylquinoline
IIId	dimethylpyrazole	VIIb	propylquinoline
IIIe	1,3,5-trimethylpyrazole	VIII	1,2,3,4-tetrahydroisoquinoline
IIIf	butylpyrazole	IXa	2-methylpyrimidine
IIIg	1-ethyl-3,5-dimethylpyrazole	IXb	3,6-dimethyl-2,4-pyrimidine-dione
IIIh	1-methylbenzopyrazole		
IVa	pyridine	X	2-methylbenzoxazole
IVb	methylpyridine	XI	benzothiazole
IVc	aminopyridine	XII	tetrahydropyridine
IVd	3-hydroxypyridine	XIII	1-piperazineethanamine
IVe	2-nitrilopyridine		

8.9
Origins of Major N Compounds Identified

As to the origins of the major N compounds identified, it is possible that at least a portion of some of these compounds are pyrolysis products of amino acids, peptides, proteins, [18] and porphyrins (a component of chlorophyll), [19] or originate from the microbial decomposition of plant lignins and other phenolics in the presence of ammonia. [20] Of considerable interest are the identifications aromatic and aliphatic nitriles. Nitriles can be formed from amines with the loss of 2 H_2, from amides with the loss of H_2O, and also by reacting n-alkanoic acid with NH_3. [21] The detection of long-chain alkyl- and dialkyl-nitriles points to the presence in the soil or SOM of long-chain amines and/or amides. On the other hand, there is considerable evidence that N-heterocyclics are significant soil-N components rather than degradation products of other molecules produced by pyrolysis. The arguments in favor of N-heterocyclics as genuine soil components are: (a) heterocyclics are formed by microbial synthesis in the soil from plant residues or remains of animals which contain phenols, proteinaceous materials, aromatic compounds and lipids; [22] (b) unsubstituted and substituted N-heterocyclics such as pyrroles, pyrrolidines, pyridines, pyrazoles, and pyrans have been identified in aquatic humic substances and dissolved organic matter at pyrolysis temperatures of only 200–300 °C; [22] (c) N-heterocyclics such as those referred to above have also been identified in soils without pyrolysis by gel chromatography-GC/MS after reductive acetylation, [15] by X-ray photoelectron spectroscopy, [23] and by spectroscopic, chemical, and isotopic methods. [24]

Further research is needed to identify additional N-heterocyclics in soils and to determine whether the N-heterocyclics are present in the soil in the forms in which they were identified or whether they originate from more complex structures. If the latter is the case, we need to isolate these complex N-containing molecules and attempt to identify them.

8.10
^{15}NMR analysis of soils

^{15}N NMR spectroscopy is widely used in organic chemistry and biochemistry for determining chemical structures of N-containing molecules. In a recent review on applications of ^{15}N NMR to soils, SOM, and composts, Preston [25] notes that in all studies done so far on these materials, the spectra recorded are very similar and remarkably simple, consisting of one major peak due to amide/peptide and a few minor signals arising from N in indoles, pyrroles, and amino acids. Similarly, Knicker et al. [26] report that 85 % of the signal intensity of ^{15}N-enriched composts and recently formed SOM is due to amide/peptide, and that no signals in the range of heteroatomic N compounds were detected. These authors conclude from their ^{15}N NMR analyses that at most 10 % of the total N in SOM or soils could be present in heteroatomic N structures or Schiff bases.

In contrast to the finding by ^{15}N NMR that 85 % of the total soil N occurs in proteins or peptides, chemical methods show that only about 40 % of the total soil N occurs in proteins. [4] What are the reasons for these wide divergencies? To provide answers to these questions, it may be useful to consider the following:

(a) Natural ^{15}N abundance levels in soils are low (0.4 %), so that direct analysis by ^{15}N NMR is difficult. Another problem is the small gyromagnetic ratio of the ^{15}N nucleus.

(b) To overcome these difficulties, ^{15}N concentrations in soils are increased by adding ^{15}N-labeled salts such as $(^{15}NH_4)_2$ SO_4 and incubating. But as has been demonstrated, [27] even incubation for 600 days does not produce the same array of ^{15}N compounds as those synthesized in the soil in the presence of reactive (catalytic) surfaces over a period of hundreds or thousands of years. It is likely that during the early stages of incubation, the microbial synthesis of proteins is the predominant reaction but that of heterocyclic N compounds may take a much longer time.

(c) Aside from amino acids and amino sugars, chemical and spectroscopic analyses of soils have led to the identification of over 120 organic N compounds; many of these are heterocyclic, [16, 17, 28]. It is likely that most of these compounds occur in low concentrations only, so that ^{15}N NMR in its current state of development is unable to detect them. One can hope that with significant improvements in instrumental design, the gulf between results obtained by ^{15}N NMR and by chemical and mass spectrometric methods will eventually narrow.

8.11
Distribution of N in Soils

From the data presented herein and in earlier publications, [16, 17, 28] it is possible to deduce the following distribution of total N in soils: proteinaceous materials (proteins, peptides, and amino acids) 40%, amino sugars 5%, heterocyclic N compounds (including purines and pyrimidines) 35%, and NH_3 20% with about 1/4 of the NH_3 fixed as NH_4^+ to clay minerals. Thus, proteinaceous materials and heterocyclic N compounds are the major soil N components.

8.12
Concluding Comments

During the past three years, the author and his collaborators have identified in soils by Py-FIMS and CpPy-GC/MS over 120 N-containing organic compounds. Pyrolysis was used to set free N-containing compounds from other organic and inorganic soil constituents and to volatilize them so that they could be transferred to the mass spectrometer in Py-FIMS and to the gas chromatograph and subsequently to the mass spectrometer in CpPy-GC/MS. Many of these N compounds were identified in soils for the first time, and many were heterocyclic. Among the heterocyclic compounds were unsubstituted and substituted pyrroles, imidazoles, pyrazoles, pyridines, pyrazines, indoles, quinolines, isoquinoline, pyrimidines, oxazoles, benzothiazole and a number of saturated N-heterocyclics. Some of the heterocyclics appear to originate from biological precursors such as proteinaceous materials, carbohydrates, phenols, chlorophyll, nucleic acids, and alkaloids, which enter the soil system via plant residues or remains of animals; others may be formed by microbial synthesis. Of special interest are the identifications of long-chain (up to n-C_{17}) aliphatic nitriles and dinitriles, which point to the presence of long-chain aliphatic amines and/or amides in soils. As to future work, research is needed on the identification of additional N-containing compounds in soils and to determine whether these are present in the soil in the forms in which they were identified or whether they originate from more complex structures. If the latter is the case, then we need to isolate these complex N-containing molecules and attempt to identify them.

The work reported in this paper is important from both the fundamental and practical standpoints. From the fundamental point of view, we need more detailed information on the major N components of soils aside from proteins and amino sugars. Incubation studies show that N-heterocyclics in soils yield bioavailable NH_3, [8] likely at a slower rate than proteinaceous substances. But we need to know more about (a) the identities of these compounds, and (b) how they interact and associate with other organic and inorganic soil constituents, so that we can obtain a better understanding of the chemical and biochemical reactions of N in soils. From the practical point of view, we may need to consider the following: Nitrogen is essential for plant production, is used in relatively large amounts, and is expensive. A better understanding of the chemistry and biochemistry of N in soils will allow us to develop management practices that will improve the economy of nitrogen use, while at the same time reduce pollution.

References

1 SCHNITZER, M. in Humic Substances in Soil, Sediment, and Water (eds. AIKEN, G.R., MCKNIGHT, D.M., WERSHAW, R.L. and MAC-CARTHY, P.) Wiley-Interscience: New York, 1984, chapter 12.

2 SMITH, D. in Experimental Microbial Ecology (eds. BURNS, R.G. and SLATER, J.H.), Blackwell: London, 1982, chapter 4.

3 ROSSWALL, T. *Plant and Soil*, **1982**, *67*, 15–34.

4 SOWDEN, F.J., CHEN, Y., SCHNITZER, M. *Geochim. Cosmochim. Acta*, **1977**, *41*, 1524–26.

5 GRIFFITH, S.M., SOWDEN, F.J., SCHNITZER, M. *Soil Biol. Biochem.*, **1976**, *8*, 529–31.

6 ANDERSON, G. *Nature*, **1957**, *180*, 287–88.

7 CORTEZ, J., SCHNITZER, M. *Soil Sci. Soc. Am. J.*, **1979**, *43*, 958–61.

8 IVARSON, K.C., SCHNITZER, M. *Can. J. Soil Sci.*, **1979**, *59*, 59–67.

9 KEENEY, D.R., BREMNER, J.M. *Soil Sci. Soc. Am. Proc.*, **1964**, *28*, 653–56.

10 MEINTS, V.W., PETERSON, G.A. *Soil Sci.*, **1977**, *124*, 334–42.

11 OTTOW, J.C.G. *Naturwissenschaften*, **1978**, *65*, 413–23.

12 ZHUO, S., WEN, Q., CHENG, L. *Pedosphere*, **1995**, *5*, 183–86.

13 SCHNITZER, M., HINDLE, D.A. *Can. J. Soil Sci.*, **1980**, *60*, 541–48.

14 SCHNITZER, M., MARSHALL, P.M., HINDLE, D.A. *Can. J. Soil Sci.*, **1983**, *63*, 425–33.

15 SCHNITZER, M., SPITELLER, M. Trans. 13th Congr. Int. Soil Sci. Soc. Hamburg, **1986**, *3*, 473–74.

16 SCHULTEN, R.-H., SORGE-LEWIN, C., SCHNITZER, M. *Biol. Fertil. Soils*, **1997**, *24*, 249–54.

17 SCHULTEN, R.-H., SCHNITZER, M. *Biol. Fertil. Soils*, **1998**, *26*, 1–15.

18 IRWIN, W.J. Analytical Pyrolysis, Dekker: New York, 1982.

19 BRACEWELL, J.M., PACEY, N., ROBERTSON, G.W. *J. Anal. Appl. Pyrolysis*, **1987**, *10*, 199–213.

20 BREMNER, J.M. *Pontif. Acad. Sci. Scr. Varia*, **1967**, *32*, 143–193.

21 EVANS, E.J., BATTS, B.D., CANT, N.W., SMITH, J.W. *Org. Geochem.*, **1985**, *8*, 367–74.

22 SCHNITZER, M., SCHULTEN, H.-R., *Adv. Agron.* **1996**, *55*, 167–217.

23 PATIENCE, R.I., BAXBY, M., BARTLE, K.D., PERRY, D.L., REES, A.G.W., ROWLAND, S.J. *Org. Geochem.* **1992**, *18*, 161–69.

24 IKEAN, R., IOSELIS, P., RUBINSZTAIN, Y., AIZENSHTAT, Z., MILOSLAVSKY, I., YARIV, S., PUGMIRE, R., ANDERSON, L.L., WOOLFENDEN, W.D., KAPLAN, I.R., DORSEY, T., PETERS, K.E., BOON, J.J., LEEUNEE, J.W., ISHIWATARI, R., MORINAGA, S., YAMAMOTO, S., MACIHARA, T., MULLER-VONMOOS, T., RUB, A. *Sci. Total Environ.* **1992**, *117/18*, 1–12.

25 PRESTON, C.M. *Soil Sci.*, **1996**, *161*, 144–66.

26 KNICKER, H., FRIEND, R., LUEDEMANN, H.-D., *Naturwissenschaften*, **1993**, *80*, 219–21.

27 KNICKER, H., LUEDEMANN, H.D. *Org. Geochem.* **1995**, *23*, 329–41

28 LEINWEBER, P., SCHULTEN, H.R. *Soil Sci. Soc. Am. J.* **1998**, *62*, 383–93.

9

Spherical Molecular Assemblies: A Class of Hosts for the Next Millennium

Leonard R. MacGillivray and Jerry L. Atwood

9.1
Introduction

During the past century, the field of X-ray crystallography has provided numerous insights into problems in chemistry. Indeed, since the discovery by Laue, in 1912, that crystals diffract X-rays, [1] chemists have used information gained from single crystal studies to formulate principles for understanding properties of inorganic, organic, and biochemical compounds and to synthesize new chemical entities. The work of Pauling, for example, in 1928, provided sets of rules for describing and predicting arrangements of ions in inorganic compounds such as minerals, [2] while structure determinations of deoxyribonucleic acid (DNA) and viruses by Watson and Crick and Klug, in 1953 [3] and 1962, [4] respectively, furnished an understanding of those processes associated with the chemistry of life.

9.1.1
Supramolecular Chemistry

With the role of crystallography in providing structural details of molecular connectivity and in "traditional" areas of chemistry thus established in the first three quarters of the century, it is perhaps not surprising that it has only been within the last two decades that chemists have, in part, used this information to turn to a multidisciplinary approach to synthesis that utilizes non-covalent bonds (e.g. hydrogen bonds, π-π interactions) for the design of highly organized chemical entities. Known as supramolecular chemistry, or chemistry beyond the molecule, this field, first conceptualized by Lehn, in 1973, [5] deals with the chemist's ability to exploit the structure-directing properties of non-covalent forces for the design of multi-component chemical species that, in a similar way to biological systems (e.g. proteins), exhibit properties (e.g. host-guest) not found in the individual components (e.g. amino acids). Ideally, such systems have been designed to gain insights into various biological phenomena. If accomplished, these insights may ultimately permit chemists to rationally design drugs, catalysts, and perhaps life itself, as well as generate new classes of materials with unique properties.

9.1.2
Towards Supramolecular Synthesis

Although it is now generally accepted that a supramolecular approach to chemistry can provide a route to achieving these goals, it has become clear that if chemists are to synthesize entities analogous to biological systems, they must face the realization that the dimensions of biological structures typically range from tens to thousands of angstroms [6] and that an approach to synthesis that depends exclusively upon the stepwise formation and breakage of covalent bonds is burdened with inherent limitations. [7] For example, in addition to requiring a large amount of time, such an approach often results in low product yields (e.g. natural product synthesis). As a result, an alternative method for designing large chemical systems (e.g. nanosystems) must be employed.

9.1.3
Self-Assembly

Within the last decade, it has become evident that self-assembly, the single-step construction of molecular architecture using non-covalent forces, has provided an attractive means for constructing large, highly organized chemical entities. Owing to the reversibility of such interactions, non-covalent bonds, upon selection of appropriate chemical subunits, can facilitate error-free generation of either discrete or infinite supramolecular species, usually in quantitative yield (Fig. 9.1). Although the idea is not entirely new (nature uses this approach in a number of multi-component systems such as hemoglobin [8] and viruses [4] while crystallization can be viewed as an example of self-assembly *par excellence* [9]), chemists are just beginning to understand those principles which govern the self-assembly process.

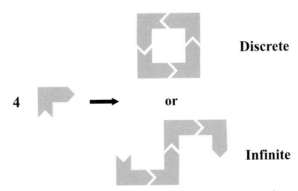

Discrete

4

or

Infinite

Fig. 9.1 Two modes of self-assembly of a complementary subunit.

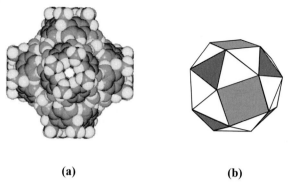

(a) **(b)**

Fig. 9.2   (a) X-ray crystal structure of **1** and (b) the snub cube, one of the 13 Archimedean solids. The square faces of the snub cube correspond to the resorcin[4]arenes; the eight shaded triangles that adjoin three squares correspond to the water molecules of **1**.

9.2
Overview

It is with these ideas in mind that we now provide an account of work performed in our laboratory during the last two years, which focuses upon the discovery, in the solid state, of a chiral, spherical host assembly held together by 60 hydrogen bonds (**1**) [10] (Fig. 9.2a) and the development of a rational design strategy for the construction of related host systems. [11] In particular, upon recognizing that the topology of **1** conforms to the structure of a snub cube, one of the 13 Archimedean solids (Fig. 9–2b), [12] we have structurally classified a variety of organic, inorganic, and biological shells and have shown that their structures may be catalogued according to principles of solid geometry in which the five Platonic and 13 Archimedean solids may be used as models for host design. This approach, which we recognize as being driven by biology, has led us to identify containers that have yet to be synthesized or discovered and we propose such systems as targets for chemical synthesis.

9.3
A Spherical Molecular Assembly Held Together by 60 Hydrogen Bonds

We have recently shown that it is possible to extend the cavities of bowl-shaped molecules known as resorcin[4]arenes (**2**) supramolecularly, using hydrogen bond acceptors such as pyridines as extender units (Fig. 9.3). [13,14] During studies aimed at co-crystallizing *C*-methylcalix[4]resorcinarene (**2a**) with hydrogen bond acceptors in aromatic solvents, we discovered the ability of **2a** to self-assemble in the crystalline state as a spherical hexamer, along with eight water molecules, to form a spherical container assembly, with idealized *432* symmetry, that is held together by 60 O–H$\cdots$O hydrogen bonds **1**. [10] The assembly, which is chiral, possesses a well-

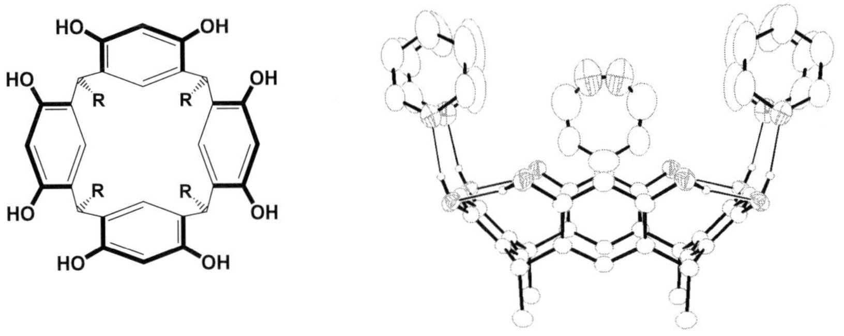

(a)   **(b)**

Fig. 9.3 (a) Iconic representation of **2a** and (b) X-ray crystal structure of 2a·4(4pyridine)·pyridine.

defined cavity with a maximum diameter of 1.8 nm and an internal volume of about 1.4 nm^3, a volume five time larger than the largest molecular capsule previously reported. [15] Although guest species could be located with the interior of **1** (i.e. electron density maxima), it was not possible to determine their identity from the X-ray experiment, presumably owing to the high symmetry of the host and high thermal motion within the host cavity. Notably, solution studies also revealed the ability of C-undecylcalix[4]resorcinarene (**2b**) to maintain the structure of **1** in apolar organic solvents such as benzene. [10]

9.3.1
Polyhedron Model – Snub Cube

Consultation of polyhedron models revealed the structure of **1** to conform to a snub cube, one of the 13 Archimedean solids, in which the vertices of the square faces correspond to the corners of the resorcin[4]arenes and the centroids of the eight triangles that adjoin three squares correspond to the eight water molecules. [12] Indeed, to us, the ability of six resorcin[4]arenes to self-assemble to form **1** was reminiscent of spherical viruses in which identical copies of proteins self-assemble, by way of non-covalent forces, to form viral capsids having icosahedral symmetry and a shell-like enclosure. In fact, owing to the fit displayed by its components, **1** exhibits a topology that agrees with the theory of virus shell structure which states that octahedral systems must contain 24 asymmetric units and possess *432* symmetry. [4] Moreover, these observations suggested that to design related spherical hosts, one must consider the limited possibilities available in space for constructing spherical frameworks based upon regular polygons, those being the five Platonic and 13 Archimedean solids.

9.4
General Principles for Spherical Host Design

The discovery that members of the resorcin[4]arene family self-assemble to form **1**, owing to its classification as an Archimedean solid, prompted us to examine the topologies of related spherical hosts with a view to understanding their structures on the basis of symmetry. In addition to providing grounds for classification, we anticipated that such an approach would allow us to identify similarities at the structural level, which, at the chemical level, may not seem obvious and may be used to design large, spherical host assemblies similar to **1**.

Thus, it is herein that we now describe the results of this analysis which we regard as the development of a general strategy for the construction of spherical molecular hosts. [11] We will begin by presenting the idea of self-assembly in the context of spherical hosts and then, after summarizing the Platonic and Archimedean solids, we will provide examples of cubic symmetry-based hosts, from both the laboratory and nature, with structures that conform to these polyhedra.

9.4.1
Spheroid Design

The strategy stems from ideas developed within host-guest [16] and supramolecular chemistry. [5] As is the case for spherical viruses [4] and fullerenes, [17] an appropriately sized, shaped, and functionalized guest is packaged within the interior of a host such that it is completely surrounded or enclosed. From a geometrical standpoint, this may be achieved by centralizing the guest within a hollow spherical shell S (Fig. 9.4). It must be noted, however, that, from a chemical standpoint, it is impossible to mimic S, since atoms and molecules are discrete entities whereas the surface of S is uniform. Thus, to design a spherical molecular host, an alternative procedure must be employed.

Fig. 9.4   Hollow spherical shell S.

9.4.2
Self-Assembly

In a paper describing the structure of regular viruses, Caspar and Klug [4] have shown that viral capsids use self-assembly to construct spherical shells up to a hundred nanometers in diameter by utilizing identical copies of proteins as chemical subunits. Indeed, such a design strategy is desirable since it employs an economy of information, [17,18] giving rise to a host whose subunits exist in identical chemical environments, exposed surface area is at a minimum, and strain energy is distributed evenly along its surface. That this process is also amenable for spherical shells at the angstrom level was later realized with the discovery of buckminsterfullerene or C_{60}. [19] Notably, in the extreme case, an infinite number of identical subunits placed along the surface of a sphere will lead to a shell topologically equivalent to S. Thus, to construct a spherical molecular host, one must ultimately consider the number of subunits n for spheroid design and their placement along the surface of the shell.

9.4.3
Subunits for Spheroid Design and Self-Assembly

To construct a spherical shell using a single subunit, $n = 1$, the only structure obtainable is S. As outlined above, it is impossible to construct a spherical molecular host using a single chemical entity, and therefore S does not represent a self-assembled spherical framework.

For $n = 2$, each subunit must cover one half of the surface of the sphere. This can only be achieved if the subunits exhibit curvature and are placed such that their centroids lie at a maximum distance from each other. These criteria place two points along the surface of a sphere separated by a distance equal to the diameter of the shell. We assign these positions the north and south poles. As a consequence of this arrangement, there exist two structure types for $n = 2$. The first belongs to the point group $D_{\infty h}$ and consists of two identical subunits attached at the equator (Fig. 9.5a). Since it is impossible to create a shell-like hemisphere that possesses ∞-fold rotation symmetry using atoms and molecules, this structure is not obtainable from a chemical standpoint. The second belongs to the point group D_{nd} and its simplest member, $n = 2$, is topologically equivalent to a tennis ball (Fig. 9.5b). Each subunit of this system is symmetrical and may be divided into four identical asymmetric units (Fig. 9.5c), which implies that eight asymmetric units are required to design the shell.

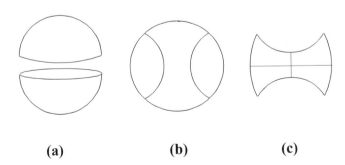

(a)   **(b)**   **(c)**

Fig. 9.5 $n = 2$ shells, (a) $D_{\infty h}$ symmetry, (b) D_{2d} symmetry (tennis ball), (c) subunit of D_{2d} shell depicting the four asymmetric units.

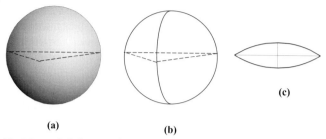

Fig. 9.6 $n = 3$ shell, (a) equilateral triangle from S, (b) D_{3h} symmetry, (c) subunit of D_{3h} shell depicting the four asymmetric units.

Thus, $n = 2$ represents the minimum number of subunits which may be used to construct a spherical molecular host via self-assembly.

For $n = 3$, each subunit must cover one third of the surface of the sphere. Following the design conditions described previously, placing three identical subunits along the surface of a sphere results in an arrangement in which their centroids (Fig. 9.6) constitute the vertices of an equilateral triangle. As a result, there is only one structure type for $n = 3$ (Fig. 9.6a). The structure belongs to the point group D_{3h} and, as for the $n = 2$ system, the subunits must exhibit curvature (Fig. 9.6b). Since each "arm" may be divided into four identical asymmetric units (Fig. 9.6c), 12 asymmetric units are required to construct the shell.

For $n = 4$, positioning four points along the surface of a sphere such that they lie a maximum distance from each other places the points at the vertices of a tetrahedron (Fig. 9.7a).

This is the first case in which joining the points via line segments gives rise to a closed-surface container. The container, a tetrahedron, is comprised of four identical subunits in the form of equilateral triangles, where surface curvature is supplied by edge sharing of regular polygons rather than the subunits themselves (Fig. 9.7b). Owing to its symmetry, each triangle may be divided into six asymmetric units (Fig. 9.7c), which implies that 24 asymmetric units (4 × 6) are required to generate the shell. The tetrahedron belongs to the point group T_d and its polygons are related by combinations of two-fold and three-fold rotation axes.

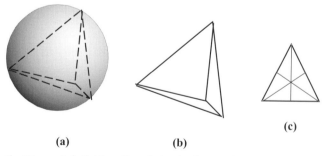

Fig. 9.7 $n = 4$ shell, (a) equilateral triangles from S, (b) T_d symmetry, (c) subunit of T_d shell depicting the six asymmetric units.

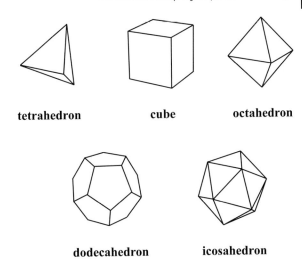

tetrahedron   cube   octahedron

dodecahedron   icosahedron

Fig. 9.8   The five Platonic solids (see also Table 9.1).

9.4.4
Platonic Solids

The Platonic solids comprise a family of five convex uniform polyhedra which possess cubic symmetry (i.e. *32, 432,* or *532* symmetry) and are made of the same regular polygons (e.g. equilateral triangle, square) arranged in space such that the vertices, edges, and three coordinate directions of each solid, are equivalent (Fig. 9.8, Table 9–1). [12] That there is a finite number of such polyhedra is due to the fact that there exists a limited number of ways in which identical regular polygons may be adjoined to construct a convex corner. Equilateral triangles may be adjoined in three ways while squares and pentagons may be adjoined in only a single manner. Moreover, it is impossible to create a convex corner using regular polygons with six or more sides since the sum of the angles around each vertex would be greater than or equal to 360 degrees. [12] These principles give rise to five isometric polyhedra which are achiral and whose polygons are related by combinations of *n*-fold rotation axes. The Platonic solids include the tetrahedron, which belongs to the point group

Tab. 9.1   Platonic solids.

Solid	Vertices	Edges	Face Type	Faces
Tetrahedron	4	6	triangle	4
Cube	8	12	square	6
Octahedron	6	12	triangle	8
Dodecahedron	20	30	pentagon	20
Icosahedron	12	30	triangle	20

T_d, possesses *32* symmetry, and requires a minimum of 12 asymmetric units; the cube and octahedron, which belong to the point group O_h, possess *432* symmetry, and require a minimum of 24 asymmetric units; and the dodecahedron and icosahedron, which belong to the point group I_h, possess *532* symmetry, and require a minimum of 60 asymmetric units. The number of asymmetric units required to generate each shell doubles if mirror planes are present in these structures.

9.4.5
Archimedean Solids

In addition to the Platonic solids, there exists a family of 13 convex uniform polyhedra known as the Archimedean solids. Each member of this family is made up of at least two different regular polygons and may be derived from at least one Platonic solid through either truncation or the twisting of faces (Fig. 9.9, Table 9.2). [12] In the case of the latter, two chiral members, the snub cube and the snub dodecahedron, are realized. The remaining Archimedean solids are achiral. Like the Platonic solids, the Archimedean solids possess identical vertices, exhibit either *32*, *432*, or *532* symmetry, and require a minimum of either 12, 24, or 60 asymmetric units, respectively. The Archimedean solids possess a wider variety of polygons than the Platonic solids. These include the equilateral triangle, square, pentagon, hexagon, octagon, and decagon.

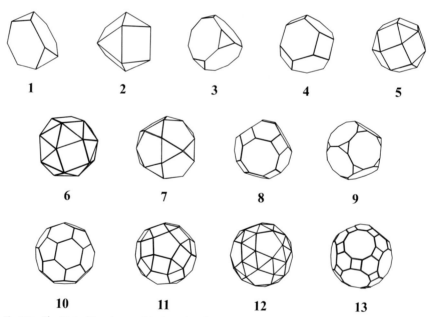

Fig. 9.9 The 13 Archimedean solids, in order of increasing number of vertices. Truncated tetrahedron (1), Cuboctahedron (2), Truncated cube (3), Truncated octahedron (4), Rhombicuboctahedron (5), Snub cube (6), Icosidodecahedron (7), Rhombitruncated cuboctahedron (8), Truncated dodecahedron (9), Truncated icosahedron (10), Rhombicosidodecahedron (11), Snub dodecahedron (12), Rhombitruncated icosidodecahedron (13) (see also Table 9.2).

Tab. 9.2 Archimedean solids.

Solid	Vertices	Edges	$f(3)^a$	$f(4)$	$f(5)$	$f(6)$	$f(8)$	$f(10)$
(1) truncated tetrahedron	12	18	4	–	–	4	–	–
(2) cuboctahedron	12	24	8	6	–	–	–	–
(3) truncated cube	24	36	32	6	–	–	6	–
(4) truncated octahedron	24	36	–	6	–	8	–	–
(5) rhombicuboctahedron	24	48	8	18	–	–	–	–
(6) snub cube	24	60	32	6	–	–	–	–
(7) icosidodecahedron	30	60	20	–	12	–	–	–
(8) rhombitruncated cuboctahedron	48	72	–	12	–	8	6	–
(9) truncated dodecahedron	60	90	20	–	–	–	–	12
(10) truncated icosahedron	60	90	–	–	12	20	–	–
(11) rhombicosidodecahedron	60	120	20	30	12	–	–	–
(12) snub dodecahedron	60	150	80	–	12	–	–	–
(13) rhombitruncated icosidodecahedron	120	180	–	30	–	20	–	12

[a] f(n) where f denotes face and n indicates the number of sides of the face.

9.4.6
Models for Spheroid Design

It is our contention here that the Platonic and Archimedean solids represent the limited ways in which $n > 3$ identical chemical subunits which correspond to regular polygons may be arranged along the surface of a sphere to approximate S. [20] As a result, these solids may be used as models for the design of spherical molecular hosts. These polyhedra provide combinations of n-fold rotation axes and subunits for self-assembly, in the form of regular polygons, which allow one to determine where chemical subunits of a host should be placed and the bonding arrangements they should adopt. In effect, the Platonic and Archimedean solids simplify the task of constructing a spherical molecular host by facilitating *a priori* spheroid design, and serving as targets in chemical synthesis.

9.5
Examples from the Laboratory and from Nature

The premise that spherical molecular hosts may be constructed according to principles of solid geometry renders both organic and inorganic components viable for their design and permits the individual subunits to be held together by covalent and/or noncovalent bonds. Indeed, a common feature displayed by organic and inorganic chemists is that they must utilize space for assembling atoms into molecular frameworks. To demonstrate the utility of this approach, we will now present selected examples of spherical hosts from the laboratory and nature. We will begin with the Platonic solids and work our way to the Archimedean polyhedra.

9.5.1
Platonic Solids

As stated, the Platonic solids constitute a family of five convex uniform polyhedra made up of the same regular polygons and possess either *32, 432,* or *532* symmetry. As a result, the three coordinate directions within each solid are equivalent, which makes these polyhedra models for spheroid design.

9.5.1.1 Tetrahedral Systems (T_d, T_h, T)

The macrotricyclic spherand designed by Lehn et al. was the first tetrahedral host (Fig. 9–10a). [21] The bridgehead nitrogen atoms, located at the corners of the tetrahedron, and ethyleneoxy units, the edges, supply the three-fold and two-fold rotation axes, respectively. As a result, the spherand is composed of 24 asymmetric units [(N/3)-(CH_2CH_2)-(O/2)]/2. Notably, this molecule and its tetraprotonated form has been shown to bind an ammonium and chloride ion, respectively. Schmidtchen et al. have introduced similar tetrahedral cages with edges comprised entirely of methylene bridges, [22] while Vögtle et al. have demonstrated the synthesis of a hollow hydrocarbon called spheriphane (Fig. 9.10b). [23]

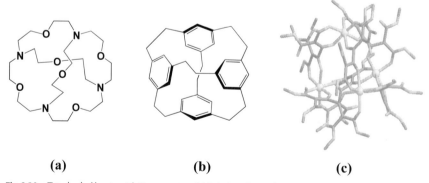

(a)   (b)   (c)

Fig. 9.10 Tetrahedral hosts with T_d symmetry, (a) Lehn's spherand, (b) Vögtle's spheraphane, (c) Saalfrank's metal-based cage.

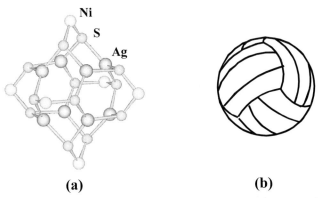

(a)　　　　　　　　**(b)**

Fig. 9.11 (a) Tetrahedral host with T_h symmetry, X-ray crystal structure of the $[Ag(I)_8Ni(II)_6\{SC(Me)_2CH(NH_2)CO_2\}_{12}Cl]^{5-}$ ion showing the positions of the Ag, Ni, and S atoms and (b) a volleyball, an object that possesses T_h symmetry.

Saalfrank et al. were the first to introduce metal-based tetrahedral cages by using metal ions as corner units and bridging malonate ligands as edges (Fig. 9.10c). [24] Owing to a bend in each ligand, these M_4L_6 cages are adamantane-like. In terms of host-guest behavior, an iron-based system has been shown to complex a single ammonium ion. [25]

Following the introduction of these metal-based systems, similar tetrahedral cages have emerged. Raymond et al. [26] and Huttner et al., [27] for example, have revealed that linear, rather than bent, bridging ligands may be used to form such cages. In the case of the former, bidentate ligands were used to completely fill the coordination sites around each metal while, in the latter, monodentate ligands along with "stopper" units were employed. These hosts were shown to form complexes with four dimethylformamide molecules and a tetrafluoroborate ion, respectively. [26, 27]

Huan et al., [28] Birker et al., [29] and Bürgi et al. [30] have described tetrahedral (Fig. 9–11a) shell-like hosts, $[H_{12}(VO_2)_{12}(C_6H_5PO_3)_8]^{4-}$, $[M(I)_8M(II)_6\{SC(Me)_2CH-(NH_2)CO_2\}_{12}Cl]^{5-}$ (where $M(I)_8M_6(II) = Cu(I)_8Ni(II)_6$, $Ag(I)_8Ni(II)_6$, $Ag(I)_8Pd(II)_6$), and $[ICd_8(HOCH_2CH_2S)_{12}]^{3+}$ that possess vanadium and sulfur atoms at the vertices of a icosahedron. These structures also possess phosphorus and metal ions at the centroids of triangular faces which correspond to the corners of a cube. As a result, these shells belong to the point group T_h, the point group of a volleyball (Fig. 9.11b).

9.5.1.2 Octahedral Systems (O_h, O)

We will now illustrate four octahedral hosts related to the Platonic solids. Three are based upon the cube while one possesses features of both a cube and an octahedron.

The first is a cyclophane-based system reported by Murakami et al. (Fig. 9.12a). [31] The sides of the host consist of tetraaza-[3.3.3.3]paracyclophane units, and its octaprotonated cation has been shown to bind anionic guests. The molecule possesses 48 asymmetric units of $[(N/3)-(CH_2)-\{(C_6H_4)/2\}]/2$.

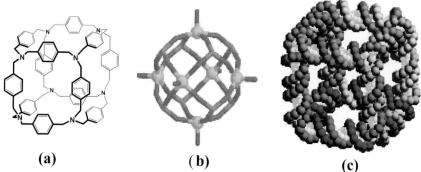

(a) **(b)** **(c)**

Fig. 9.12 Octahedral hosts, (a) Murakami's cyclophane-based cube, (b) X-ray crystal structure of the $[(VO_6)(RPO_3)_8]^+$ (R = tBu, OSiMe$_3$) ion, (c) Seeman's DNA-based cube (printed with permission by Professor Nadrian Seeman, for further details see http:\\seemanlab4.chem.nyu.edu).

The second is a polyoxovanadate, $[(VO_6)(RPO_3)_8]^+$ (R = tBu, OSiMe$_3$), reported by Zubieta et al. [32] and Thorn et al. [33] which consists of VO$_5$ pyramids and phosphonate ligands (Fig. 9.12b). The vanadium atoms of the shell are located at the vertices of an octahedron, while the phosphorus atoms are located at the corners of a cube, thus displaying the dual relationship of these polyhedra. In both cases, the host has been shown to complex a chloride ion.

The third is a gold selenide cube, $[NaAu_{12}Se_8]^{3-}$, reported by Kanatzidis et al. [34] The anion is made up of eight selenium atoms and 12 gold atoms. A sodium cation occupies the center of the complex.

The fourth is a cube synthesized by Chen and Seeman, the components of which are based upon DNA (Fig. 9.12c). [35] The directionality and ability of the double helix to form branched junctions are exploited for the edges and vertices, respectively. Interestingly, each face of this molecule forms a cyclic strand which is catenated with strands of adjacent faces. Molecular modeling experiments indicate the length of each edge to be approximately 6.8 nm.

9.5.1.3 Icosahedral Systems (I$_h$, I)

Spherical viruses are icosahedral molecular hosts related to the Platonic solids (Fig. 9–13a). [4] Consisting of identical copies of proteins which assemble by way of noncovalent forces, these hosts range from 15 to 90 nm in diameter and encapsulate strands of ribonucleic acid (RNA). Although spherical viruses require a minimum of 60 subunits, most are made up of 60n (n= 2, 3, 4...) subunits owing to a reduction in symmetry of their polygons (Fig. 9.13b). This process, known as triangulation, gives rise to quasi-equivalent positions along the surface of the shell, which enable the virus particle to cover the RNA with the largest number of subunits. Since only certain triangulations are permitted by symmetry, viruses may be classified into a coherent system. [4]

Kretschmer et al. have recently described the ability of 12 CpSmCl$_2$ units to form a neutral samarium-based shell, $[Cp_{12}Sm_{12}(\mu_3\text{-Cl})_{24}]$, in which 12 samarium atoms

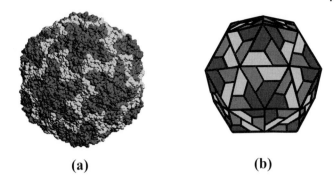

(a)   **(b)**

Fig. 9.13 An icosahedral host, (a) X-ray crystal struture of the rhino-virus, a spherical virus linked to the common cold, (b) a schematic representation of the rhinovirus displaying triangulation.

are located at the vertices of an icosahedron and 20 chloride ions are at the vertices of a dodecahedron. [36] The remaining chloride ions form a tetrahedron at the center of the shell.

9.5.2
Archimedean Solids

As stated, the Archimedean solids constitute a family of 13 convex uniform polyhedra made up of two or more regular polygons and, like the Platonic solids, possess either *32, 432,* or *532* symmetry. As a result, the three coordinate directions within each solid are equivalent, making these polyhedra, in addition to the Platonic solids, models for spheroid design.

9.5.2.1   Truncated Tetrahedron (1)

Fujita et al., [37] Stang et al., [38] and Steel et al. [39] have recently described the synthesis of M_6L_4 cages which are topologically analogous to a truncated tetrahedron (Fig. 9–14). These systems, which may be regarded as inverted M_4L_6 frame-

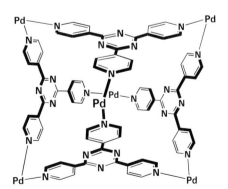

Fig. 9.14. Fujita's metal-based cage, a host based upon the truncated tetrahedron.

works, consist of metal ions and aromatic-based bridging ligands which constitute the two-fold and three-fold rotation axes, respectively. Notably, owing to the presence of a chiral stopper unit, the system reported by Stang et al. possesses T symmetry and is therefore chiral. [38] In terms of host-guest behavior, X-ray crystallographic studies have revealed the assembly reported by Fujita et al. to form a complex with four adamantyl carboxylate ions, [37] while that of Steel et al. was found to encapsulate a molecule of dimethylsulfoxide. [39] According to mass spectrometric data, the cage reported by Stang et al. associates with four triflate ions. [38]

Zubieta et al. have also demonstrated the formation of a cage topologically equivalent to a truncated tetrahedron, $[Mo_{16}(OH)_{12}O_{40}]^{8-}$. Composed of four Mo(VI) and twelve Mo(V) centers, this host possesses a central $[Mo_{12}O_{40}]^{20-}$ core which encapsulates a proton or sodium ion. [40]

9.5.2.2 Cuboctahedron (2)

Gonzàlez-Duarte et al. have recently described the ability of eight cadium ions and 16 thiolate ligands to assemble from aqueous media to form a highly charged cage, $[ClCd_8\{SCH(CH_2CH_2)_2N(H)Me\}_{16}]^{15+}$, the sulfur atoms of which sit at the vertices of a cuboctahedron. [41] The host contains a central chloride ion and an inner tetrahedral array of cadmium ions. Interestingly, Ross et al., using MM2 molecular model simulations, have considered the existence of a molecule containing eight benzene rings and either 12 oxygen or sulfur atoms, which they refer to as heterospherophane (Fig. 9–15). [42] Although it is not mentioned in the original report, the shell exhibits a topology identical to a cuboctahedron.

9.5.2.3 Truncated Octahedron (4)

Seeman et al. have constructed a DNA-based shell analogous to a truncated octahedron. The edges of this molecule, each of which contains two turns of the double helix, contain 1440 nucleotides, and the molecular weight of the structure, which is an overall 14-catenane, is 790 000 Daltons. [43] Interestingly, the design strategy relies on a solid support approach in which a net of squares is ligated to give the polyhedron. It is currently unclear what shape the molecule adopts in various media.

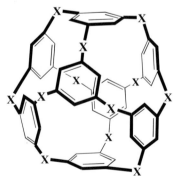

Fig. 9.15 A theoretical organic shell based upon the cuboctahedron (X = O, S).

Kretschmer et al. have described the formation of a lanthanide complex, $[Cp_6Yb_6Cl_{13}]^-$ (Cp = cyclopentadienyl), which conforms to a truncated octahedron. [36] The anion contains six ytterbium ions, located at the corners of an octahedron, and 12 bridging choride ions. A single chloride ion occupies the center of the shell.

9.5.2.4 Rhombicuboctahedron (5)

Müller et al. have shown that 24 oxygen atoms of the polyoxometalate $[As_4Mo_6V_7O_{39}]^{2-}$ may be attributed to the structure of a rhombicuboctahedron. [44] Notably, a strong "tetrahedral distortion" of each ion is required to correspond each host to the polyhedron. This shell has been shown to complex a sulfate ion in the solid state.

9.5.2.5 Snub Cube (6)

We have recently demonstrated the ability of six resorcin[4]arenes and eight water molecules to assemble in apolar media to form a spherical molecular assembly which conforms to a snub cube (Fig. 9.3). [10] The shell consists of 24 asymmetric units – each resorcin[4]arene lies on a four-fold rotation axis and each H_2O molecule on a three-fold axis – in which the vertices of the square faces of the polyhedron correspond to the corners of the resorcin[4]arenes and the centroids of the eight triangles that adjoin three squares correspond to the water molecules. The assembly, which exhibits an external diameter of 2.4 nm, possesses an internal volume of about 1.4 $Å^3$ and is held together by 60 O–H$\cdots$O hydrogen bonds.

9.5.2.6 Truncated Icosahedron (10)

Buckminsterfullerene, an allotrope of carbon, is topologically equivalent to a truncated icosahedron, an Archimedean solid that possesses 12 pentagons and 20 hexagons (Fig. 9–16). [17] Each carbon atom of this fullerene corresponds to a vertex of the polyhedron. As a result, C_{60} is held together by 90 covalent bonds, the number of edges of the solid.

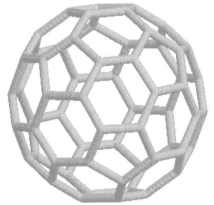

Fig. 9.16 X-ray crystal structure of buckminster-fullerene, C_{60}, a shell based upon the truncated icosahedron.

9.5.3
Archimedean Duals and Irregular Polygons

The Platonic and Archimedean solids comprise two finite families of polyhedra in which each solid consists of identical vertices, edges, and either a single or two or more regular polygons. It is of interest to note, however, that there exists a family of spherical solids which are made up of irregular polygons which may also be used as models for spheroid design. Known as Archimedean duals, [12] these polyhedra are constructed by simply connecting the midpoints of the faces of an Archimedean solid. Such a treatment gives rise to 13 polyhedra which possess two or more different vertices and identical irregular polygon faces (Fig. 9.17). As a result, chemical subunits used to construct hosts which conform to these polyhedra cannot be based upon regular polygons.

9.5.3.1 Rhombic Dodecahedron (2)

To the best of our knowledge, there is one host which conforms to the structure of an Archimedean dual. Harrison was the first to point out that the quaternary structure of ferritin, a major iron storage protein in animals, bacteria, and plants, corresponds to the structure of a rhombic dodecahedron. [45] This protein, which is approximately 12.5 nm in diameter, consists of 24 identical polypeptide subunits (Fig. 9.18), and holds up to 4500 iron atoms in the form of hydrated ferric oxide with

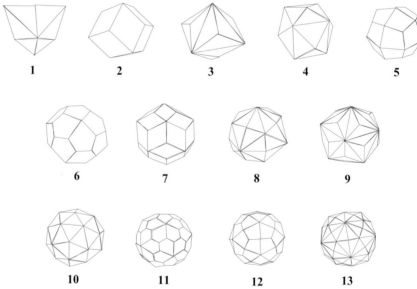

Fig. 9.17 The 13 Archimedean duals derived from corresponding Archimedean solids (see Fig. 9.9). Triakis tetrahedron (1), Rhombic dodecahedron (2), Triakis octahedron (3), Tetrakis hexahedron (4), Deltoidal icositetrahedron (5), Pentagonal icositetrahedron (6), Rhombic tri-contahedron (7), Disdyakis dodecahedron (8), Triakis icosahedron (9), Pentakis dodecahedron (10), Deltoidal hexecontahedron (11), Pentagonal hexecontahedron (12), Disdyakis triacontahedron (13).

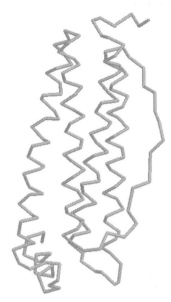

Fig. 9.18 Ferritin, a spherical host based upon the rhombic dodecahedron. X-ray structure of the polypeptide subunit.

varying amounts of phosphate $[Fe_2O_3(H_2O/H_3PO_4)_n]$. [46] The polypeptides, which consist of four helix bundles, assemble by way of non-covalent forces and form dimers which correspond to the faces of the solid

9.5.4
Irregular Polygons

It is also important to point out that if "partial" truncation is applied to the Platonic solids such that Archimedean solids are not realized, or if truncation is applied to the Archimedean solids, then the resulting polyhedra will not possess regular faces, but, like the Archimedean duals, may be used as models for spheroid design as a consequence of their cubic symmetries. Indeed, it is striking to note that of the spherical hosts synthesized to date, all have been constructed using chemical subunits which either correspond to regular polygons (e.g. calix[4]arenes, VO_4 pyramids, tridentate bridging ligands) or form regular polygons (e.g. carbon-based hexagons and pentagons). Moreover, the realization that spherical shells may be constructed using polyhedra with irregular faces, as in the case of ferritin, implies that spherical shells based upon irregular polygons may be rationally designed.

9.6
Why the Platonic and Archimedean Solids?

With interests in chemical synthesis moving towards the fabrication of nanometer scale molecular frameworks [6, 47–49] and the miniaturization of functional microstructures [50] (Scheme 9–1), it becomes apparent why the Platonic and Archimedean solids are appropriate models for shell design.

covalent synthesis
self-assembly ➡ **nanoscale architecture** ⬅ **miniaturization**

Scheme 9.1

First, for a given chemical subunit, these solids inherently give rise to larger structures. This may be illustrated by comparing three molecules which adopt the structures of two Platonic solids and an Archimedean solid: cubane, [51] dodecahedrane, [52] and C_{60}. [19] In each molecule, a carbon atom is located at the vertex of a polyhedron. Upon traversing the series, however, one observes a gradual increase in the size of these molecules. In fact, the increase is such that C_{60} is capable of entrapping a guest. [17]

Second, we note that these solids incorporate economy of design. This may be illustrated upon considering a sphere (guest) inscribed within either a tetrahedron (T_d), cube (O_h), or icosahedron (I_h) (host). Upon comparing these hosts, one realizes that the icosahedron facilitates encapsulation of the guest with the largest number of subunits (Table 9.1). Moreover, from a chemical perspective, the icosahedron allows the host to complex the guest with the smallest chemical subunits. Indeed, this feature could eliminate errors in subunit design and shell formation and, in principle, reduce certain "costs" of a given framework.

Thus, the Platonic and Archimedean solids not only provide a means for host design, but a way in which to maximize chemical information, allowing the chemist to simplify the structures of complex molecular frameworks and, in effect, engineer host-guest systems.

9.7
Conclusion

Using information obtained from X-ray crystallography, we have described the structure of a chiral, spherical molecular assembly held together by 60 hydrogen bonds. [10] The host, which conforms to the structure of a snub cube, self-assembles in apolar media and encapsulates guest species within a cavity that possesses an internal volume of approximately 1.4 nm^3.

From this information, general principles for the design of spherical molecular hosts have been developed. [11] These principles rely on the use of convex uniform polyhedra as models for spheroid design. To demonstrate the usefulness of this approach, structural classification of organic, inorganic, and biological hosts – frameworks which can be rationally compared on the basis of symmetry – has revealed an interplay between symmetry, structure, and function. [53]

Indeed, we anticipate that the Platonic and Archimedean solids may be used for the construction of hosts which conform to those solids not yet realized and additional members of each family, where supramolecular synthesis, via self-assembly, will play a major role in their design, ushering in an era of spherical host-guest chemistry.

References

1 FRIEDRICH, W.; KNIPPING, P.; VON LAUE, M. T. F. *Sitzungsber. Math. Phys. Kl. K. Bayer. Akad. Wiss. Munchen* **1912**, 303.

2 PAULING, L. *Z. Kristallogr.* **1928**, *67*, 377.

3 WATSON, J. D.; CRICK, F. H. C. *Nature* **1953**, *171*, 737.

4 CASPAR, D. L. D.; KLUG, A. *Cold Spring Harbor Symp. Quant. Biol.* **1962**, *27*, 1.

5 LEHN, J.-M. *Structure and Bonding* **1973**, *16*, 1.

6 WHITESIDES, G. M.; MATHIAS, J. P.; SETO, C. T. *Science* **1991**, *254*, 1312.

7 LAWRENCE, D. S.; JIANG, T.; LEVETT, M. *Chem. Rev.* **1995**, *95*, 2229.

8 STRYER, L., In *Biochemistry*, 3rd Edition; Freeman: New York, 1988.

9 DUNITZ, J. D. In *Perspectives in Supramolecular Chemistry. The Crystal as a Supramolecular Entity*, DESIRAJU, G. R., Ed.: Wiley: Chichester, 1996.

10 MACGILLIVRAY, L. R.; Atwood, J. L. *Nature* **1997**, *389*, 469.

11 MACGILLIVRAY, L. R.; Atwood, J. L. *Angew. Chem. Int. Ed. Engl.,* **1999**, *38*, 1018.

12 WENNINGER, M. J. *Polyhedron Models*; Cambridge Univ. Press: New York, 1971.

13 MACGILLIVRAY, L. R.; ATWOOD, J. L. *J. Am. Chem. Soc.* **1997**, *119*, 6931.

14 MACGILLIVRAY, L. R.; ATWOOD, J. L. *Chem. Commun.* **1999**, 181.

15 CONN, M. M.; REBEK, JR., J. *Chem. Rev.* **1997**, *97*, 1647.

16 CRAM, D. J. *Nature* **1992**, *356*, 29.

17 *Buckminsterfullerenes* (Eds.: BILLUPS, W. E.; CIUFOLINI, M. A.), VCH, New York, 1993.

18 PHILIP, D.; STODDART, J. F. *Angew. Chem. Int. Ed. Engl.* **1996**, *35*, 1155.

19 KROTO, H. W.; HEATH, J. R.; O'BRIEN, S. C., CURL, R. F.; SMALLEY, R. E. *Nature* **1985**, *318*, 162.

20 Hosts based upon *n* = 2 subunits possess dihedral symmetry and their structures may be considered to be based upon tennis balls, rugby balls, prisms, and antiprisms (see ref. [11]).

21 GRAF, E.; LEHN, J.-M. *J. Am. Chem. Soc.* **1975**, *97*, 5022.

22 SCHMIDTCHEN, F. P.; MÜLLER, G. *J. Chem. Soc., Chem. Commun.* **1984**, 1115.

23 VÖGTLE, F.; SEEL, C.; WINDSCHEIF, P.-M. In *Comprehensive Supramolecular Chemistry, Cyclophane Hosts: Endoacidic, Endobasic, and Endolipophilic Large Cavities, Vol. II*; ATWOOD, J. L.; DAVIES, J. E. D.; MacNICOL, D. D.; VÖGTLE, F., Ed.; Pergamon: New York, 1996.

24 SAALFRANK, R. W.; STARK, A.; PETERS, K.; VON SCHNERING, H. G. *Angew. Chem. Int. Ed. Engl.* **1988**, *27*, 851.

25 SAALFRANK, R. W.; BURAK, R.; BREIT, A.; STALKE, D.; HERBST-IRMER, R.; DAUB, J.; PORSCH, M.; BILL, E.; MÜTHER, M.; TRAUTWEIN, A. X. *Angew. Chem. Int. Ed. Engl.* **1994**, *33*, 1621.

26 BEISSEL, T.; POWERS, R. E.; RAYMOND, K. N. *Angew. Chem. Int. Ed. Engl.* **1996**, *35*, 1084.

27 MANN, S.; HUTTNER, G.; ZSOLNAI, L.; HEINZE, K. *Angew. Chem. Int. Ed. Engl.* **1996**, *35*, 2808.

28 HUAN, G.; DAY, V. W.; JACOBSON, A. J.; GOSHORN, D. P. *J. Am. Chem. Soc.* **1991**, *113*, 3188.

29 BIRKER, P. J. M. W. L.; FREEMAN, H. C. *J. Am. Chem. Soc.* **1977**, *99*, 6890.

30 BÜRGI, H. B.; GEHRER, H.; STRICKLER, P.; WINKLER, F. K. *Helv. Chim. Acta* **1976**, *59*, 2558.

31 MURAKAMI, Y.; KIKUCHI, J.; HIRAYAMA, T. *Chem. Lett.* **1987**, 161.

32 Salta, J.; Chen, Q.; Chang, Y. D.; Zubieta, J. *Angew. Chem. Int. Ed. Engl.* **1994**, *33*, 757.

33 Thorn, D. L.; Harlow, R. L.; Herron, N. *Inorg. Chem.* **1995**, *34*, 2629.

34 Huang, S. P.; Kanatzidis, M. G. *Angew. Chem. Int. Ed. Engl.* **1992**, *31*, 787.

35 Chen, J.; Seeman, N. C. *Nature*, **1991**, *350*, 631.

36 Kretscher, W. P.; Teuben, J. H.; Troyanov S. I. *Angew. Chem. Int. Ed. Engl.* **1998**, *37*, 88.

37 Fujita, M.; Oguro, D.; Miyazawa, M.; Oka, H.; Yamaguchi, K.; Ogura, K. *Nature* **1995**, *378*, 469.

38 Stang, P. J.; Olenyuk, B.; Muddiman, D. C.; Smith, R. D. *Organometallics* **1997**, *16*, 3094.

39 Hartshorn, C. M.; Steel, P. J. *Chem. Commun.* **1997**, 541.

40 Khan, M. I.; Mhller, A.; Dillinger, S.; Bögge, H.; Chen, Q.; Zubieta, J. *Angew. Chem. Int. Ed. Engl.* **1993**, *32*, 1780.

41 GonzBlez-Duarte, P.; Clegg, W.; Casals, I.; Sola, J.; Rius, J. *J. Am. Chem. Soc.* **1998**, *120*, 1260.

42 Ross, R. S.; Pincus, P.; Wudl, F. *J. Phys. Chem.* **1992**, *96*, 6169.

43 Zhang, Y.; Seeman, N. C. *J. Am.Chem. Soc.* **1994**, *116*, 1661.

44 Müller, A.; Krickemeyer, E.; Dillinger, S.; Bögge H.; Stammler, A. *J. Chem. Soc. Chem. Commun.* **1994**, 2539.

45 Smith, J. M. A.; Stansfield, R. F. D.; Ford, G. C.; White, J. L.; Harrison, P. M. *J. Chem. Edu.* **1988**, *65*, 1083.

46 Trikha, J.; Theil, E. C.; Allewell, N. M. *J. Mol. Biol. B*, **1995**, 949.

47 Fyfe, M. C. T.; Stoddart, J. F. *Acc. Chem. Res.* **1997**, *30*, 393.

48 Stang, P. J.; Olenyuk, B. *Acc. Chem. Res.* **1997**, *30*, 502.

49 Ozin, G. A. *Acc. Chem. Res.* **1997**, *30*, 17.

50 Wise, K. D.; Najafi, K. *Science* **1991**, *254*, 1335.

51 Eaton, P. E.; Cole, T. W. Jr. *J. Am. Chem. Soc.* **1964**, *86*, 3157.

52 Paquette, L. A.; Ternansky, R. J.; Balogh, D. W.; Kentgen, G. *J. Am. Chem. Soc.* **1983**, *105*, 5446.

53 Pennisi, E. *Science* **1998**, *279*, 978.

10
The Combinatorial Approach to Materials Discovery

L. F. Schneemeyer and R. B. van Dover

10.1
Introduction

The discovery of new materials has traditionally been a time-consuming and labor-intensive process. Recently, the costs of doing research and the need to reduce R&D times is driving the materials community to explore new methodologies which promise greater efficiency via a parallel approach to materials discovery and optimization. Increasing the number of materials that are studied should lead to increased understanding of composition/property relationships and increase the probability of a breakthrough materials discovery.

In the early 1990s, massively parallel (also known as "combinatorial" or "high throughput") research methods became important tools in the drug discovery process. Large numbers of discrete chemical entities, sets of molecules termed libraries, were generated using rapid synthesis techniques, and were examined by rapid screening techniques in efforts to speed the identification of new leads in the drug development process. Methodologies have advanced to the point that some companies can examine upwards of 1 million discrete compounds per year. A number of advances, including database and data mining approaches, and new synthetic, analytical and screening methodologies, have enabled these impressive numbers to be achieved. Indeed, combinatorial chemistry, including the related rapid analog synthesis, is now widely utilized as a tool of in pharmaceutical research not only for lead identification but for optimization as well.

The ability to prepare large numbers of candidate materials and efficient evaluation strategies are attractive to the inorganic materials community also. However, the problems of interest to the materials community are diverse, ranging from homogeneous and heterogeneous catalysis to identification of new electronic materials such as piezoelectrics and dielectrics and new intermetallics including soft magnetic materials. Further, many of the issues related to the utilization of inorganic materials are really questions of process optimization. Combinatorial-type, parallel approaches to such problems can be valuable. However, significant challenges exist in terms of the scalability of advanced materials from the microscale used for exploration to the macroscale scale necessary for production, since the properties of these materials often vary greatly with processing. It is unlikely that any single experimental methodology or platform will be relevant to all of the various materials research problems. Still, approaches to the inorganic materials problems considered so far share common themes. In all cases, experimenters envision a parallel synthesis step that produces samples with attributes similar to those of interest for the particular application of interest, a characterization step that is similarly rapid and that provides relevant evaluation of samples, and a data evaluation and handling mechanism.

In this article, we review the new field of combinatorial materials science. We describe in detail our particular materials exploration approach that we call the continuous compositional spread (CCS) approach. This CCS approach allows the synthesis and characterization of a major fraction of a ternary or pseudo-ternary phase spread in about 24 hours. We compare the CCS approach with other schemes for rapid materials investigations. We also provide an example of the successful application of the CCS approach in the discovery of a new high dielectric constant thin-film material, an amorphous zirconium tin titanate, aZTT. New combinatorial approaches to materials research can allow us to do studies that would not have been possible using conventional techniques.

10.2
History of Rapid Synthesis Approaches in Materials Research

10.2.1
Early Work

High-throughput approaches to the investigation of materials systems have been proposed and executed, with varying degrees of success, for about thirty years. Sawatzky and Kay, [1] working at IBM, reported in 1969 the use of a combinatorial-type approach to prepare a series of rf-sputtered gadolinium iron garnet films with varying defect concentrations. The aim of the study was to determine the effects of cation vacancies on magneto-optic properties of the films. Sample preparation for this study involved the use of a two target sputtering system. A polycrystalline garnet target was placed on one rf diode sputtering cathode and a metallic iron target was placed on a second cathode. Sawatzky and Kay chose to use one rf power supply and one dc supply because of concerns about the interactions between the power supplies to the two targets if both were operated in rf mode at the same frequency. The iron target was therefore operated in a dc reactive sputtering mode. The two target regions were separated by a grounded stainless steel shield which served to minimize the interaction between the two glow discharges and acted to prevent cross contamination.

The design of the Sawatzky-Kay apparatus produced an as-grown film with a continuously varying concentration gradient across the sample surface as well as a continuously varying film thickness. Samples were characterized by standard multiple-beam interferometer techniques to measure the film thickness as a function of position along the sample. The compensation temperature of the sample was also measured along the length of the sample using a Faraday-effect hysterisigraph. Results obtained were in good agreement with those expected for the observed Gd/Fe ratios. Conventional synthetic approaches would have required that a series of ceramic targets be prepared with varying Gd/Fe ratios, and thin films deposited individually from each. Each of those films would then have required characterization, including verification of the Gd/Fe ratio, measurement of the film thickness and measurement of magnetic properties. The two-target approach produced an entire series of doped samples in a single run. Additionally, the range of dopants incorporated could be shifted by varying the potential on the target providing the dopant ion.

Hanak, at RCA Labs, promoted what he termed the "multisample approach in materials science". [2] His approach, first described in a 1970 paper, allowed him to investigate the synthesis of entire multicomponent systems. The innovation that Hanak contributed was the use of a single multi-sectioned cathode for rf cosputtering along with a simple compositional analysis method for these cosputtered films that was based on film thickness measurements. Samples were sputtered from a sectioned target as indicated in Fig. 10.1, producing cosputtered, intimately mixed, samples having compositional spreads. Hanak prepared both two-component and three-component compositional spreads, and investigated systems including alloy superconductors and magnetic materials.

While producing samples more efficiently is one aspect which can lead to more efficient materials discovery efforts, efficient characterization is also needed. In his 1970 paper, [1] Hanak spoke to the issue of materials testing and evaluation. He discussed advances in the measurement of a number of chemical, physical and mechanical properties, measurements which must be tailored to the specific materials problem under investigation. Ultimately it was difficulties in characterization that limited the impact of these approaches. Computers were not yet commonly available, and automated sample evaluation methods remained to be developed.

Cosputtering is particularly suited to phase-spread studies because sputter deposition rates are usually quite reproducible and independent of time. Therefore it is

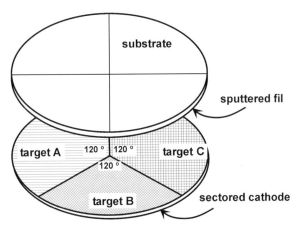

Fig. 10.1 Sputter target layout used for ternary phase spread deposition in the Hanak approach.

simple to set up a multisector target system like that of Hanak, or a system with independent power supplies like that of Sawatzky and Kay. Variations of this technique have been used routinely since 1970, typically to form two-component spreads to explore the detailed effects of composition on properties. For example, the amorphous Tb-Fe magnetic system was explored by van Dover et al. using independent magnetron sputtering guns. [3] The advantage of using magnetron guns is that the glow discharge associated with each gun is almost completely independent of the other, allowing great versatility and simplicity in setting relative deposition rates. The technique was also used to discover a new ternary compound in the Gd-Cr-N system, [4] the first time a ternary compound had been discovered using a combinatorial-type synthetic approach. The difficulties of phase identification using thin films did not allow the precise composition or structure of the new compound to be identified.

It is much more challenging to synthesize phase spreads using co-evaporation than cosputtering, since evaporation rates are very sensitive to input power and condition of the evaporant source. Of course, for low-boiling-point materials, temperature stabilized crucibles can provide very steady evaporation rates, but for refractory materials electron beam evaporation is required, and feedback control of the evaporation rate is also required. Hammond and coworkers developed a scheme for implementing this feedback using a chopped ionization gauge and dual-time-constant feedback scheme. [5] Using this technique, a phase spread of V-Al was prepared, allowing the discovery [6] of A15-structure V_3Al. This was probably the first new compound discovered using a parallel synthesis technique. Use of a codeposition technique was particularly essential in this case, as the new phase was found to be unstable at normal reaction temperatures, and could only be synthesized below 600 °C.

Automated systems for bulk ceramic sample preparation are straightforward to envision and have been described in the literature. These were especially popular in Japan in the late nineteen eighties during the intense discovery phase for new copper oxide-based superconductors. Ceramic synthesis was accomplished using computer-controlled powder delivery systems together with robotic systems for mixing/grinding the starting materials together and then transferring them to a suitable reaction vessel or crucible. In this case, the emphasis was on speeding the synthesis step since moderately rapid assessment of superconductivity in a sample was fairly well developed.

While these more efficient materials synthesis approaches were exploited in some laboratories, traditional one-sample-at-a-time methods have persisted generally. Often this is appropriate because the intense testing regimens that the samples are being subjected to take much longer than the synthetic step. Improving the speed of the synthetic step alone would have little effect on the overall progress of materials discovery.

10.2.3
Recent Innovations

More recently, inspired by the impact of combinatorial chemistry on the drug development process and the desire to lower research costs, interest in approaches to efficient materials investigations has enjoyed a resurgence. [7] Xiang and coworkers [8] described a new, highly flexible approach to materials investigations in which thin-film layers are deposited sequentially through masks to achieve desired compositions and then are subsequently subjected to a high-temperature reaction step. The high-temperature treatment is necessary to obtain a uniform film by interdiffusion of the layers. This discrete combinatorial synthesis (DCS) approach is a process with enormous versatility in selecting compositions. However, the requirement of a relatively high temperature diffusion step means that it is not suitable for many low-temperature or metastable materials, including many amorphous materials.

A proof-of-concept experiment utilizing the DCS approach involved the preparation of a 16-member library in the Bi-Pb-Cu-Ca-Sr-O composition space, with identical molar compositions for each cation. As expected, $BiCuCaSrO_x$ and $BiPb-CuCaSrO_x$ were found to be superconducting. While these materials were already known, the value of the approach was demonstrated. In a subsequent experiment aimed at further demonstrating the power of the approach, a 128-combination manganate-based materials array was prepared to identify phases displaying "colossal magnetoresistance". In both of these cases, an easily measured materials property, diamagnetism and changes in electric field upon application of a magnetic field, respectively, allowed straightforward albeit tedious evaluation of sample arrays.

Experiments using the DCC approach aimed at the discovery of improved phosphor materials have also been described. [9] In this case, samples are evaluated optically, an approach well suited to direct comparisons of large numbers of samples, although it is somewhat difficult to compare the results to the optical properties of bulk materials. Further spectroscopic evaluations of individual elements of the sample array are also easily accomplished by a variety of approaches including scanning fiber techniques. One concern in studies of phosphors is the sensitivity of the optical behavior including fluorescence intensity to processing effects such as details of the microstructure or surface preparation.

Films for the DCC approach can be deposited by any conventional film deposition technique including CVD, evaporation, PVD, sol-gel, etc. By monitoring the rates and the deposition time for each of the constituents in a given sample, approximate compositions of the various samples can be tracked. However, in any thin-film sample the direct structural and compositional evaluation is problematic.

Ink jet printers or similar precision solution handling equipment can be used to deliver accurate amounts of solution species, including ceramic precursors such as citrates or nitrates. This approach has the potential to produce large array of samples efficiently and at low cost. Again, there is no inherent limitation to the number of constituent elements that can be incorporated and sample compositions can readily be tracked. Because the samples are prepared in solution form, for example, as solutions of nitrates, mixing issues are not a severe problem. After all of the elements have been added, samples are heated to dryness. High-temperature calcining or heating under inert or reducing atmospheres is straightforward. One challenge limiting widespread utilization of this approach at present is the lack of commercial refractory, multiple-well sample plates. Some researchers have described homemade sample plates in which an array of holes are drilled in a high-purity Al_2O_3 plate. To allow reuse, the plate is cleaned, for instance by boiling in acid, to remove all traces of previous samples.

Mallouk and coworkers recently exploited ink jet printers as a synthesis tool in an investigation of precious metal catalysts for fuel cell applications. [10] The study identified a high-efficiency catalyst for methanol electro-oxidation in previously unexplored ternary and quaternary platinum alloy systems. In this study, the authors explored various mixtures of five elements: platinum, ruthenium, osmium, iridium, and rhodium. A critical factor in the success of this catalyst study was the development of a novel and rapid evaluation procedure suited to evaluation of the relatively small samples that had been prepared. The oxidation of the methanol on the array electrode results in the generation of acid. A fluorescent acid-base indicator in the solution above the array pinpoints the most active dots, where the concentration of acid is highest. Thus, to determine the catalytic activity of each dot, the researchers chemically converted the electrical current to an optical signal so that the good catalysts lit up, making them easy to identify. This work provided a convenient and efficient study of previously unexplored phase space. The time and cost involved in preparing a large set of these noble metal alloys would likely have precluded this discovery had conventional metallurgy approaches been the only alternative.

10.2.4
The Continuous Compositional Spread (CCS) Approach

We recently described the materials exploration approach which we call the continuous compositional spread (CCS) approach. [11] This approach exploits the versatility of off-axis reactive sputtering, which is a low temperature deposition technique, although the use of a sample heater or post-deposition anneals can allow investigations of higher temperature materials. A schematic diagram of the experimental set-up is shown in Fig. 10.2. In the CCS approach, phase spreads are produced by reactive cosputtering using independently-controlled rf sputtering guns. For material deposited from a given gun, the rate is greatest close to the gun and drops off approximately exponentially with distance from the gun. Because the power to each gun is controlled independently, the deposition rate of each component in a phase spread that is being produced can be adjusted to shift the portion of ternary phase space that is examined. Substrate bias is controlled independently so that samples can be prepared under high or low ion bombardment which can affect the film microstructure.

Films grown by codeposition inherently have atomic-level mixing of the constituent atoms in the film. In the case of reactive deposition in oxygen to form oxide films, the sputtered species are typically dimers. The technique results in a fully oxidized film with atomic level mixing being produced directly, in contrast with a sequential deposition approach. Because independent guns are used, a wide range of compositions can be accessed, and constituent elements are easily changed. Also, substrate bias can be controlled in the CCS approach allowing us to adjust the degree of ion bombardment which affects the microstructure of the resulting film. Finally, as with on-axis sputtering, the parameters of sputtering ambient such as the Ar/O_2 ratio and total pressure, are important variables that can strongly influence the properties of resulting films because of defects such as oxygen vacancies that

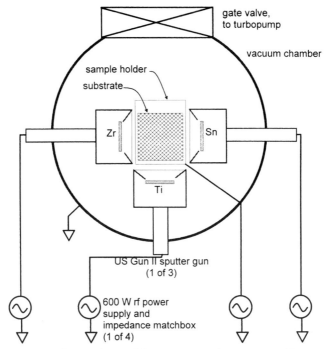

Fig. 10.2 Schematic layout of the ternary sputtering system used for Continuous Composition Spread depositions.

may be more or less prevalent under different conditions. Nitrogen can also be introduced as a dopant into the films during deposition through the use of an $Ar/O_2/N_2$ atmosphere or investigations of pure nitrides carried out through the use of an Ar/N_2 atmosphere.

The anneal step required by the DCS multilayered technique necessarily drives a film towards a thermodynamic limit. The CCS approach, by contrast, allows the preparation of low temperature, amorphous materials. Phase space at higher temperatures can also easily be explored using CCS films deposited on a heated substrate. Additionally, CCS films deposited at lower temperatures can be subjected to high temperature anneals, for example, to understand the crystallization behavior of amorphous films in a given phase system. Another advantage of the CCS approach is that there is less opportunity for contamination of samples relative to the DCS approach. In the DCS approach it is possible to get backsputtering off of the mask resulting in unexpected ingredients in a sample and therefore properties that may not be those of the system of interest.

A final advantage of the cosputtered CCS approach is that rf bias, typically 10 W, under independent control can be supplied to the substrate as indicated in the schematic diagram (Fig. 10.2). We speculate that the role of substrate bias is to increase surface mobility during film growth. This changes the microstructure and decreases the defect density of the as-grown film, which is related to the leakage behavior of that film. We have found that the use of substrate bias during off-axis deposition of dielectric materials is critical for the growth of low-leakage, high breakdown field dielectric thin films in the off-axis geometry.

10.3
Systematized Search for a New High-ε Thin-film Material

10.3.1
General Considerations for Investigating New Materials Systems

It is nearly impossible to carry out exhaustive studies of materials systems involving even a few elements. The periodic table currently has 111 elements. If we consider only those elements that are not gases or liquids under ambient conditions as well as those that are neither radioactive nor extremely reactive in air (because we want to be able to make a target for sputtering), then we are left with about 50 elements as indicated in the periodic table in Fig. 10.3. Taken three at a time, there are over 100 000 possible ternary systems that can be synthesized. In addition, there are usually a variety of processing parameters that can influence the properties of the samples that are grown. These include substrate, substrate temperature, and oxygen fugacity during sputtering. Thus, even the efficiency provided by the CCS approach is easily overwhelmed by the sheer numbers of possible samples. Systems with more than three cations are obviously even less amenable to exhaustive searches.

In combinatorial-type materials studies – as in all research problems – suitable constraints must be identified to yield a tractable experimental space for investigation. These constraints can include a processing window and limitations of the elements investigated. For example, some metals might be too expensive for a particular end use. By reducing the number of experiments to a suitably small number, then exhaustive understanding of a particular system becomes feasible. In any proposed combinatorial study it is therefore critical to first establish a clear vision of the goal of the study, so that parameters of the study can be defined and constrained.

1A	2A	3B	4B	5B	6B	7B	8B	8B	8B	1B	2B	3A	4A	5A	6A	7A	8A
H																	He
Li	Be											B	C	N	O	F	Ne
Na	Mg											Al	Si	P	S	Cl	Ar
K	Ca	Sc	Ti	V	Cr	Mn	Fe	Co	Ni	Cu	Zn	Ga	Ge	As	Se	Br	Kr
Rb	Sr	Y	Zr	Nb	Mo	Tc	Ru	Rh	Pd	Ag	Cd	In	Sn	Sb	Te	I	Xe
Cs	Ba	La	Hf	Ta	W	Re	Os	Ir	Pt	Au	Hg	Tl	Pb	Bi	Po	At	Rn
Fr	Ra	Ac															

			Ce	Pr	Nd	Pm	Sm	Eu	Gd	Tb	Dy	Ho	Er	Tm	Yb	Lu
			Th	Pa	U	Np	Pu	Am	Cm	Bk	Cf	Es	Fm	Md	No	Lr

Fig. 10.3 Periodic table of the elements, with elements most suitable for reactive sputtering in *bold face*.

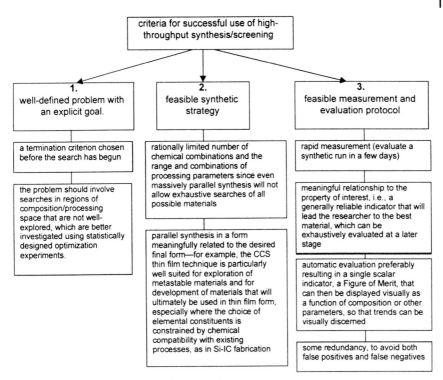

Fig. 10.4 Diagram outlining criteria for successful use of high-throughput synthesis/screening approaches to materials discovery and exploration.

Figure 10.4 outlines some of the criteria which, if met, would lead to successful application of the combinatorial approach to materials exploration.

We anticipated that the CCS technique could prove an effective research tool for solving an imminent technological problem: the need for an alternative to SiO_2 for capacitors in Integrated Circuit (IC) technology. In the following discussion, we begin by motivating the goal of the research and defining the problem we aim to solve. We then discuss the application of the CCS approach for the materials synthesis, including constraints and limitations. Finally, we discuss our measurement protocol and the definition of a Figure of Merit, FOM, that can identify the most promising candidate materials. This approach was successful: using it we discovered new high dielectric constant materials in the Zr-Sn-Ti-O psuedo-ternary phase system.

10.3.2
The Problem: Finding New High Dielectric-Constant Materials

Capacitors are charge storage devices that are essential in many circuit families, including dynamic random access memory, DRAM, and RF chips. For example, in RF chips, capacitors occupy a large fraction (at present about 50 %) of the area of the

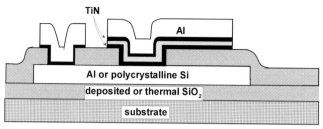

Fig. 10.5 Schematic illustration of an integrated-circuit capacitor structure.

chip, even despite the fact that complicated folded geometries are fabricated to conserve space. A schematic drawing of a metal-oxide-metal, MOM, capacitor is shown in Fig. 10.5. If capacitors with a higher specific capacitance (C/A, where C is the capacitance and A is the area) were used, a greater number of chips could be fabricated at the same cost, leading to lower prices and/or greater profit. The key to higher specific capacitance is finding a material with a higher dielectric constant, since for a parallel-plate capacitor $C/A = \varepsilon_r \varepsilon_0 / d$, where $\varepsilon_r \equiv \varepsilon$ is the relative dielectric constant of the insulating film, ε_0 is the permittivity of free space, and d is the thickness of the insulating film. In the case of DRAM as well, the properties of SiO_x ($\varepsilon = 3.9$), Table 10.1 will not meet the needs of future generations of integrated circuits. Basic circuit considerations in DRAM require a minimum capacitance of 20–40 fF/cell. As the area of a cell has shrunk, designers have maintained the cell capacitance by resorting to extremely thin a-SiO_x films, limited by the decreasing reliability of thinner films due to the finite breakdown fields, and exotic trenches, limited by the ability to obtain reliable high-aspect-ratio trenches and contacts. [12] Utilizing a different material with improved properties, e.g. a material with a higher dielectric constant and/or breakdown field, could avoid the limitations of a-SiO_x.

Thus the goal for our combinatorial-type materials search was the identification of a superior thin-film dielectric for embedded DRAM applications. We addressed the specific problem of embedded DRAM as a back-end process, which means that all of the transistors and conductors have been built into the integrated circuit structure before the DRAM is fabricated. This constrains the processing conditions that the dielectric film can be subject to. In particular, deposition and processing temperatures must be limited to 400 °C or less. Because integrated circuits are fabricated in extremely expensive facilities and because chemical contamination is a threat to the integrity of established fabrication processes, there are serious constraints on the elements that can be used in the film. A small set of elements are presently used in silicon fabrication. These include Si, O, N, Al, Ti and W. Particularly unwelcome are elements that can diffuse rapidly in silicon. However, where a new element provides necessary functionality that cannot be obtained using existing elements, then the barriers to entry can be relaxed. Such was the case recently for Cu, which forms deep traps and dramatically decreases the recombination lifetime in Si. However, the advantage of its much higher conductivity compared with that of Al and the development of barrier layers such as TiN that prevent contamination of the silicon

Tab. 10.1 Comparison of the properties of leading candidates for DRAM capacitors.

Material	ε_r	E_{br} MV/cm	FOM mC/cm^2	J_{leak} A/cm^2	Comments
SiO_2	4.0	10	3.5	6×10^{-12} (typical)	Standard material ε_r is low
TaO_x	23	4	8.1	6×10^{-10} (best)	Moderate ε_r, good for one generation?
$(Ba,Sr)TiO_3$	200	1	18	1×10^{-8} (~typical)	Etching, exotic electrodes Ref. [22].
Zr-Sn-Ti-O a-ZTT	62	4.4	24.3	2×10^{-9}	Ref. [11].

layers by copper have led to its recent adoption in high-performance integrated circuits for microprocessors. A further consideration in the selection of candidate elements is the likely need to deposit dielectric film using CVD approaches, because of
their superior step coverage abilities. Finally, it will be necessary to pattern any thin
film that is deposited, thus the etching behavior of any constituent is of interest.

10.3.3
Measurement Strategy and Figure of Merit

The dielectric constant and breakdown field of a material are not the only considerations. Because a capacitor is used for charge storage, the leakage properties of the
film are also critical. These are largely controlled by the defects in the films which
can include oxygen (anion) vacancies or mixed valence effects in the cations. It is
important that films that are as fully oxidized as possible be studied.

The CCS approach is well matched to the synthesis part of the problem of the
identification of new thin-film high dielectric constant materials for embedded
DRAM applications. As noted above, low-temperature deposition is essential and
the CCS approach provides in situ mixing at low temperatures. Samples are
obtained in thin-film form and can be made in thicknesses that are similar to those
that will actually be used.

In a materials search it is essential to distinguish interesting samples from mun-
dane samples. Two considerations are important. First, the measurement technique
chosen must provide data that is meaningfully related to the problem under study.
Second, the measurement time should be about as rapid as the synthesis step in
order for meaningful progress to occur.

In the case of defining a measurement strategy that bears a meaningful relationship to the property of interest, the aim of the evaluation is to guide the researcher
to materials with the most interesting behavior. The initial screen need not be thorough nor definitive but must be indicative. A limited subset of candidate materials

can be subject to much more thorough or exhaustive evaluations following the initial screening step. The most valuable indicator allowing a researcher to sort among candidate materials is a Figure of Merit (FOM), preferably a scalar. This FOM can be displayed visually as a function of composition or other variables so that trends can be ascertained. Because combinatorial approaches provide an abundant supply of samples, it can be helpful for the data to contain some redundancy. This extra information can provide a means to avoid false positives and negatives.

In the case of new high dielectric constant thin-film materials for use in embedded DRAM applications, we have sought materials having both high dielectric constants and high breakdown fields as well as low leakage currents. Because the CCS approach yields films that have thickness variations, it is convenient if the FOM is independent of thickness. Neither the dielectric constant nor the breakdown field alone is a useful FOM. The product, $\varepsilon E_{br} = (Ct/A) \times (V_{br}/t) = CV_{br}/A$, is the most useful and relevant FOM. This value physically corresponds to the maximum charge that can be stored on the capacitor. For a-SiO_x, a typical value is 3.5 mC/cm^2, and typical values for conventional candidates for embedded DRAM applications, including a-TaO_x and $BaSrTiO_x$, BST, are in the range of 10 – 20 mC/cm^2. This FOM, εE_{br}, is crude. It neglects any possible thickness dependence of ε and contains no direct information about the leakage current, J. However, it is a suitable indicator to provide reliable feedback and is easy to measure. The capacitance value can be measured in about a second and the breakdown field, V_{br}, in a minute or less, so that an entire ternary spread containing 5000 points can be measured overnight. After using this FOM to distinguish the interesting materials in an efficient manner, we can take time to go back and explore the best candidates in greater detail. We note that a more refined FOM which does include a direct measurement of leakage currents might be defined, but we find that it offers little advantage over the simple scalar FOM described above.

10.3.4
Electrical and Compositional Evaluation

To evaluate new thin-film dielectric materials, measurements of capacitance, breakdown voltage, and leakage current, are needed. There are three approaches to obtaining data on our thin films. For the first two approaches, we create capacitor structures to obtain these quantities. The measurement therefore is directly related to the manner in which the films are intended to be utilized. Our materials are deposited on single-crystal silicon substrates coated with layers of Ti and TiN. The Ti/TiN bottom layer is a metallic conductor and is a material that is currently used in the fabrication of integrated circuits. This layer forms the bottom electrode of a metal-oxide-metal, MOM, capacitor structure. To form the top electrode, we can deposit top counterelectrodes onto our dielectric thin film. The top electrodes can be deposited through a shadow mask to create the structures of interest, or the top metal can be blanket deposited and a lithography step used to pattern the structures. The top electrodes are then addressed one at a time using an automated probe station using a standard LCR meter to measure capacitance and leakage current. We obtain information on the breakdown field by measuring a current-voltage, I-V, curve at each point using a current source and electrometer. Breakdown is taken to occur when the current rises abruptly by a factor of ten or more, or when the current density rises above 2×10^{-4} A/cm^2; when breakdown is observed the measurement is terminated. As an alternative to preparing patterned top electrodes, we have used a scanned mercury, Hg, electrode to form the top contact. This approach has the advantage of requiring minimal sample preparation. The mercury is confined within a capillary which defines the surface area addressed by the electrode as shown schematically in Fig. 10.6. The electrical measurements are made as described above.

This scanned mercury electrode technique is not a conventional measurement approach. We have compared Hg as a top electrode to other thin-film counterelectrodes, particularly platinum and aluminum, using known dielectric materials including a-SiO$_x$, a-TaO$_x$ and a-Al$_2$O$_3$ thin films. Current-voltage curves measured using mercury top electrodes are comparable to those measured using Pt or Al top electrodes.

Information on the dielectric properties of a material might also be obtained using noncontact scanning techniques. These can include rf or microwave measurements made with an electrode that is held in the vicinity of the surface. This approach has the advantage that samples remain undamaged by the measurement and thus can be reevaluated or subjected to other studies as needed. However, this approach cannot provide accurate values for the capacitance, in very thin films such that $d/\varepsilon < 4$ Å. Moreover, the current-voltage measurements are still needed in order

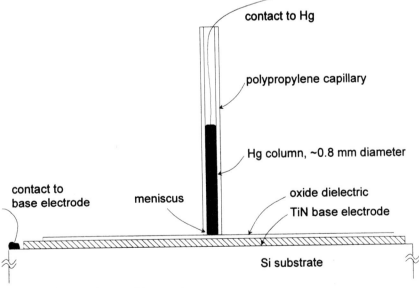

Fig. 10.6 Schematic drawing of the scanning mercury counterelectrode technique.

to obtain values for E_{br} and J_{leak}, which are necessary to evaluate the materials, so a top contact is still needed.

Knowledge of the composition of samples in an off-axis deposited phase spread is important. Calibration runs were used to determine the thickness vs position for single components deposited in our system at a given power setting. The drop-off in thickness as a function of distance from the gun is approximately exponential and can be modeled accurately to about 5 at%. We find that the deposition rates are additive for the three guns. RBS measurements have been made on a number of samples to validate our assumptions. In addition, we routinely use RBS measurements as a double check on the specific compositions that we identify as having superior properties.

10.4
Identification of a Promising Candidate and Discussion of Trends

10.4.1
Initial Survey

To discover new high dielectric constant thin-film materials, we began by surveying about 30 multinary phase spreads. We focussed particularly on spreads containing Ti, since TiO_x is a high dielectric constant material. It is generally found that TiO_x films have a low breakdown field and high leakage current, rendering the undoped material unsuitable for capacitor applications. Our hope was to find amorphous pseudo-binary or -ternary oxides in which the desirable high dielectric constant of titania was retained while the leakage and breakdown behavior have improved characteristics. The initial survey revealed several systems that warranted further investigation. While screening quickly identified systems of interest, the more detailed studies required to understand and verify film characteristics consumed much more time.

10.4.2
The Zr-Sn-Ti-O System

Based on our initial review of the data, the Zr-Ti-Sn-O pseudo-ternary phase system was chosen for a more detailed investigation. Figure 10.7 shows the experimentally obtained capacitance values and breakdown voltages displayed as a function of position. The location of the guns with respect to the substrate are indicated for the capacitance (see also Fig. 10.2). Raw capacitance data alone can be misleading because they depend both on the dielectric constant of the material and on the thickness. Thus, very thin regions, at the corners or the substrate, have high capacitance values. Usually, these thin regions do not have excellent breakdown behavior, and thus the FOM values in such regions tend to be low. The capacitance and breakdown voltage are combined at each point to obtain the Figure of Merit (FOM = εE_{br}) data shown in Fig. 10.8. To obtain a useful indication of the leakage current, we analyzed the I-V curves taken at each spatial location, and identified the leakage current at a fixed

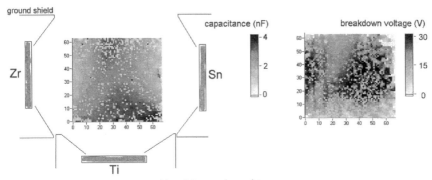

Fig. 10.7 aZTT raw capacitance and breakdown voltage data, as a function of position on the substrate, with guns shown schematically.

value of stored charge. For the right hand plot of Fig. 10.8, we chose $Q = 7$ mC/cm^2 and plotted the leakage current on a logarithmic scale as shown. With knowledge of the composition, as described above, we can then plot this data onto a ternary phase diagram as shown in Fig. 10.9. In the raw data, a high FOM region, black points can be seen in the data toward the bottom of the sample, i.e. relatively close to the Ti gun.

Regions of high FOM or low leakage, "sweet spots", can be characterized schematically by 95th percentile contours calculated from the full data sets as plotted in Fig. 10.10. In this figure, we show the 95th percentile FOM regions for thin films of amorphous Zr-Ti-Sn-O and also the Hf-based analog, [13] Hf-Ti-Sn-O, discussed below. Also shown is the location of the single-phase crystalline low loss, dielectric ceramic of composition, $Zr_{1-x}Sn_xTiO_4$, which is used for filter elements in wireless applications. [14] This material has a near-zero temperature coefficient of resonant frequency ($\tau f = \pm 4$ ppm/C). The region of this phase space that was previously explored in thin-film deposition experiments is indicated by the black line and dot. A film with the composition $Zr_{0.8}Sn_{0.2}TiO_4$ was prepared by sputtering from a single target by Nakagawara et al., [15] and several samples with varying Zr/Ti ratios

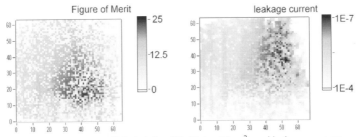

Fig. 10.8 aZTT Figure of Merit ($\varepsilon E_{br} = CV_{br}/A$, in μC/cm^2) and leakage currents (measured at a stored charge of 7 μC/cm^2, current density measured in A/cm^2) as a function of position on the substrate.

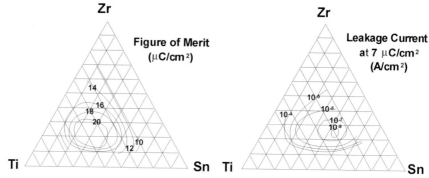

Fig. 10.9 Figure of Merit and leakage current data of Figure 10.8, replotted on a conventional ternary composition diagram.

were prepared and studied by Ramakrishnan, et al. [16] These studies were clearly inspired by the crystalline $Zr_{1-x}Sn_xTiO_4$ material and were aimed at producing films displaying the behavior of the crystalline ceramic including its near-zero temperature coefficient of resonant frequency. In the case of a-Zr-Ti-Sn-O films, neither the high FOM region nor the low leakage region correspond to the area of phase space occupied by the dielectric ceramic.

A powerful feature of the combinatorial approach as applied to the problem of high dielectric constant thin-film materials relative to a point-by-point investigation is its propensity to discovery interesting properties in unexpected regions of phase space. Conventional approaches would necessarily be guided by prior studies, often of crystalline phases, because of the time and expense involved in preparing multiple targets and examining the resulting films. Our high-throughput search approach allowed us to investigate broad areas of phase space with fewer preconceived constraints and has led to several cases of materials with excellent properties being found in unexpected regions of phase space.

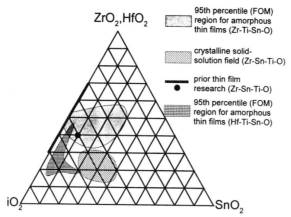

Fig. 10.10 Schematic illustration of the regions of composition space that yield optimum properties for various systems.

10.4.3
Single-Target Synthesis and Detailed Electrical Characterization

Deposition of thin films of new Zr-Ti-Sn-O high dielectric constant materials using a conventional deposition process provides important verification of the potential utility of these materials. Conventional techniques are those that are more suitable for large-scale processing, such as conventional on-axis sputtering, CVD, etc. We studied the deposition of aZTT dielectric thin films using conventional on-axis reactive sputtering [17] to show that film with properties comparable to or better than those obtained by off-axis sputtering could be grown. A composite $Zr_{0.2}Sn_{0.2}Ti_{0.6}$ metal target was fabricated to our specifications, and films were grown by on-axis sputtering using one of the 2-inch diameter sputter guns in the system. Approximately thirty runs were performed in which the process conditions were varied to find a rough optimum. The location of this optimum was not unexpected: a total pressure of 4 mTorr, 50±10 % O_2 in Ar and 200 ± 25 °C, although the best properties were obtained with the target-substrate distance no greater than 3 cm in our system, a surprisingly small distance.

Using these roughly optimal conditions, a 40–50 nm thick film was obtained with a uniform region about 25 mm in diameter. This is quite reasonable, and suggests that scale-up to a deposition system with a 12-inch diameter target would give acceptable uniformity over an 8-inch diameter Si substrate. Capacitors were prepared on this film by evaporating Pt through a metal shadow mask, giving circular counterelectrodes 200 μm in diameter. Figure 10.11 shows a J-E curve for a typical capacitor, for which the specific capacitance is 11 fF/μm^2, the breakdown field is 3.4 MV/cm, and the dielectric constant (ε) is 62. Leakage currents, measured at 1.0 MV/cm, were in the range 10^{-9} to 10^{-7} A/cm^2. These films have properties that suggest utility for use in future generations of Si-IC device technology, including DRAM.

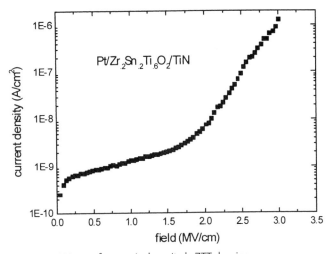

Fig. 10.11 I-V curve for on-axis deposited aZTT showing excellent leakage behavior.

10.4.4
HfTT Analog

The chemistries of hafnium and zirconium, two group IV early transition metals, are nearly identical. The lanthanide contraction leads to essentially identical atomic and ionic radii for these two atoms. Therefore, having observed the excellent properties in the aZTT system, we explored the Hf-Sn-Ti-O system using the CCS approach. [13] We have found that the properties of thin films in the a-Hf-Sn-Ti-O system depended strongly on deposition conditions, particularly substrate temperature. Phase spreads were prepared at three different substrate temperatures, 185 °C, 260 °C, and 305 °C, with all other conditions including sputtering atmosphere and powers to the sputtering guns held constant. In all cases, the high dielectric region lies in the TiO_x-rich region of the phase diagram. However, the maximum dielectric constant measured changes with deposition temperature with the best properties obtained in films deposited at 260 °C. In the region that had FOM > 18 mC/cm^2, the highest dielectric constant measured was 58.

This region is roughly triangular, with the bulk of the region lying along the Ti–Hf line as shown in Fig. 10.10. The current-voltage behavior of a sample having the nominal composition $Hf_{0.2}Sn_{0.05}Ti_{0.75}O_2$ shows reasonable leakage, ca. 1×10^{-7} A/cm^2, and a breakdown field of 4 MV/cm. The film measured in this experiment does not represent an optimized sample. Further work on this material might be expected to result in films with reduced leakage currents and still higher breakdown fields.

10.4.5
Other Systems

The CCS approaches also lends itself conveniently to the study of pseudo-binary systems. Van Dover recently showed [18] that lanthanide additions to amorphous titania produces films of composition $Ti_{1-x}Ln_xO_y$, which have high dielectric constants (ε50) and greatly improved leakage characteristics relative to undoped $aTiO_x$ films ($J_{leak}10^{-8}$ A/cm^2 at 1 MV/cm, compared to $\sim 10^{-5}$ A/cm^2 for undoped TiO_x). The CCS approach was used to prepare phase spreads that allowed van Dover to chart the dielectric constant, breakdown field and leakage behavior as a function of lanthanide concentration. The particular cases Ln=Nd, Tb, and Dy were examined, of which the $Ti_{1-x}Dy_xO_y$ films were found to be slightly superior. Figure 10.12 compares the I-V curves for Ti-O and Ti-Dy-O films deposited in the same sputtering system. Incorporation of the lanthanide, in this case Dy, produces dramatically improved leakage currents. The advantage of $Ti_{1-x}Ln_xO_y$ films compared to aZTT or BST films is that they have only two cation components, which may simplify process development for integration into a conventional Si IC process flow.

Recently, much attention has focused on amorphous tantalum oxide ($aTaO_x$) films for DRAM applications. This material has a higher FOM than SiO_x, with ε23, $E_{br}4$ MV/cm, FOM8.1 μC/cm^2, and is of particular interest for embedded DRAM applications, for the reasons mentioned above in the context of aZTT films. State-of-the-art films are believed to have adequate electrical properties, at least for present

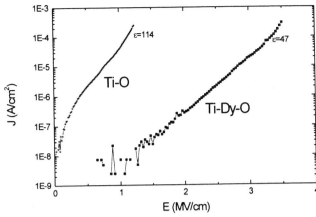

Fig. 10.12   I-V for Ti-Ln-O compared to Ti-O, showing the dramatic improvement in leakage current (for films deposited in same sputtering system).

DRAM cell designs. But, any improvements in the leakage behavior of films would have positive design implications. Leakage currents in dielectric films are believed to be controlled largely by the defect structure, in particular, the number, type and distribution of oxygen vacancies. Removing or passivating those vacancies might be expected to result in films with lower leakage currents. Little detailed knowledge of defects in a-TaO_x is available, so efforts to improve films have centered on two empirical approaches: post-deposition annealing and incorporation into the film of a second component – a metal oxide such as AlO_x, TiO_x, WO_x, SiO_x, or ZrO_x, or an anion such as nitrogen.

We used the CCS approach to explore the possibility that simultaneous incorporation of cation and anion dopants into an a-TaO_x film would result in further improvements of film properties. [19] We prepared films having continuously varying aluminum contents using the CCS approach. This allowed us to prepare films with Al/(Ta+Al) atomic ratios ranging from 0.1 to 0.6 in a single run, eliminating run-to-run variations and improving the sensitivity and confidence regarding systematic composition/property effects. Films were prepared with and without N_2 included in the sputtering atmosphere. The simultaneous inclusion of Al and N into a-TaO_x films was found to improve the leakage properties of films with only a small reduction in ε, as shown in Fig. 10.13, which compares I-V curves for Ta-Al-O-N with Ta-O. About 10–20 at% Al along with 5 vol% N_2 in the sputter gas gave excellent results. Further, the simultaneous addition of both Al and N to TaO_x appears to provide an improvement in statistical measures of quality for Ta-Al-O-N compared to Ta-O, Ta-Al-O and Ta-O-N (750 capacitors in each experiment) as shown in Table 10.2.

Materials with better properties are also of interest as possible replacements for aSiO_x for gate oxide applications, since the ultrathin SiO_x layers required for low-voltage, high-drive logic is inherently too leaky to be employed. Instead, a thicker layer of a material with a higher dielectric constant is needed. As in capacitor appli-

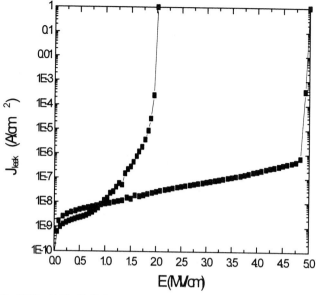

Fig. 10.13 I-V for Ta-Al-O-N compared to Ta-O, showing that Al improves the electrical properties (based on depositions in the same sputtering system). The composition of this film was about $Ta_{0.8}Al_{0.2}O_{1.45}N_{0.05}$.

cations, the basic Figure of Merit is still εE_{br}. However, in contrast to the "back end" processing described previously in the context of DRAM capacitors, the gate oxide is inherently subject to front end processing, which means that materials must survive thermal treatments including various annealing steps. For $aTaO_x$ to be a candidate gate oxide material, it must be able to survive a short anneal at roughtly 850 °C without substantial degradation of properties. However, $aTaO_x$ crystallizes at 650 °C and the crystallized film is much leakier and physically rougher than the initial amorphous material. We used the CCS approach to allow us to examine the effect of dopants on the annealing behavior of $aTaO_x$. We recently reported [20] that aluminum additions raise the crystallization temperature substantially, as shown in Fig. 10.14, which shows roughness, a measure of the degre of crystallinity of the film,

Tab. 10.2 Improvement in statistical measures of quality for Ta-Al-O-N compared to Ta-O, Ta-Al-O and Ta-O-N (750 capacitors in each experiment).

Material	Fraction (fraction of capacitors with $\varepsilon E_{br} > 3$ $\mu C/cm^2$)	Median (median value of εE_{br}, in $\mu C/cm^2$)
Ta-O	0.66	4.5
Ta-Al-O	0.27	5.8
Ta-O-N	0.61	4.8
Ta-Al-O-N	0.88	5.9

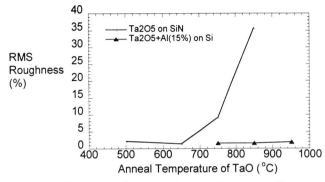

Fig. 10.14 Effect of anneal and crystallization of 10 nm Ta_2O_5 film on Si on roughness as measured by AFM. The addition of 15 % Al in the 20 nm films suppressed crystallization.

with anneal temperature for Ta_2O_5 on SiN and for Ta_2O_5 + 15 at% Al in Si. The addition of 15 % Al in these films suppressed crystallization while causing only a small reduction in dielectric constant. Thus aluminum-doped tantalum oxide films, a-$Ta_{1-x}Al_xO_y$ may be useful in gate oxide applications.

10.4.6
Other Problems for Which a Combinatorial Approach is Well Suited

Our research using the CCS approach has focused on the discovery of new dielectric thin films. This example illustrates the power of this approach and key considerations in utilizing high throughput synthesis and screening approaches. A variety of other problems appear well matched to such combinatorial-style methods. A few of these are discussed briefly below, touching on both experimental challenges and research opportunities.

10.4.7
New Magnetic Materials

Many of the technological advances made in the magnetic data storage industry have depended directly on innovations in magnetic materials. The read/write heads in magnetic hard disks utilize advanced magnetic thin films and multilayers. For example, the write heads require soft magnetic materials with a very high saturation magnetization in order to generate the gap fields needed to reverse domains on high-performance media. Nanocrystalline materials such as FeTaN have been recently developed and appear to be adequate, but alternative materials might be identified by a combinatorial-type approach. One specific goal might be the identification of magnetic materials with a resistivity higher than that of FeTaN, ca. 100 $\mu\Omega$ cm. Materials with higher and/or more controllable anisotropies might also be useful, but anisotropies generally depend strongly on microstructure, which is not easy to control or vary in a combinatorial-type study.

The identification and study of colossal magnetoresistance (CMR) materials could be amenable to combinatorial-style searches. Such approaches could be useful for elucidating the structural and electronic phase diagrams of these fascinating materials. Ultimately this could help sort out the apparently complex physics that underlies their behavior. An indication of the value of these approaches was recently provided by Xiang et al. [21]

10.4.8
Superconductors

The search for new superconductors is a daunting challenge since these materials have been extensively investigated for over fifty years and were the subject of a massive search initiated by the discovery of "high-temperature" superconductivity in 1986. While it is true that a number of new superconductors with higher and higher transition temperatures have been found, it also is clear that most of these require very specific and narrow processing windows for phase formation, rendering them extremely difficult to capture using a combinatorial-style approach. Yet the payoff in terms of basic understanding as well as possible commercially important materials can be large. Convenient and rapid methods to evaluate superconducting properties including scanning approaches are well understood. Combinatorial-type approaches may be useful in elucidating phase diagrams and well as providing new insights into the underlying physics of superconductivity.

10.4.9
Thermoelectric materials

Thermoelectric materials have the potential to revolutionize a variety of heating and cooling technologies provided a breakthrough material can be identified. Again, many studies of candidate materials have already been done, so the likelihood of discovering a breakthrough seems low. A well-accepted Figure of Merit for thermoelectric materials exists already. This FOM may be difficult to measure in a thin-film geometry, however, because of the difficulty of measuring thermal conductivity in a thin film. It is possible that a more crude (and easier to measure) yet indicative FOM might be used to identify promising candidates.

10.4.10
Piezoelectric materials

Piezoelectric materials have a variety of commercial applications, and a large number of such materials are known. To examine such materials in thin-film form, highly textured crystalline thin films are likely to be needed, so microstructure control is therefore an issue. Screening approaches should be straightforward.

10.4.11
Ferroelectric Materials

As with piezoelectric materials, ferroelectric materials have a variety of applications and have been intensively investigated. Again, textured crystalline thin films are likely to be needed, which implies that microstructure control is again an issue. Screening and measurement approaches are straightforward.

10.4.12
Optical Materials

A variety of problems fall under the heading of optical materials including new glass compositions, transparent conductors, phosphors and amplifier materials. Screening approaches might be straightforward, but a candidate material must then be examined in detail to understand the match between performance and application requirements. Also, it is helpful if the form that the material will be used in is related to the form in which it is screened. Thus, where bulk single crystals are required for implementation, then evaluation of a film sample may miss significant and relevant information.

10.4.13
Catalysts

Catalysts are so important in chemical industry that combinatorial-type approaches will certainly be utilized as tools in the identification and optimization of both heterogeneous and homogeneous catalysts. These are, in many cases, high value-added, commercially important materials. In any research activity it makes sense to remove any choke points that impede progress. However, many engineering considerations are important in producing commercial catalytic processes; thus, combinatorial-type approaches are unlikely to be a panacea. Still, a variety of unanticipated discoveries may result from various efforts that are under way.

10.5
Concluding Comments

A powerful feature of the combinatorial approach as applied to the problem of high dielectric constant thin-film materials relative to a point-by-point investigation is its propensity to discovery interesting properties in unexpected regions of phase space. Conventional approaches would necessarily be guided by prior studies, often of crystalline phases, because of the time and expense involved in preparing multiple targets and examining the resulting films. Our high-throughput search approach allowed us to investigate broad areas of phase space with fewer preconceived constraints and has led to several cases of materials with excellent properties found in unexpected regions of phase space.

References

1 SAWATZKY, E., and KAY, E., *IBM J. Res. Dev.* **13**, 696 (1969)

2 HANAK, J. J., *J. Mater. Sci.* **5**, 964–971 (1970)

3 VAN DOVER, R. B., HONG, M., GYORGY, E. M., DILLON, J.F., and ALBISTON, S.D., *J. Appl. Phys.* **57** 3897–9 (1985)

4 VAN DOVER, R. B., HESSEN. B., and WERDER D., *Chem Mater.* **5**, 32–35 (1993)

5 HAMMOND, R. H. et al., *J. Appl. Phys.* **40**, 2010 (1969)

6 HARTSOUGH, L. D., and HAMMOND, R. H., *Sol. St. Comm.* **9**, 885 (1971)

7 GALLOP, M. A., BARRETT, R. W., DOWER, W. J., FODOR, S. P. A., and GORDON, E. M., *J. Med. Chem.* **37**, 1233–1251 (1994); GORDON, E. M., BARRETT, R. W., DOWER, W.J., FODOR, S. P. A., and GALLOP, M. A., *J. Med. Chem.* **37**, 1385–1401 (1994); DOYLE, P. M., *J. Chem. Tech.* **64**, 317–324 (1995)

8 XIANG, X.-D. et al., *Science* **268**, 1738–1740 (1995)

9 SUN, X.-D., GAO C., WANG J., and XIANG, X.-D., *Appl. Phys. Lett.* **70**, 3353 (1997); DANIELSON, E., DEVENNEY, M., GIAQUINTA, D. M., GOLDEN, J. H., HAUSHALTER, R. C., MCFARLAND E. W., POOJARY, D. M., REAVES, C. M., WEINBERG, W. H., and WU, X. D., *Science* **279**, 837 (1998).

10 REDDINGTON, E., SAPIENZA, A., GURAU, B., VISWANATHAN, R., SARANGAPANI, S., SMOTKIN, E. S., and MALLOUK, T. E., *Science* **280**, 1735 (1998).

11 VAN DOVER, R. B., SCHNEEMEYER, L. F., and FLEMING, R. M., *Nature,* **392**, 162 (1998)

12 EL-KAREH, B., BRONNER, G. B., and SCHUSTER, S. E., *Solid State Technology,* May 1997, pp.89–101.

13 SCHNEEMEYER, L. F., VAN DOVER, R. B., and FLEMING, R. M., *Ferroelectric thin films VII,* (Pittsburgh, PA, USA: Mater. Res. Soc, 1999); YOUNG, P. L., FEHLNER, F. P., and WHITMAN, A. J., *J. Vac. Sci. Tech.* **14**, 174 (1977)

14 WOLFRAM, G., and GÖBEL, H., *Mat. Res. Bull.* **16**, 1455 (1981); Iddles, D. M., Bell, A. J., and Moulson A. J., *J. Mater. Sci.* **27**, 6303 (1992)

15 NAKAGAWARA, O., TOYODA, Y., KOBAYASHI, M., YOSHINO, Y. KATAYAMA, Y., TABATA, H., and KAWAI, T., *J. Appl. Phys.* **80**, 388 (1996)

16 RAMAKRISHNAN, E. S., CORNETT, K. D., SHAPIRO, G. H., and HOWNG, W.-Y., *J. Electrochem. Soc.* **145**, 358 (1998)

17 VAN DOVER, R. B., and SCHNEEMEYER, L. F., *IEEE Electron Dev. Lett.* **19**, 329 (1998).

18 R. B. VAN DOVER, *Appl. Phys. Lett.* **74**, 3041 (1999).

19 VAN DOVER, R. B., SCHNEEMEYER, L. F., FLEMING, R. M., and WERDER, D. J., *Ferroelectric thin films VII,* (Pittsburgh, PA, USA: Mater. Res. Soc., 1999); YOUNG, P. L., FEHLNER, F. P., and WHITMAN, A. J., *J. Vac. Sci. Tech.* **14**, 174 (1977)

20 VAN DOVER, R. B., FLEMING, R. M., SCHNEE-MEYER, L. F., ALERS G. B., and WERDER, D. W., 1998 IEDM Technical Digest, p. 823.

21 X.-D. XIANG, *Bull. Am. Phys. Soc.* **44**, 103 (1999).

22 PARK, S. O. et al., *Jpn. J. Appl. Phys.* **35** Part 1, 1548–1552 (1996)

11

On One Hand But Not The Other: The Challenge of the Origin and Survival of Homochirality in Prebiotic Chemistry

Patrick Frank, William A. Bonner, and Richard N. Zare

What are the facts of life? One of the most striking is that all known living systems involve the same types of polymers, i.e., three varieties of homochiral biopolymers. That is, each variety is composed of unique molecular building blocks having the same three-dimensional handedness. Thus, with rare exceptions, the proteins found in cells are composed exclusively of the l-enantiomers of 19 optically active amino acids (Fig. 11.1).[*] Similarly, only D-ribose and 2-deoxy-D-ribose sugars are found in the nucleic acid polymers that make up the RNAs and DNAs, which are essential for protein synthesis in the cell and for the transmission of genetic information from one generation to the next.

What is the origin of biological chirality and how did it successfully propagate itself to dominance? The riddle is at once intriguing, central, and profound. It has been persuasively argued and is now widely accepted that homochirality of biopolymers is essential for the existence of life, and that life would be impossible in its absence. [4] Thus the solution to the riddle of terrestrial homochirality may well be intimately related to how life on Earth originated.

Was Nature's selection of these stereochemical structures a random event, or was it the consequence of determinate processes? Although the presence of amino acids and other organic molecules on primitive Earth has been attributed to both endogenous and exogenous sources, [5,6] the origin of the chiral bias of prebiotic molecules remains a fundamental mystery. A mystery perhaps even more central to the origin of life is the mechanism that brought into dominance and survival any putative small chiral molecular excess on early Earth.

---

[*] The prefixes D and L refer to absolute spatial configurations, which do not necessarily designate how plane polarized light is rotated or how circularly polarized light is preferentially absorbed. The latter are denoted by the prefixes d (dextrorotatory) and l (levorotatory). The observation that a chiral compound can be assigned to d or l can vary with wavelength and other factors. For example, at the sodium D line, eight of the naturally occurring L-amino acids are d and ten are l at 25°C in water solution. At shorter wavelengths, near 200 nm, all nineteen L-amino acids absorb preferentially left-handed versus right-handed circularly polarized radiation. See: L. Fowden, P. M. Scopes, and R. N. Thomas, *J. Chem. Soc.* (C) **1971**, 833–840; and L. I. Katzin and E. Gulyas, *J. Am. Chem. Soc.* **1968**, *90*, 247–251.

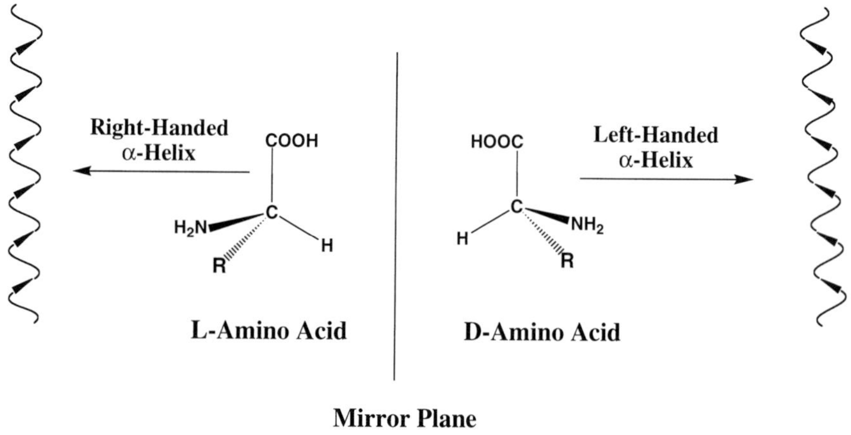

Fig. 11.1 Mirror-image relationship of the amino acids. All amino acids (except glycine) can exist in one of two mirror-symmetric forms, called enantiomers. Proteinogenic amino acids are almost exclusively the L-enantiomer. With some exceptions, biopolypeptides fold into right-handed α-helices. [1,2] In all cases, polypeptides composed of D-amino acids fold into helices of opposing handedness. [3] The *bolded* and *hashed* bonds are above and below the plane of the drawing, respectively. "R" represents the pendant group that distinguishes one amino acid from another.

The traditional starting point for a solution to the mystery of terrestrial homochirality has been a search for factors capable of spontaneously breaking molecular symmetry, thus causing a stereochemical preference. One possibility is that Nature has a built-in preference for one handedness over another and that this preference is reflected in the homochirality of biopolymers. Indeed, we shall see that such an energetic preference does exist, although it is so small compared to the thermal energy of the ambient surroundings that its presence may not be determining. Another possibility is that some statistical fluctuation led to the preferential selection of one enantiomer over the other in a racemic system, that is, a system composed of equal amounts of left-handed and right-handed molecules. A third possibility is that prebiotic chemical systems have been exposed to some external set of conditions that have broken their chiral symmetry in such a manner as to favor particular stereochemical structures over their mirror images. Suggested examples [4] of such symmetry-breaking external conditions involve exposure to circularly polarized light of just one handedness, exposure to polarized β-radiation, or subjection to crossed force fields (electric and magnetic, or gravitational and rotational) of one origin or another. Once some form of primitive self-replicating chiral molecular system was established, its ability to reproduce itself and to metabolize its surroundings caused one type of handedness to become dominant over the other.

Almost 140 years ago Pasteur showed how a racemic mixture could be separated into its chiral constituents. Ever since, theories such as the three possibilities above have been proposed to explain an abiotic origin for molecular chirality in living systems. At the present time, however, no agreement exists about which explanation is best. In each of these scenarios, we can imagine production of some initial enantiomeric excess (e.e.).

One crucial question is not whether some small initial e.e. can be generated, but how such a small e.e. can be amplified so that one enantiomer eventually dominates completely over the other. The second crucial question is how a homochiral dominance, once established, can survive. These are the questions that challenge us in our search to explain the homochirality of terrestrial biopolymers.

In what follows we first review some of the mechanisms that have been proposed for establishing initial enantiomeric excesses in a racemic mixture of optically active stereoisomers. Next, we summarize experiments that have been performed to find ways in which such small enantiomeric excesses can become amplified to afford highly enantiomerically enriched yields. Then we discuss ways in which homochirality could have persisted in the challenging environment of early Earth. Finally, we synthesize several lines of evidence into a research outline directed toward developing a self-consistent theory for the origin and survival of homochirality, and thus of life. The end of the 20th century marks nearly 100 years since the first critical comments on the chemical origin of life were made by Chamberlin and Chamberlin. [7] It seems fitting, after the intense, brilliant, and optimistic work of so many people, that the end of the century that witnessed such work might see the emergence of a scientifically credible research outline. Such a research outline should promise all the elements of a viable, testable, and self-consistent theory for the origin of life.

11.1
Symmetry Breaking and Chiral Induction

Chiral molecules have a nonsuperposable mirror image (see Fig. 11.1) and so possess intrinsic handedness in three-dimensional space. In a perfectly symmetric, chirally unbiased world, each handed version (enantiomer) must exist with an equal probability. The observed preference of one enantiomer over the other in biomolecules implies that this symmetry has been broken. The initial induction of a symmetry-breaking chiral excess is the sine qua non of eventual chiral dominance. Several mechanisms have been proposed to bring about chiral symmetry breaking.

11.1.1
Is it Intrinsic?

More than forty years ago, Lee and Yang [8] observed anomalies in the decay patterns of theta and tau mesons, which suggested to them that parity was not conserved for certain weak interactions involved in the β-decay of radioactive nuclei. This Nobel-prize-winning prediction was experimentally validated by Wu et al., [9] who found that the longitudinally polarized electrons emitted during the β-decay of ^{60}Co nuclei had a notable (40 %) left-handed bias, i.e., their spins were predominantly antiparallel to their directions of motion. These experiments established that parity violation and symmetry breaking occurred at the nuclear level.

In 1957 Vester and Ulbricht attempted to couple this parity violation on the nuclear level to events at the molecular level. Vester et al. [10,11] suggested that circularly polarized gamma rays produced when longitudinally polarized β-decay electrons impinge on matter might, in the presence of suitable organic substrates, engender stereoselective photochemical reactions, synthetic or degradative, that would lead to chiral products having an enantiomeric excess.

A more subtle aspect of this broken symmetry, however, is its effect on the energies of D- and L- enantiomers, which, previous to the work of Lee and Yang, were assumed to be energetically identical. In 1966 Yamagata [12] proposed what might be called the Yamagata PVED-induced chirality hypothesis. He argued that the D- and L-forms of chiral molecules would have slightly different energies and thus different probabilities for chemical reaction because of parity violation. Yamagata argued further that this parity violation energy difference (PVED) would account for the asymmetric appearance of biomolecules on Earth.

In 1974 Garay and Hrasko [13] contended that PVEDs between enantiomers would, in the course of millions of years of evolution, lead to almost complete selection of one isomer, and in 1975 Letokhov [14] asserted that PVED-generated rate differences as small as 10^{-16} would over 10^8 to 10^9 years be "quite sufficient for full selection of either of the two stereoisomeric forms of all the amino acids that occur in animate nature." In 1983 Kondepudi and Nelson [15] claimed that a value of $\Delta E/kT$ of 10^{-17} to 10^{-15} is "sufficient to have a strong chiral selectivity."

Building on an early theoretical scheme by Frank [16] (see below) for the spontaneous autocatalytic symmetry breaking in which enantiomers act as catalysts for their own production, Kondepudi and Nelson [17–20] further suggested a generalized scheme by which the minor excess of one enantiomer, caused by PVED, might lead to a state of enantiomeric homogeneity. They then calculated that a period as short as 15 000 years could cause chiral domination. [19] In 1989 MacDermott and Tranter [21] maintained that "this amplification time would be reduced from 10^4 years to just one year if the PVED were increased to 10^{-16} kT.", i.e., by a factor of only ten.

The traditional treatment of molecules relies upon a molecular Hamiltonian that is invariant under inversion of all particle coordinates through the center of mass. For such a molecular Hamiltonian, the energy levels possess a well-defined parity. Time-dependent states conserve their parity in time provided that the parity is well defined initially. Such states cannot be chiral. Nevertheless, chiral states can be defined as time-dependent states that change so slowly, owing to tunneling processes, that they are stationary on the time scale of normal chemical events. [22] The discovery of parity violation in weak nuclear interactions drastically changes this simple picture, [14, 23–28] For a recent review, see Bouchiat and Bouchiat. [29]

A proper treatment of molecular chirality must include parity-violating forces in the quantum chemistry calculations, even though the corresponding parity-violating molecular potentials are extremely small – on the order of $10^{-18\pm3}$ kJ/mol for molecules composed of light elements. Such forces may be neglected for almost all purposes except the treatment of chiral molecules. This computational task, however, is not readily carried out. Among the best calculations to date appears to be the recent work of Bakasov, Ha, and Quack, [30] who report that they are unable to calculate the energy difference at the stage of refinement they have reached, although their calculations do show that the difference between the parity-violating potentials is more than an order of magnitude greater than previously estimated. Lazzeretti and coworkers [31, 32] have reached the same conclusion. Even so, Bakasov et al. find that the values of the PVEDs for sugars and amino acids are uncertain even as to their sign. Therefore, at the present level of theory, it is far from established that the energy differences of enantiomers arising from parity-violating weak nuclear forces can account for the homochirality of biopolymers. Recent highly precise PVED calculations comparinh D- and L-alanine indicate the continued force of this conclusion. [30a]

11.1.2
Is it Fluctuational?

In 1981 Nicolis and Prigogine [33] considered the behavior of far-from-equilibrium racemic mixtures that can bifurcate randomly into a direction of an enantiomeric singularity on passing through a critical point. They showed that such systems are profoundly sensitive to environmental asymmetries that at the critical point can cause the system to adopt one handedness over the other.

Are examples of such systems to be found in nature? Indeed, crystallization of silica within cooling magma tubes exhibits just this behavior. Although silica is not chiral on the atomic scale, it can crystallize from a melt into one of two helical structures that are morphologically chiral. [34] The choice of which of the two crystallographic helices is expressed is purely a fluctuational episode producing a nucleation event within the cooling, saturated melt. Figure 11.2 shows an example of single crystals of left- and right-handed quartz.

The observed distribution of chirality among quartz crystals within the Earth is extremely close to 50:50, as would be expected from a randomly selected sample of

Fig. 11.2 Single crystals of morphologically enantiomeric quartz. Note the mirror symmetry of the facets on the respective crystals. Unlike amino acids, the component silicon dioxide molecules have no chirality. The spontaneous resolution of quartz into crystals of opposite morphological handedness is an example of local symmetry breaking in the environment.

limited size if terrestrial quartz had no net chiral excess. [35] For example, in the largest study to date (of 27 053 naturally occurring quartz crystals), 49.83 % of these (13 481) were found to be left-handed, and 50.17 % (13,572) were found to be right-handed. [36]

Spontaneous resolution seems to be a general phenomenon, because it is also observed to occur from solutions of achiral sodium chlorate, which crystallizes into morphologically chiral crystals. In the same way as quartz, however, any given spontaneous crystallization produces nearly equal numbers of left- and right-handed sodium chlorate crystals. The explanation for this behavior is that the primary crystallization nuclei that form throughout a cooling concentrated solution must occur with a random distribution of the two morphological chiralities.

In an interesting finding, however, Kondepudi, Kaufman, and Singh [37] reported that, when stirred, aqueous solutions of sodium chlorate would deposit crystals that were overwhelmingly of one chirality or the other. The rationale for this was that stirring dispersed the initial crystallization nucleus throughout the solution into a multitude of secondary crystallization nuclei. These secondary nuclei, of the same handedness as the primary nucleus, induced rapid and complete crystallization of the solute into a single chiral form. Although the chirality of the initial nucleus was randomly selected, the choice, once made, determined the chiral fate of the process. Recently, Kondepudi, Laudadio, and Asakura [38] have extended these findings to 1,1′-binaphthyl.

As found for quartz, however, such processes occurring across the Earth will necessarily sum to zero net chirality. That is, because the choice of nucleating chirality is random, the chirality of the crystals deposited from disturbed solutions is also random, and over a large number of such events, the outcomes must sum to equality.

In the organic regime, Niori et al. [39] observed that certain large self-associating molecules that in solution have a shape like a recurved bow can spontaneously assemble into macroscopic chiral regions of one handedness or the other. This work has recently been extended by Walba and coworkers, [40] who showed that similar molecules can form observable chiral domains despite the lack of any molecular chiral center. The chirality of the domains alternate in the fluid, and so there is no net chiral induction. Likewise, racemic mixtures of chiral amphiphiles dissolved in lipid monolayers can spontaneously segregate into D- and L-enriched regions. [41, 42] Similar behavior occurs in monolayers of racemic (R)(S)-2-bromohexadecanoic acid on graphite. The racemic monolayer was observed to spontaneously segregate into alternating chiral domains that were directly visualized using scanning tunneling microscopy. [43]

From these examples, it becomes clear that molecular symmetry can spontaneously break into chiral domains in the absence of any external force or seed. However, in every known case, the net symmetry remains intact, and the overall chirality sums to zero in the environment. Nevertheless, the spontaneous formation of macroscopic chiral regions in systems of associating achiral molecules is of interest to those who contemplate induction of molecular chirality in the context of prebiotic reaction chemistry.

11.1.3
Is it Extrinsic?

Other means exist to break symmetry than a way intrinsic to weak nuclear forces. For example, it is well known that chiral molecules are characterized by circular dichroism (CD), that is, left circularly polarized light (left CPL) and right circularly polarized light (right CPL) are not absorbed equally by the two enantiomers. Because of CD, CPL from an external source is able to engender asymmetric stereo-selection in the photodegradations, photochemical syntheses, or photochemically induced interconversions of chiral molecules, thereby leading to small e.e.s. Such processes and the theory behind them, which stem from the pioneering asymmetric photolyses reported by Kuhn and Braun [44] over 70 years ago, have been more recently reviewed in detail. [45]

On Earth, light from the sun can exhibit weak (ca. 0.2 %) left circular polarization at sunrise and right circular polarization at sunset. Light scattering from atmospheric aerosols and particulates induce this polarization difference. These differences, as discussed by Bonner, [46] would sum to zero over a completely flat Earth. Where the terrain is tilted, however, a net circular polarization of light could result. Thus, photolysis of racemic terrestrial organics exposed on some eastward-facing slope could be invoked to rationalize the appearance of chirality on Earth. This explanation, although it cannot be ruled out, does seem to be a form of special pleading after the fact.

Because solar CPL is weak and variable, Rubenstein and coworkers [47] looked further afield and suggested that the discriminating CPL could have originated from the polarized synchrotron radiation of neutron stars. In this mechanism, further extended by Bonner and Rubenstein [48] and Greenberg, [49] irradiation of interstellar grain mantles by circularly polarized ultraviolet starlight would produce an enantiomeric excess in the organic materials of interstellar dust clouds. This handedness would reverse along with the hemispheric polarity of the CPL field across the ecliptic of the neutron star. Planets within solar systems newly formed by condensation of these irradiated dust clouds would then be seeded with molecules of one e.e or the other by meteoric infalls of dust, comets, and carbonaceous chondrites. This process is thus envisioned to have provided the initial chiral impetus toward a general planetary chiral excess. Which chirality dominates life is, in this theory, a matter of happenstance.

11.2
Experimental Studies of Chiral Induction

Clearly what we need are verifying experimental demonstrations to help us select among the various mechanisms outlined above for the appearance of a small enantiomeric excess (e.e.) within a mixture of two enantiomers. Only after such an initiating event can an e.e. be amplified into the state of homochirality and enantiomeric purity necessary to permit the emergence of self-replicating biopolymers. In what follows we review briefly a number of experimental systems that have focused on abiotic e.e. induction. We consider both their generality and their inherent plausibility in the light of what we know about early Earth.

11.2.1
Intrinsic Mechanisms

The hypothesis that polarized β-emission could induce a chiral excess in racemic mixtures was tested experimentally by its originators in the presence of several β-emitting nuclides, [10, 11] but no polarimetric evidence was found for an enantiomeric excess of chiral products. Nevertheless, this work provided the impetus for numerous other investigations, which have been recently reviewed elsewhere. [50] There is some experimental evidence demonstrating that a preferred molecular chirality results from spin-polarized subatomic particles through induced chemical transformations. Tokay et al. [51] reported that after a year of autoradiolysis the β-decay of solid racemic mixtures of ^{14}C-enriched D,L-leucine exhibited an excess of the L-isomer. This result was suggested to indicate preferential decarboxylation of the D-isomer due to the polarization of the emitted β-particle. A more recent report suggests that spin-polarized radiation can influence the outcome of the crystallization of aqueous sodium chlorate. [52] In the described experiments, β-radiation produced an excess of $NaClO_3$ crystals exhibiting clockwise rotation, whereas excess crystals of opposite habit were produced using positrons. These results could indicate an alternative route to the appearance of a chiral excess on early Earth, that is, $β^{\pm}$-induced phase transformations.

Despite the theoretical difficulties outlined above, some small PVED between enantiomers does exist, on the order of $10^{-18\pm3}$ times the average thermal energy (kT) at room temperature per light-atom molecule. In a mole of a racemic mixture of amino acids, for example, this energy difference leads to an excess of approximately a million molecules of the more energetically stable enantiomer. Thus, we are led to search experimentally for how such minuscule excesses could be translated into a macroscale preference. As yet, another challenge, the measurement of the energy differences associated with the different enantiomers (PVEDs) so far eludes our detection abilities.

In his 1966 paper discussing PVEDs, Yamagata [12] postulated that the small reaction rate differences between enantiomers might be enhanced by "an accumulation principle" in a sequential series of polymerization steps. In 1974, Thiemann and Darge [53] investigated the possibility that PVED effects might be observable during polymerization reactions. They conducted polymerizations of the N-carboxyanhydrides of scrupulously racemized samples of alanine, α-aminobutyric acid, and lysine, and then measured the optical activity of each resulting polymer polarimetrically. All of the polymers showed rotations at 310 nm that ranged between −0.00025° and −0.00084°. These rotations were claimed to indicate a relative difference in the polymerization rates for L- and D-amino acids of about 8×10^{-6}. From these results it is clear that the evidence for PVED accumulation is small to nonexistent.

In 1974 Wagener [54] suggested a number of physical methods that might be exploited to detect any inherent PVED-induced chiral excess. These included chromatography, ion exchange, electrophoresis, polymerization, and crystallization. Analyzing the suitability of each, he concluded that fractional precipitation was the most promising procedure. Immediately thereafter, Thiemann [55] tested this conclusion by examining the fractional precipitation of D,L-asparagine. The results were ambiguous in that repeated recrystallizations below 7.5° yielded a small excess of the D-isomer in the precipitate, whereas above 8° the reverse was true.

The significance of a PVED in influencing the outcome of a chemical reaction remains controversial, and Bonner [56] has recently reviewed experimental work directed toward demonstrating PVED effects. In 1970 Thiemann and Wagener [57] studied the fractional crystallization of sodium ammonium D,L-tartrate, examining the initial fractions for optical activity. In ten such experiments they observed a rotation of approximately −0.001° at 280 nm, and concluded that the effect was caused by a difference of 10^{-5} in the lattice energies of the enantiomorphic crystals. In 1973 Yamagata [58] proposed refinements of such crystallization experiments, but unfortunately these were not implemented. Two years later Kovács and Garay [59] reexamined the crystallization of sodium ammonium D,L-tartrate, but now in the presence of the β-rays from codissolved ^{32}P-phosphate. They observed a slight preference for crystallization of the L-enantiomer in the presence of the radioactive ^{32}P, but control experiments lacking the ^{32}P showed no selectivity whatsoever, and offered "No support ... in favor of any measurable energy content difference between optical isomers." [60] These negative results may reflect, however, the use of light atoms in the search.

PVED scales strongly with the nuclear charge Z. Recently PVED -derived atomic chirality has been directly measured in the gas phase for cesium ($Z = 55$) and thallium ($Z = 81$). [61, 62] Relativistic calculations indicate that PVED effects produced in molecules containing multiple heavy atoms should be several orders of magnitude larger than PVEDs of molecules containing only light atoms. [63] Enzymes containing transition metal sulfide clusters at the catalytic site are candidates for such chemistry. Possibly of significance is that such enzyme systems are considered to be among the first that were utilized within the earliest living systems.

11.2.2
Fluctuational Mechanisms

In 1969 Calvin [64] proposed a scheme for autocatalytic symmetry breaking, which he called "stereospecific autocatalysis". Calvin's mechanism has been validated experimentally in the context of the total spontaneous resolution during the crystallization of racemic mixtures. During crystallization, crystals of one enantiomer may spontaneously separate, leaving the other enantiomer in solution. If the possibility of the equilibration of the enantiomers in solution exists and if the enantiomer in solution can convert rapidly to the enantiomer that is crystallizing before crystallization is complete, then the entire racemate may deposit as a single enantiomer. At least half a dozen examples of Calvin's stereospecific autocatalysis involving such

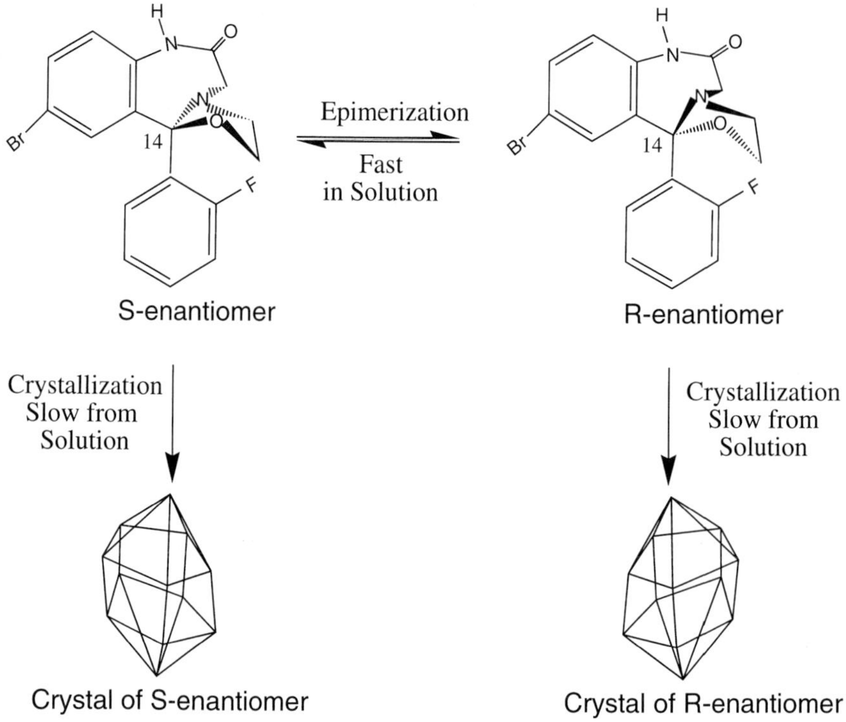

Fig. 11.3 In solution, the molecule bromo-fluoro-1,4-benzodiazepinooxazole rapidly epimerizes between two enantiomers, denoted by R and S, caused by bond rearrangement at the carbon atom of position 14. Epimerization is fast relative to crystallization. If one enantiomer is removed from solution, Le Chatelier's principle will cause the equilibrium to shift producing more of that enantiomer in solution. Thus, fluctuational nucleation of one enantiomer can produce a spontaneous resolution by crystallization. The *bolded* and *hashed* bonds are above and below the plane of the drawing, respectively.

spontaneous resolution under racemizing conditions (SRURC) have substantiated the mechanism. [65]

SRURC is such an interesting example of the facile formation of chiral induction from racemic mixtures in the absence of any external symmetry-breaking agent that it deserves special attention. One of the best studied examples is the crystallization of bromofluoro-1,4-benzodiazepinooxazole (Fig. 11.3), which possesses a single asymmetric carbon atom at C14 and a potentially asymmetric bridgehead nitrogen atom at N4.

Okada and coworkers [66] discovered that the crystallization of this molecule from methanol yielded optically active crystals that were sometimes dextrorotatory and sometimes levorotatory, and that underwent rapid racemization upon redissolving in methanol. They went on to study the kinetics of this process as well as the nature of the different crystalline forms. [67, 69] Bonner [50, 65, 70] has extended and confirmed these observations and conclusions, with special attention to the possibility that the crystallization might provide the means for studying the efficacy of

PVED effects. After much study, Bonner [50] concluded that "the data offer no evidence whatsoever for the efficacy or intervention of PVEDs."

SRURC has an additional drawback, however, for realistic prebiotic symmetry breaking. A hypothetical solution of equilibrating racemates undergoing total spontaneous resolution by crystallization – for example, from some evaporating tide pool on the primitive Earth – would inevitably racemize on redissolving prior to undergoing the next stage of its prebiotic chemical evolution, barring prior implausible solid-state reactions.

11.2.3
Extrinsic Mechanisms

That chiral molecules can be produced in a CPL field, either from achiral precursors by photo-activated synthesis or by preferential chiral photodestruction of a racemic mixture, is now well demonstrated and has been reviewed. [46] In all cases currently known, however, such processes have proved very inefficient. For example, asymmetric photochemical ring-closures of achiral helicene precursors induced by CPL have produced only about 0.2 % e.e. in the products. Likewise, the CPL-induced photolysis of racemic camphor produced about 20 % e.e., but only after 99 % photodestruction, and photolysis of D,L-glutamic acid produced only 0.22 % e.e. after 52 % photodecomposition. [71]

Recently, the theory that a terrestrial e.e. could arise from seeding by chiral molecules produced extraterrestrially by stellar CPL, discussed in Section 11.1.3, received strong and direct experimental support. Cronin and Pizzarello [72] resolved the earlier [73] contended [74] claims that a chiral excess of L-amino acids could be extracted from the Murchison meteorite. They did so by looking for, and finding in the meteorite, a chiral excess of several nonbiogenic L-β-methyl-amino acids, including L,L-2-amino-2,3-dimethylpentanoic acid (2-α-2,3-dmpa). These chiral amino acids, undoubtedly of interstellar origin, were further deduced to have originated by way of CPL photolysis of precursor racemic mixtures because no other mechanism could convincingly be adduced to account for the formation of the two chiral centers of 2-α-2,3-dmpa. Bailey et al. [75] have recently disputed the possibility that CPL from neutron stars could be the source of extraterrestrial molecular handedness, but they further reported detecting up to 17 % circular polarization in the infrared light emitted from the OMC-1 star-forming region of the Orion nebula. Although not produced by neutron stars, the inferred ultraviolet CPL would be similarly effective in inducing a net e.e. in photolyzed racemic organic precursors within dusty interstellar clouds.

The possible photoresolution of racemic mixtures of chiral molecules by irradiation with CPL under interstellar conditions has also been examined experimentally and discussed critically. [76] These experiments have proven interesting, and significant quantities of biologically familiar amino acids as well as other molecules such as glycerol have been produced. [76, 77] In addition, irradiation of racemic tryptophan with ultraviolet CPL at 10 K has produced sufficient chiral products to allow the estimate that as much as 50 % e.e. may be generated in irradiated regions of interstellar molecular clouds. [35] Thus, although the exact source of the extraterrestrial CPL remains disputed, [79, 80] the likelihood of chiral induction in interstellar molecular clouds by some sort of stellar CPL irradiation field remains very real.

More general experimental support for this idea comes from laboratory studies [81] that have shown that irradiation of realistic models of interstellar and cometary ices containing methanol and formaldehyde, with lesser quantities of ammonia, produced notable amounts of hexamethylenetetramine (HMT). HMT, in turn, is known to produce several amino acids on acid hydrolysis. [82] In addition, irradiation under similar icy conditions of complex polycyclic aromatic hydrocarbons (PAHs), which are believed to make up about 20 % of the cosmic carbon budget, [83, 84] produced an extensive variety of organic products. [85] Therefore it can be justifiably imagined that the infall to Earth of organic material from meteoric flux might have delivered not only an e.e. of amino acids but also a wide diversity of chiral organic chemicals. These chemicals were possibly enriched in the L-enantiomer selected by the polarization of the CPL field of the extraterrestrial environment, which would have provided the seed for L-homochirality in proteins.

Crick [86, 87] has conjectured that the early genetic code was G-C rich. This conjecture is based on the enhanced stability of G-C rich duplexes compared to A-T rich duplexes. If this conjecture is correct, several authors [88, 90] have proposed that the earliest genomes coded primarily for the amino acids having G or C in their codon. Together, these authors agree that these amino acids include alanine, aspartic acid, proline and glycine. Interestingly, these four amino acids include one that is hydrophobic (alanine), another that is hydrophilic (aspartic acid), and two that permit (glycine) or require (proline) folding of protein structures. Moreover, three of these amino acids are known to form polypeptides under simple conditions of high salt concentration and dehydration in which metal ions may serve as a catalyst. [91] These four amino acids are aliphatic, and they resist photodegradation compared to other (nonaliphatic or hetero-atom containing) amino acids. A consequence of the resistance to photodegradation is that these amino acids are more likely to persist in an enantiomerically enriched state in the presence of short-wavelength circularly polarized radiation. Thus, the impact of an extrinsic CPL field on racemic mixtures of amino acids rationalizes why the earliest forms of primitive life appear to have been based predominantly on homochiral proteins made up of these four amino acids.

11.3
Chiral Amplification and Takeover

How are the small-to-microscale excesses of one enantiomer over the other, produced by any of the scenarios outlined above, capable of generating a final state of enantiomeric purity? In 1953 Frank [16] developed a mathematical model for the "autocatalytic" random symmetry breaking of a racemic system. He proposed that the reaction of one enantiomer yielded a product that acted as a catalyst for the further production of more of itself and as an inhibitor for the production of its antipode. He showed that such a system is kinetically unstable, which implies that any random fluctuation producing a transient e.e. in the 50:50 population of the racemic reactant would result in the eventual domination of the chiral product from that reactant. This kinetic amplification mechanism has been critically reviewed by Nordén, [92] who corroborated this conclusion. Thus a theoretical system is available in support of the search for an experimental mechanism for amplification of a small enantiomeric excess.

11.3.1
Autoamplification by Polymerization/Depolymerization

In 1957 Wald [93] first proposed that the secondary α-helix structure of a polypeptide chain should bias the selection of amino acid enantiomers toward homochirality as polymerization progressed. In other words, the particular chirality of the α-helix should dictate that the chirality of the new monomers adding to it be of the same handedness as those already making up the helix. This seminal suggestion soon received abundant verification in a variety of prototypical experiments involving base-catalyzed conversions of monomeric amino acid N-carboxyanhydrides (NCAs) into polypeptides. These studies led Bonner and coworkers in the mid-1970s to initiate a series of model experiments designed to explore the further possibility that Wald's mechanism might also be applicable for the e.e. amplification of enantiomerically impure amino acids.

Their studies involved the partial polymerization of NCAs of mixtures of specific amino acids having known e.e.s, followed by determination of the e.e.s of the amino acids in both the resulting polypeptides and in the residual unreacted NCA monomers. [94] In a typical experiment it was found that when an optically impure leucine NCA monomer having an L > D e.e. of 31.2 % was polymerized to the extent of 52 % to the helical polyleucine peptide, the e.e. of the polymer was enhanced to 45.4 %, an increase of 14.2 %. In the same experiment the e.e. of the unreacted leucine NCA monomer was depleted to a similar extent. Analogous experiments with valine NCAs of known e.e.s, however, led to a reverse effect, namely, the preferential incorporation of the racemate rather than one enantiomer into the growing polyvaline peptide. This finding was interpreted to be the result of the fact that polyvaline consists of β-sheets rather than α-helices, emphasizing that the Wald mechanism applies only to α-helix polymers. At about the same time Brach and Spach [95] showed that, under proper conditions, β-sheet polymers could also be implicated in the amplification of amino acid e.e.s.

In the above experiments with polyleucines it was also observed that the partial hydrolysis of mixtures of nonhomochiral leucine peptides led to the preferential hydrolysis of those components of the mixtures that were enantiomerically more random. This produced a corresponding e.e. enrichment in the unhydrolyzed polymer. [96] For example, when the above polyleucine sample having an L > D e.e. of 45.4 % was 27 % hydrolyzed, the residual unhydrolyzed polymer had an e.e. of 55.0 %, representing an e.e. increase of 9.6 %.

Thus in the two typical experiments cited, a sequential combination of partial polymerization, with its 14.2 % e.e. increase, and partial hydrolysis, with its 9.6 % e.e. increase, resulted in an impressive overall L > D e.e. enhancement of 23.8 %.

This enhancement required just two steps from the initial leucine NCA monomer to the final polyleucine peptide.

The success of these model experiments led Bonner and coworkers to propose a mechanism involving repetitive cyclic sequences of partial polymerization followed by partial depolymerizaton in which the latter is caused by hydrolysis. Thus, this process is driven by environmental dry and wet cycles that could ultimately have led to homochiral polypeptides on early Earth. [97] Brach and Spach [95] have also proposed a mechanism involving partial hydrolysis for the enantiomeric enrichment of polypeptides having β-sheet secondary structures.

These model experiments involving e.e. amplification of amino acids during polymerization admittedly need prebiotically "unrealistic" substrates as well as carefully contrived experimental conditions. Nevertheless, it is noteworthy that both secondary structures of proteins, α-helices, and β-sheets have been found capable of acting stereoselectively to provide e.e. enhancements during these model polymerizations.

Analogous amplification experiments embodying prebiotically realistic environmental conditions have yet to be reported. Even so, the chiral amplification of non-homochiral peptides could be tested in a biogenically reasonable way by application of a very relevant method recently developed and reported by Imai et al. [98] These workers simulated a hydrothermal submarine vent within a single reactor by continuously circulating an aqueous solution between a high-pressure (24 MPa), elevated-temperature (200–250 K) region and an equally pressurized rapid cold quench. With glycine as solute in pure deionized water, this reactor produced oligomers up to the triglycine stage after only 30 min and up to the hexaglycine stage when Cu^{2+} was present. Diketopiperazine was also produced, and the glycine oligomers were apparently produced by amination of this compound. No other amino acids were reported tested. Extension of similar experiments to racemic amino acid mixtures, however, would directly test the hypothesis that homochiral peptide oligomers might be induced and/or amplified under conditions of dynamic polymerization-depolymerization that reproduce plausible prebiotic conditions.

Recently, Eschenmoser and coworkers [98] have suggested a more elaborate mechanism for spontaneous symmetry breaking. Accompanying their description of the self-assembly of higher oligomers of pyranosyl-RNA by ligative oligomerization of tetranucleotide-2′, 3′-cyclophosphates, they postulated that starting with a racemic mixture containing all possible diastereomers sets of such "tetramers with different but mutually fitting base sequences can be expected to co-oligomerize stochastically and generate sequence libraries consisting predominantly of homochiral (D)- and (L)-oligomers." They then showed how, after reaching a sufficient chain length, the oligomers are "driven beyond a critical level of constitutional complexity," such that a true racemic mixture of the oligomers is no longer possible. Then, with "internal and external selection pressures" either the homochiral (D)- or (L)-oligomer must eventually predominate. Although these novel suggestions are extrapolated from the authors' solid laboratory observations, they have yet to be demonstrated experimentally. However, these results extend to polynucleotides the findings noted above, that polymerization reactions of racemic materials could produce chiral symmetry breaking and amplification in a way arising naturally out of simple chemical processes. The same concept might apply to polymerization of a racemic mixture of amino acids.

11.3.2
Enantiomeric Amplification by Change of Phase

In the 45 years since its proposal, Frank's autocatalytic mechanism (Section 11.3, above) has spawned numerous theoretical refinements including consideration of such factors as reversibility, racemization, environmental "noise," and parity-violating energy differences. [100, 101] In contrast to the above examples of stereospecific autocatalysis by the SRURC , however, none of these theoretical refinements is supported by experimental evidence. While earlier attempts to validate the Frank mechanism for the autocatalytic amplification of small e.e.s in other experimental systems have generally been unsuccessful, several recent attempts have shown more promising results. [102, 104]

These schemes have been frequently suggested [105–107] as possible mechanisms to achieve the chirally pure "starting point" for prebiotic molecular evolution toward our present homochiral biopolymers. Demonstrably successful amplification mechanisms are the spontaneous resolution of enantiomeric mixtures under racemizing conditions, [509 lattice-controlled solid-state asymmetric reactions, [108] and other autocatalytic processes. [103, 104] Other experimentally successful mechanisms that have been proposed for chirality amplification are those involving "kinetic resolutions" [109] enantioselective occlusions of enantiomers on opposite crystal faces, [110] and lyotropic liquid crystals. [111] These systems are interesting in themselves but are not of direct prebiotic relevance because of their limited scope and the specialized experimental conditions needed for their implementation.

Alternative mechanisms that have been suggested have involved the partial evaporations of [112] or partial precipitations from [55] solutions in which there exist solubility differences between a racemate and its individual enantiomeric constituents. Such studies have given positive results in a few systems and might intuitively be considered plausible in a primitive Earth setting, but their generality at the present level of knowledge unfortunately appears limited. These experiments and those described earlier validating the mechanism of stereospecific autocatalysis have all involved rather exotic chemical systems that are of interest in connection with the Frank hypothesis but seem implausible in any realistic prebiotic environment. All known examples, however, do demonstrate the principle that autoamplification of small spontaneous e.e.s is chemically possible. The experimental charge, therefore, is to seek generalizing examples of the reactions that seem most plausible in a putative prebiotic environment.

11.3.3
Metal-Assisted Enantiomeric Amplification

A recent interesting example of the chiral amplification of a small initial e.e. has been reported by Soai et al. [113, 114] involving the induction of a chiral center in an achiral aldehyde using diisopropylzinc as an alkylating reductant and a very small e.e. (0.1 %) of a chiral effector molecule. The effector molecule is almost certainly a ligand for the zinc. Whichever enantiomer of the effector is in small excess overwhelmingly determines the chirality of the product.

Although the mechanism of this reaction remains unknown, we can visualize how such amplification could occur by assuming that the chirality-amplifying molecule is a tetrahedral zincate that includes two of the chiral effector molecules as ligands. The scheme is illustrated by the reactions depicted in Fig. 11.4.

Chiral effector ligands can be a variety of molecules, including amino acids. The chiral alcohol products of the chiral-transfer reactions in Fig. 11.4 cycle back into these reactions by subsequent complexation with the pool of diisopropyl zinc.

Such a system could proceed to chiral takeover, for example, if the active alkylating agent is the tetrahedral zinc complex (Fig. 11.4), if the initial total concentration of the chiral zinc complexes is small compared to that of the aldehyde reactant, if the tetrahedral D,L-zinc complex is more stable thermodynamically than are the tetrahedral D,D- and L,L-zinc complexes, and if the reaction of the D,L- complex with the pyrimidyl aldehyde is kinetically sluggish compared to the reaction rate of the

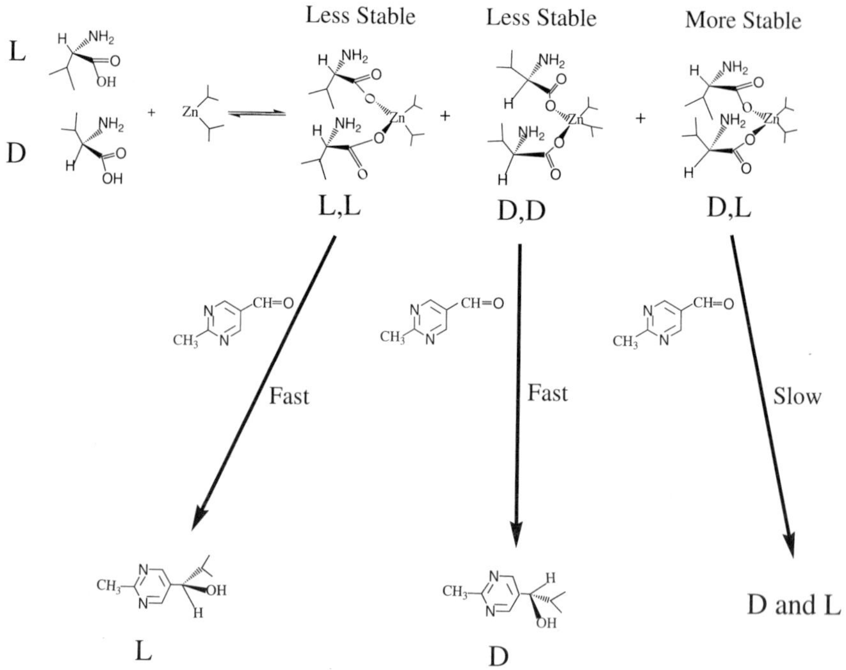

Fig. 11.4 Reaction sequence showing how a small e.e. of the reagent is converted to a large e.e. of the product by a metal-assisted reaction. In the case shown, a 1 % e.e. of L-valine produced a 51 % e.e. of L-pyrimidine alcohol by the reductive asymmetric transfer of an isopropyl group from zinc to the carbonyl carbon of the aldehyde. The chiral alcohol product can react with the di-isopropyl zinc in excess to produce additional chirally selective zinc complexes, thus catalyzing its own chiral dominance. The *bolded* and *hashed* bonds are above and below the plane of the drawing, respectively.

homochiral complexes. The last condition would be reasonable if the alkylation involved a rate-determining dissociative step. We can then see that a small initial enantiomeric excess of one chiral effector ligand will eventually dominate because the slightly smaller concentration of the opposite-handed ligand will over time come to be restricted by equilibration into the kinetically sluggish D,L-zinc complex. Although the reaction system is prebiotically unrealistic, it nevertheless demonstrates that the principle of runaway autocatalytic amplification of a small chiral excess is not chemically unreasonable.

This reaction has recently been extended into a more realistic prebiotic regime, however, by the finding that an initial symmetry-breaking event can be induced by the surface of a morphologically chiral material such as quartz. Thus, when the reaction in Fig. 11.4 was carried out without any added chiral effector ligand, but in the presence of powdered optically pure crystalline quartz [115] or $NaClO_3$ crystals, [116] the chirality of the product alcohol was overwhelmingly determined by the morphological symmetry of the mineral. In this reaction, a small chiral excess of the product pyridine alcohol must be produced very early in the reaction. This small initial e.e. could then rapidly dominate the chirality of the final product by the amplification mechanism described qualitatively above.

This sort of analysis could be extended to any metal-catalyzed chemistry in which a large runaway chiral excess is induced in the product by way of a small chiral excess of the molecules that serve as ligands to the metal. It is only necessary that the D,L-metal center be kinetically slower and thermodynamically more stable than the D,D- or L,L-complexes in order that any small e.e. of a chiral ligand be translated into chiral dominance of the reaction product. That the initial e.e. resulting in chiral takeover within a reacting system can be induced by asymmetric mineral surfaces indicates that a general chemical route to the asymmetry of life may exist.

11.3.4
Amplification by Molecular Propagation from a Chiral Center

In 1994 Wittung et al. [117] reported a provocative mechanism of chiral amplification involving achiral polymers of peptide nucleic acids (PNAs). PNAs are nonribose-containing polymers consisting of chains of poly-[N_α-(carboxymethylnucleotide)-N_α-(2 ethylamido)glycinamide] (Fig. 11.5). These molecules can form complementary base-paired helical duplexes completely analogous to those of DNA and RNA. Using circular dichroism (CD) spectroscopy these workers showed that a preferred helical handedness could be induced into an achiral PNA duplex polymer by appending a single chiral amino acid to the glycinamide terminus of each chain. Development of a CD spectrum, and thus molecular chirality, only occurred when PNA polymers with complementary nucleotides were mixed. Unmodified PNA polymers formed duplexes that did not develop a preferred helical handedness in solution. The helicity of the PNA duplex was completely determined by the chiral handedness of the amino acid tail. Although only PNA decamers were studied, there is no reason to think that longer PNA polymer chains would not behave similarly. Recent synthetic results [118] have indicated a plausible prebiotic route to the

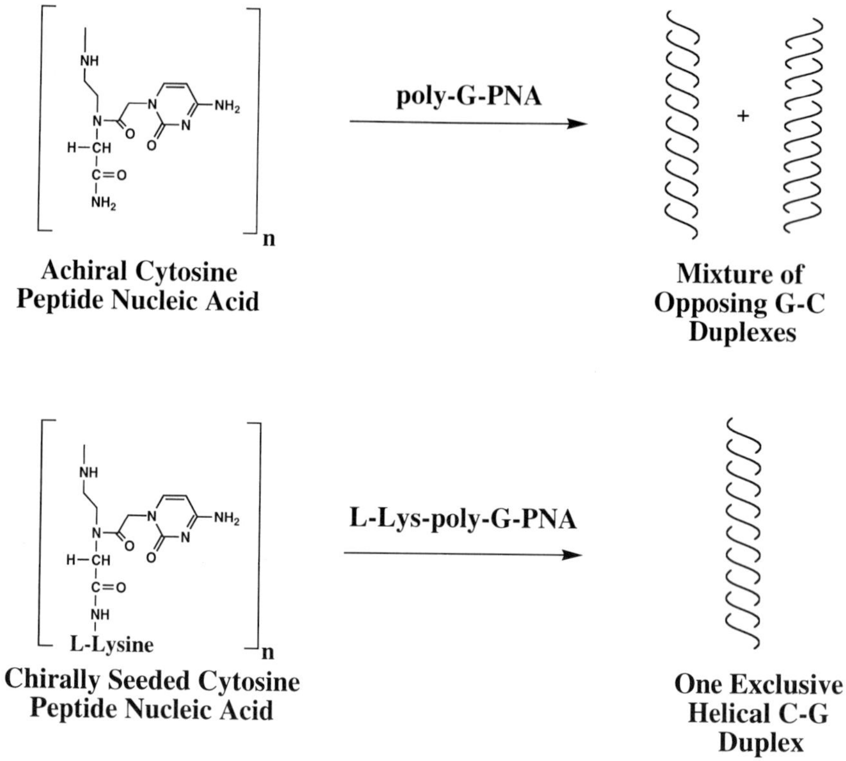

Achiral Cytosine Peptide Nucleic Acid

poly-G-PNA

Mixture of Opposing G-C Duplexes

Chirally Seeded Cytosine Peptide Nucleic Acid

L-Lys-poly-G-PNA

One Exclusive Helical C-G Duplex

Fig. 11.5 Induction and amplification of chirality in an achiral polymer. Peptide nucleic acids are DNA analogs that can fold into helical duplexes that mimic biogenic nucleic acid biopolymers. With an appended chiral amino acid, a PNA decamer folded into only (or predominantly) one of the two helical possibilities. This behavior demonstrates that a localized chiral center can induce global chirality in a polymer. The example shown is a simplification of the actual experiment.

production and polymerization of the PNA backbone, raising the possibility that such molecules were involved in early pre-RNA information storage.

11.3.5
Amplification by CPL Photoinduction

Another mechanism of chiral amplification that extends over an even larger scale has been reported by Huck et al. [119] The molecule 12-(9′H-thioxanthene-9′-yli-dene-12H-benzo[a]xanthene (Fig. 11.6), which has no chiral center, nevertheless exists, like the helicenes, in two chiral forms defined by their enantiomeric config-urations. Consistent with the discussion in Section 11.2.3, a small net handedness (ca. 0.7 %) could be induced in racemic solutions of this molecule by use of ultravio-let CPL. However, introducing 20 wt% of this molecule, which contained a 1.5 % chiral excess of one roto-enantiomer, into a nematic phase of liquid crystals pro-duced macroscopic (100 μm) regions of a chiral cholesteric liquid crystal phase. The

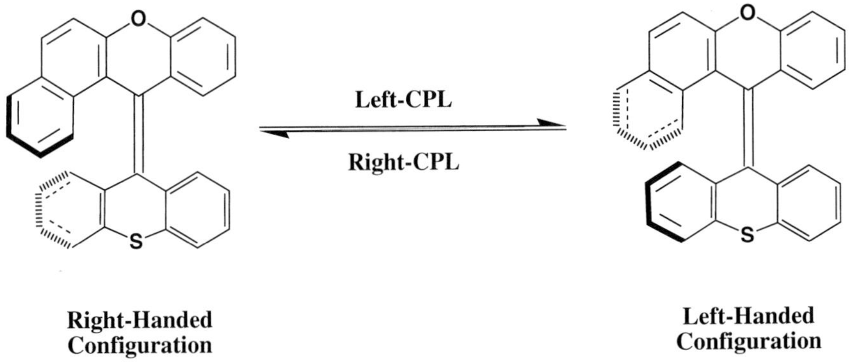

**Right-Handed
Configuration**

**Left-Handed
Configuration**

Fig. 11.6 Conversion of one enantiomer to the other by CPL. The molecule 12-(9′H-thioxanthene-9′-ylidene)-12H-benzo[a]xanthene exists in one of two chiral forms that can be intercon- verted through irradiation by CPL of the appropriate handedness. The *bolded* and *hashed* rings are above and below the plane of the drawing, respectively.

chirality of the cholesteric phase was completely determined by the handedness of the dissolved molecule in chiral excess. Similar transformations to a cholesteric phase could be produced by irradiation with ultraviolet CPL of nematic phase liquid crystals containing 20 % of a racemic mixture of the xanthene. Related work has been carried out by Burnham and Schuster, [120] who induced reversibly the cholesteric phase in a nematic liquid crystal with CPL.

11.4
The Sequestration of Chirality

Once chirality is induced and amplified by some mechanism, the excess must first persist and then propagate in order to survive. A distinctive characteristic of homochiral protein and nucleic acid biopolymers is that they function within the enclosed environment of cells, which provide a membranous boundary structure that separates the intracellular components from the external environment. It has accordingly been postulated frequently that analogous but simpler enclosed environments must have been available and operative on the primitive Earth.

In the early bombardment phase of terrestrial evolution, there would have been no retreat remote enough to allow membrane-enclosed chiral organic molecules to survive the impact of very large bolides. [121] It is only after this stage and after the end of ocean-evaporating impacts that the products of any relevant prebiotic chemistry could have persisted. Even so, the post-bombardment early Earth is thought to have been geologically much more active than it is at present. [122] Erosion from possibly fierce meteorological disturbances, intense ultraviolet irradiation from the early sun, [122] powerful volcanism, and the strong convection and diffusion from tidal churning in possibly shallow seas [124, 125] would have to be accommodated by any prebiotic molecular system. In short, any array of essentially homochiral amino acids produced by e.e. amplification processes, such as those outlined above, would not be expected to undergo further chemical evolution while floating freely in the oceans. Instead, they must have evolved only after their concentration and protection by sequestration into indigenous and robust precellular structures. We examine briefly some of the suggestions that have been offered to accomplish this sheltering.

11.4.1
Porous Minerals

In 1967, J. D. Bernal [126] elaborated his earlier ideas that life could have originated in estuarine or shoreline clays by way of the adsorption, and concentration therein, of prebiotic organic molecules. The chemistry and porous morphology of clays have been reviewed in the context of prebiotic chemistry. [127] In particular, the porosity of clays and the consequent high availability of catalytic sites that can adsorb organic molecules and activate them through complexation by metal ions, or by protonation, is discussed. These ideas led to considerable experimentation, which has been reviewed. [47] Although no chiral preference by clays has been reliably detected, the concentration of organics by these materials and the catalytic transformation of adsorbed organics are well verified.

Ideas regarding involvement of clays in the origin of life were explicitly advanced by Cairns-Smith, [128, 129] who suggested that life emerged from an association of organic molecules adsorbed into the interlayers of clays such as montmorillonite or kaolin. In his scheme, the original self-replicating elements were mineral systems derived from the clays themselves, a concept for which there is some experimental support. [130] These were eventually to be replaced, first by clay-organic systems, and then by purely organic autocatalytic reaction systems, from which later life emerged. Cairns-Smith also proposed that the metallic sites found on the edges and corners of clay materials could have promoted the emergence of organic self-replicating systems by way of selection among the products and processes of a multitudinous catalytic chemistry.

These schemes expressly included the idea that clays and porous minerals adsorbed, absorbed, and ultimately concentrated any extant organics from a dilute oceanic broth on the early Earth. This idea is critical because it is difficult to imagine that the kind of polymerization and chiral amplification processes discussed above could or would have occurred in a water solution not much more than millimolar in organics. [131]

Smith and coworkers recently proposed a specific and novel mineral-based solution to the problem of dilution and diffusion of prebiotic reactants. They have suggested [132–134] the uptake of organics within the micron-sized three-dimensional cross-linked network of pores found to exist within the top 50 μm, or so, of alumina-depleted, silica-rich weathered feldspar surfaces. These surfaces incorporate cavities typically about 0.5 μm in diameter along with cross inter-connections of about 0.2 μm. The nominal area of the weathered feldspar surface is apparently multiplied by a factor of about 130 arising from this network. The similarity of these pores to the catalytic sites in zeolite-type materials is pointedly mentioned.

The weathering of feldspars produces silica-rich surfaces that are naturally lipophilic. Such surfaces are ideal for the uptake and concentration of amino acids, polymers, and amphiphilic (surface-active) organic molecules. Amphiphiles, including fatty acids, phospholipids, and sterols, are large nonpolar hydrocarbons that include one or more small polar groups and are surface-active at an air-water interface. That is, they self-organize into monolayers with their hydrophilic polar groups extending into, and with their hydrophobic nonpolar groups remaining out of, the water surface. Thus lipophilic silica-rich surfaces may not actually be necessary for uptake of amphiphiles because the surface-active nature of amphiphiles, peptides, carboxylic acids, and amino acids may be sufficient to transform even a polar alumina-rich surface into one that is lipophilic. [135, 136]

The point is also made [134] that the very high surface areas and the richly interconnected three-dimensional networks of these micron-sized spaces, coupled with periods of desiccation, could together have produced microenvironments rich in catalytically produced complex chemicals and possibly membrane-enclosed vesicles of bacterial size. These processes would provide the proximate concatenation of lipid vesicular precursors with the complex chemicals that would ultimately produce the autocatalytic and self-replicating chiral systems. A 2.5 km^2 granite reef is estimated to contain possibly 10^{18} "catalytic microreactors, open by diffusion to the dynamic reservoir of organic molecules ... but protected from the dispersive effects of flow and convection" [134] as well as protected from the high flux of ultraviolet radiation impinging on the early Earth. [123, 137]

11.4.2
Amphiphilic Vesicles

Above a critical concentration certain amphiphiles may aggregate into membranous micelles, which are globular structures with a nonpolar interior and with a polar surface that interfaces the aqueous environment. At still higher concentrations micelles may fuse into multilamellar structures composed of bilayers. [138]

Micelles are capable of self-replication if an appropriate chemical reaction occurs within the micelle itself that produces more of the same amphiphile that forms the micelle. Such self-replication has been demonstrated for both ordinary micelles in an aqueous medium [139] as well as for "reverse micelles," [140] which are spherules of water stabilized by an amphiphile in an organic solvent. Some of the prebiotic potentialities of replicating membranous vesicles have been investigated, [141] and they have been characterized as "minimum protocells." [142]

A critical property of minimum protocells in the prebiotic environment would be their ability to sequester other molecules, including macromolecules. [142] In 1982, Deamer and Barchfeld [143] subjected phospholipid vesicles to dehydration-rehydration cycles in the presence of either monomeric 6-carboxyfluorescein molecules or polymeric salmon sperm DNA molecules as extraneous solutes. The experiment modeled a prebiotic tidal pool containing dilute dispersions of phospholipids in the presence of external solutes, with the dehydration-rehydration cycles representing episodic dry and wet eras. They found that the vesicles formed after rehydration encapsulated both the small fluorescein molecule and the polymer, with the extent of encapsulation following the amount of original liposome.

Deamer and Barchfield suggested that during the dehydration stage the original vesicles shrank and fused together, trapping the solute molecules between the alternating lipid bilayers of a multilamellar sandwich. On rehydration the lamellae swelled, reconstituting larger vesicles that encapsulated a significant percentage of the original monomeric or polymeric solute. They also suggested that if both a template and a monomer were sequestered together, the close contact in the enclosed microenvironment might enhance the efficiency of template-directed replication. This idea anticipates the later suggestion that membranous amphiphile vesicles might provide "an appropriate microenvironment for incorporation and evolution of polymer synthesis systems." [142, 144]

We offer a further suggestion. The dehydration-rehydration encapsulation of macromolecules within vesicles might occur simultaneously with the amplification of an initial e.e. in polypeptides by a partial-polymerization step (during dehydration) and a partial-hydrolysis step (during rehydration). The joint action would permit the evolution from amino acids of low e.e.s into homochiral polypeptides sequestered in the protected interiors of the vesicles. In other words, such encapsulation might augment the above e.e. amplification based on Wald's hypothesis. Although this possibility has not been investigated experimentally, the formation of membranous vesicle protocells along with their ability to self-replicate and to encapsulate exogenous molecules has been convincingly demonstrated experimentally. The conjoint mechanism of encapsulation followed by e.e. amplification could thus constitute a realistic and plausible prebiotic process on the primordial Earth.

The question as to the potential availability of the requisite amphiphilic precursors in the prebiotic environment has been addressed experimentally by Deamer and coworkers, [143, 145] who looked into the uncontaminated Murchison chondrite for the presence of such amphiphilic constituents. Samples of the meteorite were extracted with chloroform-methanol and the extracts were fractionated by thin-layer chromatography, with the finding that some of the fractions afforded components that formed monomolecular films at air-water interfaces, and that were also able to self-assemble into membranous vesicles able to encapsulate polar solutes. These observations clearly demonstrated that amphiphiles plausibly available on the primitive Earth by meteoritic infall have the ability to self-assemble into the membranous vesicles of "minimum protocells."

11.5
Setting the Scene for Life

The circumstances producing an e.e. on the early Earth could well have been capture and amplification of one of the fluctuational mechanisms outlined above, although present evidence favors the meteoric infall of chiral molecules synthesized in interstellar dust clouds. The total flux of organics on the early Earth, delivered to Earth both by way of comets and other bolides, and synthesized by terrestrial mechanisms is estimated [131] to have been on the order of 7×10^8 kg/yr for a modestly reducing atmosphere composed mostly of CO_2 and dinitrogen. The terrestrial carbon inventory would have included the organics and amphiphiles described above as originating in interstellar molecular clouds and discovered in the Murchison meteorite, and the chiral excesses suggested by the amino acids found within the Murchison meteorite.

The metal-catalyzed amplification of e.e. in small molecules, demonstrated by Soai and coworkers, along with the chiral enrichment of amino acid polymers by sequential polymerization/depolymerization steps, have shown that small enantiomeric excesses in nearly racemic mixtures can be reactively amplified to produce chiral dominance. These real chemical systems, which include plausible prebiotic reactions, experimentally demonstrate the principle of the chiral amplification of a spontaneously broken chiral symmetry in a dynamic and authentic chemical milieu. Therefore amplification to dominance of a small chiral excess of both small and polymeric molecules can be credibly incorporated into an origin-of-life model.

It is presently known that the regions surrounding extant submarine and terrestrial hydrothermal vents are rich in ammonia, hydrogen sulfide, dihydrogen, CO, and the other simple molecules [122, 146–150] thought necessary for biogenic chemistry. [151–155] The extensive organic chemistry accessible in hydrothermal vent systems has been comprehensively reviewed in monographs by several authors in a special issue (Vol. 22, 1992) of "Origins of Life and Evolution of the Biosphere." These circumstances solve the problem that a weakly reducing N_2/CO_2 atmosphere would not have provided the reactive chemicals and reducing environment necessary to produce Miller-Urey chemistry. [131] Under reducing hydrothermal conditions, a rich array of amino acids and organic molecules are expected to be efficiently synthesized. [156, 157] That is, hydrothermal activity on the archaic Earth could have provided a rich variety of chemical "feedstocks" to any conjoining catalytic porous mineral assemblage, wherein stochastic experimentation in biogenic chemistry could occur.

The production and stabilization of long polypeptides on clay surfaces [158–163] from simple evaporation cycles of dilute solutions of amino acids, [164, 165] and from the facile production of amino acids and other organics from simple precursors under a variety of conditions [151, 166–168] is well established. Therefore it is reasonable to extrapolate these laboratory results to prebiotic polymerizations of indigenous amino acids within proposed mineral dikes in a hydrothermal vent-field on early Earth. Libraries of random polypeptides, presumably existent in the critical time just prior to the emergence of the first self-replicating protocell, are known to have members that organize into robustly folded protein-like molecules that are high in α-helical content. [169] Some of these might be expected to have useful catalytic properties.

Similar efficient abiotic syntheses of nucleotide precursors and their polymerization on clays have also been reported [166, 168, 170–174] and reviewed, [175] which allows a provisional extension of the above comments concerning amino acids to these critical molecules as well. Such molecules also have interesting catalytic activity, [176–179] which would be subject to selection. The uptake and ligation of metal ions such as Mg^{2+} is known to confer catalytic activity on even random RNA oligomers. [177, 180]

The early Earth was probably much more geothermally active than the Earth is today. [122] The continents of early Earth were most likely small with vertical profiles at, or near, sea-level. [125, 181] Subaerial (in addition to submarine) tectonic rift zones and hydrothermal vent fields were probably ubiquitous and vigorous. [182] Extensive continental hydrothermal zones enriched in reducing inorganic and organic vent products and subject to wet/dry cycles were probably common. The low-profile shield continents of the early Earth must have been extensively surfaced with basalt and rhyolite, [183] some of which could have been porphyritic.

The partial pressure of CO_2 in the Earth's atmosphere during the early Archaean probably was 10 bars or more (1 bar is approximately 1 atmosphere). [182, 184] The consequent levels of carbonic acid produced acid rain (pH ca. 3.9) and a world ocean with mildly acidic (pH ca. 6) surface waters. [185] The extensive volcanism of the day would have released significant quantities of HCl vapor into the atmosphere, further acidifying rain. [186, 187] Under conditions of heat and acidic waters, which thus probably typified the hydrosphere of the early Earth, chemical weathering of even highly calcic basalt and rhyolite surfaces must have been rapid. [137] Given these circumstances, feldspar inclusions and other phenocrystic mineral assemblies could have rapidly developed the sort of porous networks described above. It is worth observing that once covered with adsorbed organics, internal rock pores would resist further aqueous chemical weathering.

11.6
The Rocky Road to Life?

We are now in a position to offer a speculation on what type of prebiotic chemistry might have occurred, informed by the ideas and observations discussed above. Failing either faster-than-light travel or spectacular advances in optics, we will almost certainly never know by exactly what process life began on Earth. The scientific charge, however, is to derive a plausible theory based upon demonstrable chemistry and verified observations.

The recent developments and ideas in the field of prebiotic chemistry can be combined with the concepts noted here to produce what we regard as a research outline, rather than a detailed hypothesis, directed toward a coherent theory of the origin of complex self-contained, self-replicating chiral assemblies. In what follows we present one possible scenario that is consistent with our current knowledge of chiral induction and amplification and with the nature of early Earth as well as early life. It is exciting that this fundamental question can be formulated in a way that allows systematic experimental testing as we enter the next century.

Credible mechanisms have been identified for the extraterrestrial production and delivery to early Earth of organic molecules and amino acids containing a small e.e. The subsequent terrestrial sequestration of an initial e.e. and its amplification into dominance are processes for which biogenically credible mechanisms exist. The production and propagation of biomimetic chiral organic polymers has also been shown, though only in part plausible in terms of prebiotic mechanisms.

Over the 200 million years or so between the end of the Hadean Age and the first appearance of biological metabolism, [121, 188] organic molecules and amino acids, including a seeding of chiral amino acids, were delivered to Earth by way of carbonaceous chondrites, comets, and interplanetary dust particles. Additional terrestrial organics and amino acids were most likely copiously synthesized in hydrothermal vent systems. The organic materials are suggested to have been absorbed and concentrated by various clays and minerals, with special attention paid to adsorption and uptake by weathered porous feldspars. These feldspar and other porous mineral regimes are considered to have been located in the vicinity of extensive near-shore and subaerial hydrothermal vent fields and thus exposed both to a fluence of reducing and reactive molecules and to thermal flux.

The continuous availability of trillions of independent microreactors greatly multiplied the initial mixture of extraterrestrial organics and hydrothermal vent-produced chemicals into a rich variety of adsorbed and transformed materials, including lipids, amphiphiles, chiral metal complexes, amino acid polymers, and nucleotide bases. Production and chiral amplification of polypeptides and other polymeric molecules would be induced by exposure of absorbed amino acids and organics to dehydration/rehydration cycles promoted by heat-flows beneath a sea-level hydrothermal field or by sporadic subaerial exposure of near-shore vents and surfaces. In this environment the e.e. of chiral amino acids could have provided the ligands required for any metal centers capable of catalyzing enantiomeric dominance. The auto-amplification of a small e.e. of L-amino acids, whether extraterrestrially delivered or fluctuationally induced, thus becomes conceptually reasonable.

Amplification of any enantiomeric excess could rapidly propagate an enantiomeric takeover within the micron-sized cavity of a weathered feldspar microreactor. We can even imagine this process assisted by the spontaneous appearance of an excess of chiral liquid crystal domains of preferred handedness, composed of membranous associations of achiral complex amphiphiles seeded with an e.e. of dissolved chiral molecules, in analogy to the laboratory observations described in Section 11.3. The porous mineral interiors also allow the arguably high likelihood of contingent metal ion-enhanced catalytic reactivity of polypeptides as progenitors to the now ubiquitous metalloproteins, and the likelihood of evolutionary selection among the produced reactivities. This massively parallel process of natural experimentation, as encouraged by networked microporous minerals, implies the likelihood of coevolution [189] and subsequent conjoining of protobiochemical pathways.

These possibilities rectify the proposed subsequent appearance and amplification of chiral autocatalytic molecules and hypercycles. [190] Any autocatalytic systems would propagate [191] throughout an extensive adjoining hydrated porous network already rich in layered amphiphiles, lipids, polymeric materials, amino acids, thiols, and so forth. In addition, amphiphiles are known to be organized into lipid membranes by interaction with the inner surfaces of porous minerals. [136] It is a small organizational jump from these membranes to fully formed lipid vesicles.

As noted above, the occurrence of such chemistry within porous feldspars or other mineral assemblages near chemically rich and strongly reducing hydrothermal vent sites in the early oceans solves two problems. First, it solves the problem of rationalizing polymerization reactions in dilute solutions. Second, it solves the problem of assuming a prolific organic chemistry in the absence of reduced molecules and ammonia in a neutral (though nonoxidizing) [131] early terrestrial atmosphere. This chemistry, occurring both in the vicinity of and within the ~0.5 μm cavities of weathered feldspar-like materials in the early Earth, provides the unifying framework for synthesis and containment of the products to critical concentrations.

Prior sequestration of the prebiotic reactions within the micropores of weathered feldspars or other porous rock matrices also avoids many of the other problems of catalysis and dilution encountered by models of chemical biogenesis. That is, this mechanism attains viable evolutionary chemical selection among spatially discrete systems *without* the need to assume an unlikely capture-and-enclosure event involving a pre-existing lipid membrane. [192] Thus autocatalysis of chiral molecules could evolve before the actual appearance of free-floating lipid vesicles.

At the point where amphiphiles were recruited to provide the precursors to cell membranes, stable lipid vesicles could have evolved [141] to enclose autocatalytic chiral hypercycles. Credible models for the subsequent evolution of vesicles containing self-replicating chiral molecules have appeared in the literature. [193, 194] These vesicles could then emerge from the feldspar spaces [134, 192] as micron-sized self-reproducing, energy-metabolizing vesicular systems: protobacteria ready to face the hydrothermal world on their own terms.

Westheimer [195] has stressed the critical importance of phosphorus-containing molecules in promoting biological processes. The cosmic origins and terrestrial geochemistry of phosphorus has been reviewed in the context of biogenic chemistry. [196] The hydrothermal origination of life within porous weathered minerals potentially solves the apparent contradiction of the ubiquity of polyphosphates in biochemistry as contrasted to both the relative rarity of crustal phosphate [197] and the insolubility of native calcium phosphate [apatite: $Ca_5(PO_4)_3X$, (X=F,Cl,OH)] and consequent low average concentration (ca. 2 μM) of dissolved oceanic phosphate. [198] Phosphate is known to be adsorbed by siliceous clays [199–201] and to have specific affinity for the surface hydroxyls of silica, [202] and is found to be concentrated by several orders of magnitude in the sedimentary and clay deposits of lakes, estuaries, and oceans, over the level of dissolved phosphate. [201, 203] Apatite inclusions have been found within some feldspars [204] and must have been deposited there by absorption and recrystallization of dissolved phosphate. These findings support the idea that primordial phosphate could have been taken up and concentrated by the same silica-rich porous minerals as the primordial organic molecules and amino acids.

As is true today, most phosphate in the primordial crust must have been sequestered in nearly insoluble calcium phosphates and carbonates or in basalts, and only dissolved monomeric phosphate was produced by weathering. [201] However, the volatile polyphosphate P_4O_{10} is known to be a component of volcanic gases. [205] This material originates from the polymerization of phosphate minerals in magmatic melts. Tetrapolyphosphate P_4O_{10} hydrolyzes to polyphosphates and trimeta-phosphate. Further, aqueous phosphoric acid is known to efficiently polymerize into polyphosphates under conditions of temperature (250–350°C) well within those attained within hydrothermal vent systems. [206] Therefore, the same heat flows inducing the dehydration/hydration cycles that might produce polypeptides at the surface of a hydrothermal system could have also produced polyphosphates at sub-terranean depths. [207] Delivery of the soluble polyphosphate fraction to the surface by the water flux of this system provides the necessary (and continuous) source of polyphosphates to the prebiotic mix. It is interesting to note that the high CO_2 atmo-sphere of the early Earth, described above, would have enhanced the solubility of apatite [203] by mild acidification of oceanic waters. The concentration of dissolved phosphate would also have been increased in the oceans of early Earth, relative to the modern pelagic level, because of the comparative scarcity of iron(III). Thus, a relatively rich phosphate-polyphosphate cycle can be plausibly argued to have existed on an early, thermally vigorous, Earth. Once present, the known reactivity of poly-phosphates can be utilized in the evolution of any self-replicating system.

These ideas are readily testable and therefore can constitute a research enterprise. For example, the hydrolysis of polyphosphates to monomeric phosphate is relatively fast in weak aqueous acid and warm temperatures. [129, 205] Curiously, however, the same acidity would promote thermal polyphosphate synthesis under the dehy-drating conditions of the heated subterranean mineral pores within a hydrothermal system. Measurement of the stability of polyphosphates in microscopic environ-ments where the activity of water might be low would test the phosphate-polypho-sphate conjecture made above. If polyphosphates are not stabilized relative to bulk water solutions, as dissolved within micron-sized mineral pores or within vesicles, perhaps in the presence of dissolved organics, then this hypothesis is in danger. Additionally, the proposed production and delivery of polyphosphates by hydrother-mal systems can be tested in an appropriate laboratory setting.

Artificial hydrothermal vents might be constructed and supplied with plausible concentrations of simple reactants such as CO, H_2, NH_3, and H_2S. Appropriate lev-els of amino acids including a small chiral excess, along with the sorts of amphiphi-lic molecules described above, can be rationalized by the findings from the Murchi-son meteorite. Organic molecules such as found in irradiated interstellar ice models, including HMT, can also be included. The system should include weathered feld-spars, which can be modified to include the reduced transition-metal minerals that they are known to contain. [134] Such minerals as Fe,Ni sulfides are likely to have been both present and stable in the environment of early Earth and are known [153, 155] to catalyze formation of organic molecules from simpler precursors.

The entire system, tested first in part and perhaps eventually in full, can subse-quently be subjected to cycles of heat and cold, and hydration/desiccation. We might then look for the production of new chiral molecules, the appearance of homochiral dominance among amino acids and their polymers, and perhaps even the eventual emergence of micron-sized lipid vesicles containing catalytic and autocatalytic reac-tion cycles.

11.7
Concluding Remarks

The universe appears to be self-organizing across all length scales. Evidence of self-organization is obvious in the self-assembly of atoms from quarks and leptons and in the building up of the periodic table by the organization of electrons in shells and subshells. It is apparent from the nature of biopolymers in cells, to neural networks in brains, to organismal development in evolutionary biology and ecology, to the organization of stars in galaxies and even to the scale of the universe itself. Since the pioneering work of Oparin, Eigen, Kaufmann, and others, the origin of life has been regarded as emerging from chemical self-organization of one type or another. Therefore to find an answer to the question "What is life?" it is imperative to answer the question "From whence life?"

At the heart of life, as we know it, are homochiral biopolymers of which RNA, DNA, and the proteins are archetypal. This chapter has considered the question of how homochiral biopolymers can possibly arise from a nearly equal mixture of the chiral monomers. That is, not only how an initiating enantiomeric excess was generated on Earth but, more importantly, what sort of amplification and survival mechanisms might have been able to transform that small enantiomeric excess into the homochiral arrays found in contemporary biomolecules. In the course of exploring this question, we have been led to develop the above outline of what may have happened. We hope that this outline might inspire a series of experiments and eventually evolve into a workable model for this process. This research outline potentially includes all the elements of a viable, testable, and self-consistent theory for the origin of life. Clearly, the question of the origin of life is among the most profound questions asked by human minds. We dare to imagine that during the 21st century a coherent and credible theory of chemical biogenesis will emerge.

11.8
Acknowledgments

This topic draws from a variety of fields of study. We are grateful for critical comments and thoughtful suggestions on earlier drafts of this chapter from Steven Benner, Peter R. Buseck, Kenso Soai, and Frank H. Westheimer. Any mistakes and omissions remain the responsibility of the authors.

References

1 SEN, A. C.; KEIDERLING, T. A. *Biopolymers* **1984**, *23*, 1519–1532.

2 YODER, G.; KEIDERLING, T. A.; FORMAGGIO, F.; CRISMA, M.; TONIOLO, C.; KAMPHIUS, J. *Tetrahedron: Asymmetry* **1995**, *6*, 687–690.

3 LAL, B. B.; NAFIE, L. A. *Biopolymers* **1982**, *21*, 2161–2183.

4 BONNER, W. A. *The Quest for Chirality* in *Physical Origin of Homochirality in Life*; CLINE, D. B., Ed.; American Institute of Physics: Woodbury, New York, 1996; Vol. 379, pp 17–49.

5 CHYBA, C.; SAGAN, C. *Nature* **1992**, *355*, 125–132.

6 SORRELL, W. H. *Astrophysics and Space Science* **1997**, *253*, 27–41.

7 CHAMBERLIN, T. C.; CHAMBERLIN, R. T. *Science* **1908**, *28*, 897–911.

8 LEE, T. D.; YANG, C. N. *Phys. Rev.* **1956**, *104*, 254–257.

9 WU, C. S.; AMBLER, E.; HAYWARD, R. W.; HOPPES, D. D.; HUDSON, R. P. *Phys. Rev.* **1957**, *105*, 1413–1415.

10 VESTER, F.; ULBRICHT, T. L. V.; KRAUCH, H. *Naturwiss.* **1959**, *46*, 68.

11 ULBRICHT, T. L. V.; VESTER, F. *Tetrahedron* **1962**, *18*, 629–637.

12 YAMAGATA, Y. *J. Theor. Biol.* **1966**, *11*, 495–498.

13 GARAY, A. S.; HRASKO, P. *Neutral currents un weak interactions and molecular asymmetry* in *International symposium on the generation and amplification of asymmetry in chemical systems* THIEMANN, W., Ed.: Juelich, Germany, 1974, pp 449–469.

14 LETOKHOV, V. S. *Phys. Lett.* **1975**, *A53*, 275–276.

15 KONDEPUDI, D. K.; NELSON, G. W. *Phys. Rev. Lett.* **1983**, *50*, 1023–1026.

16 FRANK, F. C. *Biochim. Biophys. Acta* **1953**, *11*, 459–463.

17 KONDEPUDI, D. K.; NELSON, G. W. *Phys. Lett.* **1984**, *A106*, 203–206.

18 KONDEPUDI, D. K.; NELSON, G. W. *Physica A* **1984**, *125*, 465–496.

19 KONDEPUDI, D. K.; NELSON, G. W. *Nature* **1985**, *314*, 438–441.

20 KONDEPUDI, D. K. *Biosystems* **1987**, *20*, 75–83.

21 MacDERMOTT, A. J.; TRANTER, G. E. *Croat. Chem. Acta* **1989**, *62*, 165–187.

22 HUND, F. *Z. Phys.* **1927**, *43*, 805–826.

23 REIN, D. W. *J. Mol. Evol.* **1974**, *4*, 15–22.

24 ZEL'DOVICH, B. Y.; SAAKYAN, D. B.; SOBEL'-MAN, I. I. *JETP Lett.* **1977**, *25*, 94–97.

25 HARRIS, R. A.; STODOLSKI, L. *Phys. Lett.* **1978**, *B78*, 313–317.

26 REIN, D. W.; HEGSTROM, R. A.; SANDARS, P. G. H. *Phys. Lett.* **1979**, *A71*, 499–501.

27 HEGSTROM, R. A.; REIN, D. W.; SANDARS, P. G. H. *J. Chem. Phys.* **1980**, *73*, 2329–2340.

28 HARRIS, R. A.; STODOLSKI, L. *J. Chem. Phys.* **1980**, *73*, 3862–3863.

29 BOUCHIAT, M.-A.; BOUCHIAT, C. *Rep. Prog. Phys.* **1997**, *60*, 1351–1396.

30 BAKASOV, A.; HA, T.-K.; QUACK, M. *J. Chem. Phys.* **1998**, *109*, 7263–7285.

30a BERGER, A.; QUACK, M. *Chem. Eur. J. Chem. Phys.* **2000**, *1*, 57–60, and LAERDHAL, J. K, WESENDRUB, R. SCHERDTFEGER, R. *Chem. Eur. J. Chem. Phys.* **2000**, *1*, 60–62.

31 ZANASI, R.; LAZZERETTI, P.; LIGABUE, A.; SONCINI, A. *Phys. Rev. E.* **1999**, *59*, 3382–3385.

32 LAZZERETTI, P.; ZANASI, R.; FAGLIONI, F. *Phys. Rev. E* **1999**, *60*, 871–874.

33 NICOLIS, G.; PRIGOGINE, I. *Proc. Natl. Acad. Sci. U.S.A.* **1981**, *78*, 659–663.

34 FRONDEL, C. *The System of Mineralogy,* 7 ed.; John Wiley & Sons, Inc.: New York, 1962; Vol. 3.

35 BONNER, W. A. *Origins of Chiral Homogeneity in Nature* in D-*Amino Acids in Sequences of Secreted Peptides of Multicellular Organisms*; JOLLÉS, P., Ed.; Birkhäuser: Basel, Switzerland, 1998, 159–188.

36 FRONDEL, C. *Amer. Mineral.* **1978**, *63*, 17–27.

37 KONDEPUDI, D. K.; KAUFMAN, R. J.; SINGH, N. *Science* **1990**, *250*, 975–976.

38 KONDEPUDI, D. K.; LAUDADIO, J.; ASAKURA, K. *J. Am. Chem. Soc.* **1999**, *121*, 1448–1451.

39 NIORI, T.; SEKINE, T.; WATANABE, J.; FURUKAWA, T.; TAKEZOE, H. *J. Mater. Chem.* **1996**, *6*, 1231–1233.

40 LINK, D. R.; NATALE, G.; SHAO, R.; MACLENNAN, J. E.; CLARK, N. A.; KÖRBLOVA, E.; WALBA, D. M. *Science* **1997**, *278*, 1924–1927.

41 SEKINGER, J. V.; WANG, Z.-G.; BRUINSMA, R. F.; KNOBLER, C. M. *Phys. Rev. Lett.* **1993**, *70*, 1139–1142.

42 NASSOY, P.; GOLDMANN, M.; BOULOUSSA, O.; RONDELEZ, F. *Phys. Rev. Lett.* **1995**, *75*, 457–460.

43 FANG, H.; GIANCARLO, L. C.; FLYNN, G. W. *J. Phys. Chem. B* **1998**, *102*, 7311–7315.

44 KUHN, W.; BRAUN, E. *Naturwiss.* **1929**, *17*, 227–228.

45 RAU, H. *Chem. Rev.* **1983**, *83*, 535–547.

46 BONNER, W. A. *Homochirality and Life* in *Topics in Stereochemistry* ELIEL, E. L. and WILEN, S. H., Ed.; John Wiley & Sons, Inc.: New York, 1988; Vol. 18, pp 1–96.

47 RUBENSTEIN, E.; BONNER, W. A.; NOYES, H. P.; BROWN, G. S. *Nature* **1983**, *306*, 118.

48 BONNER, W. A.; RUBENSTEIN, E. *BioSystems* **1987**, *20*, 99–111.

49 GREENBERG, J. M.; MENDOZA-GÓMEZ, C. X. *Interstellar dust evolution: a reservoir of prebiotic molecules* in *Chemistry of Life's Origins*; GREENBERG, J. M., MENDOZA-GÓMEZ, C. X. and PIRRONELLO, V., Ed.; Kluwer: Dordrecht, Netherlands, 1993; Vol. 1, pp 1–32.

50 BONNER, W. A. *Origins Life Evol. Biosphere* **1996**, *26*, 27–46.

51 TOKAY, R. K.; NORDÉN, B.; LILJENZIN, J.-O.; ANDERSSON, S. *J. Radioanal. Nucl. Chem. Lett.* **1986**, *104*, 337–340.

52 McGINNIS, M. B.; COMPTON, R.; PAGNI, R. M.; MAHURN, S.; BOGARD, J. S.; HULETT, L. E.; HANEY, J. D. *SERMACS '99*: Knoxville, TN, 1999.

53 THIEMANN, W.; DARGE, W. *Orig. Life* **1974**, *5*, 263–283.

54 WAGENER, K. *J. Mol. Evol.* **1974**, *4*, 77–84.

55 THIEMANN, W. *J. Mol. Evol.* **1974**, *4*, 85–97.

56 BONNER, W. A. *Chirality* **2000**, *12* 114-126.

57 THIEMANN, W.; WAGENER, K. *Angew. Chem. Int. Ed. Engl.* (1970) *9*, 740–741.

58 YAMAGATA, Y. *Proposal for Mirror Symmetry Test in Molecules* in *International Symposium on Generation and Amplification of Asymmetry in Chemical Systems, Sept. 24–26. Kernforschungsanlage, Juelich* **1973**, 233–255.

59 KOVÁCS, K. L.; GARAY, A. S. *Nature* **1975**, *254*, 538.

60 KOVÁCS, K. L. *Origins of Optical Activity in Nature*; WALKER, D. C., Ed.; Elsevier: New York, 1979, pp 213–227.

61 VETTER, P. A.; MEEKHOF, D. M.; MAJUMDER, P. K.; LAMOREAUX, S. K.; FORSTEN, E. N. *Phys. Rev. Lett.* **1995**, *74*, 2658–2661.

62 WOOD, C. S.; BENNET, C. S.; CHO, D.; MASTERSON, B. P.; ROBERTS, J. L.; TANNER, C. E.; WIEMAN, C. E. *Science* **1997**, *275*, 1759–1763.

63 LAERDAHL, J. K.; SCHWERDTFEGER, P. *Phys. Rev. A.* **1999** *60*, 4439–4453.

64 CALVIN, M. *Chemical Evolution*; Oxford University Press: Oxford, U.K., 1969.

65 BONNER, W. A. *Origins Life Evol. Biosphere* **1995**, *25*, 401–414.

66 OKADA, Y.; TAKEBAYASHI, T.; HASHIMOTO, M.; KASUGA, S.; SATO, S.; TAMURA, C. *J. Chem. Soc. Chem. Commun.* **1983**, 784.

67 OKADA, Y.; TAKEBAYASHI, T. *Chem. Pharm. Bull.* **1988**, *36*, 3787–3792.

68 OKADA, Y.; TAKEBAYASHI, T.; SATO, S. *Chem. Pharm. Bull.* **1989**, *37*, 5–8.

69 OKADA, Y.; TAKEBAYASHI, T.; SEIICHI, A.; SATO, S. *Heterocycles* **1990**, *31*, 1923–1926.

70 BONNER, W. A. *Origins Life Evol. Biosphere* **1994**, *24*, 63–78.

71 NORDÉN, B. *Nature* **1977**, *266*, 567–568.

72 CRONIN, J. R.; PIZZARELLO, S. *Science* **1997**, *275*, 951–955.

73 ENGEL, M. H.; MACKO, S. H. *Nature* **1997**, *389*, 265–268.

74 MASON, S. F. *Nature* **1997**, *389*, 804.

75 BAILEY, J.; CHRYSTOSPOMOU, A.; HOUGH, J. H.; GLEDHILL, T. M.; McCALL, A.; CLARK, S.; MÉNARD, F.; TAMURA, M. *Science* 1998, *281*, 672–674.

76 CERF, C.; JORISSEN, A. *Space Science Reviews* 2000, *(in press)*

77 BRIGGS, R.; ERTEM, G.; FERRIS, J. P.; GREEN-BERG, J. M.; McCAIN, P. J.; MENDOZA-GÓMEZ, C. X.; SCHUTTE, W. *Orig. Life Evol. Biosphere* 1992, *22*, 287–307.

78 GREENBERG, J. M. *Chirality in interstellar dust and in comets: Life from dead stars?* in *Physical Origin of Homochirality in Life*; CLINE, D. B., Ed.; American Institute of Physics: Wood-bury, NY, 1996, pp 185–210.

79 RUBENSTEIN, E.; BONNER, W. A.; BROWN, G. S. *Science* 1999, *283*, 1415.

80 BAILEY, J. *Science* 1999, *283*, 1415.

81 BERNSTEIN, M. P.; SANDFORD, S. A.; ALLA-MANDOLA, L. J.; CHANG, S. *Astrophys. J.* 1995, *454*, 327–344.

82 WOLLMAN, Y.; MILLER, S. L.; IBANEZ, J.; ORÓ, J. *Science* 1971, *174*, 1039–1040.

83 ALLAMANDOLA, L. J.; TIELENS, A. G. G. M.; BARKER, J. R. *Astrophys. J. Suppl. Ser.* 1989, *71*, 733–775.

84 PUGET, J. L.; LEGER, A. *Annu. Rev. Astron. Astrophys.* 1989, *27*, 161–198.

85 BERNSTEIN, M. P.; SANDFORD, S. A.; ALLA-MANDOLA, L. J.; GILLETTE, S. J.; CLEMETT, S. J.; ZARE, R. N. *Science* 1999, *283*, 1135–1139.

86 CRICK, F. H. C. *Cold Spring Harbor Symp. Quant. Biol.* 1966, *31*, 3–9.

87 CRICK, F. H. C. *J. Mol. Biol.* 1968, *38*, 367–379.

88 BAUMANN, U.; ORO, J. *BioSystems* 1993, *29*, 133–141.

89 TRIFONOV, E. N.; BETTECKEN, T. *Gene* 1997, *205*, 1–6.

90 FERREIRA, R.; CAVALCANTI, A. R. D. O. *Origins Life Evol. Biosphere* 1997, *27*, 397–403.

91 EDER, A. H.; SAETIA, S.; RODE, B. M. *Inorg. Chim. Acta* 1993, *207*, 3–10.

92 NORDÉN, B. *J. Mol. Evol.* 1978, *11*, 313–332.

93 WALD, G. *Ann. NY Acad. Sci.* 1957, *69*, 352–368.

94 BLAIR, N. E.; BONNER, W. A. *Orig. Life* 1980, *10*, 225–263.

95 BRACH, A.; SPACH, G. *J. Mol. Evol.* 1979, *13*, 35–46.

96 BLAIR, N. E.; DIRBAS, F. M.; BONNER, W. A. *Tetrahedron* 1981, *37*, 27–29.

97 BLAIR, N. E.; BONNER, W. A. *Orig. Life* 1981, *11*, 331–335.

98 IMAI, E.-I.; HONDA, H.; HATORI, K.; BRACK, A.; MATSUNO, K. *Science* 1999, *283*, 831–833.

99 BOLLI, M.; MICURA, R.; ESCHENMOSER, A. *Chem. Biol.* 1997, *4*, 309–320.

100 AVETISOV, V. A. *Spontaneous mirror symmetry breaking via enantioselective autocatalysis* in *Physical Origin of Homochirality in Life*; CLINE, D. B., Ed.; American Institute of Physics: Woodbury, New York, 1996, pp 141–150.

101 LIPPMAN, D.; PATTANAYAK, A. *Symmetry breaking by autocatalysis* in *Physical Origin of Homochirality in Life*; CLINE, D. B., Ed.; American Institute of Physics: Woodbury, New York, 1996, pp 151–155.

102 BUHSE, T.; LAVABRE, D.; MICHEAU, J.; THIE-MANN, W. *Chirality* 1993, *5*, 341–345.

103 SHIBATA, T.; MONOKA, H.; HAYASE, T.; CHOJI, K.; SOAI, K. *J. Am. Chem. Soc.* 1996, *118*, 471–472.

104 BOLM, C.; BIENWALD, F.; SEGER, A. *Angew. Chem. Int. Ed. Engl.* (1996) *35*, 1657–1659.

105 MacDERMOTT, A. J. *The weak force and SETH: The search for extraterrestrial homochirality* in *Physical Origin of Homochirality in Life*; CLINE, D. B., Ed.; American Institute of Physics: Woodbury, New York, 1996, pp 241–254.

106 AVETISOV, V.; GOLDANSKII, V. *Proc. Natl. Acad. Sci. USA* 1997, *93*, 11435–11442.

107 KONDEPUDI, D. K. *Selection of handedness in prebiotic chemical processes* in *Discovery of Weak Neutral Currents*; AIP Conference Proceed-ings, 1994; Vol. 300, pp 491–498.

108 VAIDA, M.; POPOVITZ-BIRO, R.; LEISEROWITZ, L.; LAHAV, M. *Probing reaction pathways via asymmetric transformations in chiral and centro-symmetric crystals* in *Photochemistry in Orga-nized and Constrained Media*; RAMAMURTHY, V., Ed.; VCH Publishers: New York, New York, 1991, pp 248–302.

109 KAGAN, H. P.; FIAUD, J. C. *Kinetic Resolutions* in *Topics in Stereochemistry*; ELIEL, E. L. and WILEN, S. H., Ed.; John Wiley & Sons: New York, New York, 1988; Vol. 18, pp 249–330.

110 ADDADI, L.; BERKOVITCH-YELLIN, Z.; WEISS-BUCH, I.; VAN MIL, J.; SHIMON, L. J. W.; LAHAV, M.; LEISEROWITZ, L. *Angew. Chem. Intl. Ed. Engl.* (1985), *24*, 466–485.

111 THIEMANN, W.; TEUTSCH *Origins. Life Evol. Biosphere* **1990**, *20*, 121–126.

112 MOROWITZ, H. J. *J. Theor. Biol.* **1969**, *25*, 491–494.

113 SOAI, K.; SHIBATA, T.; MORIOKA, H.; CHOJI, K. *Nature* **1995**, *378*, 767–768.

114 SHIBATA, T.; YAMAMOTO, J.; YONEKUBO, S.; OSANAI, S.; SOAI, K. *J. Am. Chem. Soc.* **1998**, *120*, 12157–12158.

115 SOAI, K.; ASANAI, S.; KADOWAKI, K.; YONE-KUBO, S.; SHIBUTA, T.; SANO, I. *J. Am. Chem. Soc.* **1999**, *121*, 11235–11236.

116 SATO, I.; KADOWAKI, K.; SOAI, K. *Angew. Chem. Int. Ed.* **2000**, *39*, 1510–1512.

117 WITTUNG, P.; NIELSEN, P. E.; BUCHARDT, O.; EGHOLM, M.; NORDEN, B. *Nature* **1994**, *368*, 651–653.

118 NELSON, K. E.; LEVY, M.; MILLER, S. L. *Proc. Natl. Acad. Sci. USA* **2000**, *97*, 3868–3871.

119 HUCK, N. P. M.; JAGER, W. F.; DE LANGE, B.; L., F. B. *Science* **1999**, *273*, 1686–1691.

120 BURNHAM, K. S.; SCHUSTER, G. B. *J. Am. Chem. Soc.* **1999**, *121*, 10243–10248.

121 ZAHNLE, K. J.; SLEEP, N. H. *Impacts and the Early Evolution of Life* in *Comets and the Origin and Evolution of Life*; THOMAS, P. J., CHYBA, C. F. and McKAY, C. P., Ed.; Springer: New York, New York, 1997, pp 175–208.

122 NISBET, E. G.; FOWLER, C. M. R. *Geol. Soc. Spec. Publ.* **1996**, *118*, 239–251.

123 ZAHNLE, K.; WALKER, J. C. G. *Rev. Geophys. Space Phys.* **1982**, *20*, 280–292.

124 WALKER, J. C. G.; LOHMAN, K. C. *Geophys. Res. Lett.* **1989**, *16*, 323–326.

125 LOWE, D. R. *Early environments: Constraints and opportunities for early evolution* in *Early Life on Earth*; BENGTSON, S., Ed.; Columbia University Press: New York, NY, 1994, pp 24–35.

126 BERNAL, J. D. *The Origin of Life*; The World Publishing Co.: Cleveland, 1967.

127 CAIRNS-SMITH, A. G.; HARTMAN, H. *Clay Minerals and the Origin of Life*; Cambridge University Press: Cambridge, 1986.

128 CAIRNS-SMITH, A. G. *The Life Puzzle*; The University of Toronto Press: Toronto, 1971.

129 CAIRNS-SMITH, A. G. *Genetic Takeover and the Mineral Origins of Life*; Cambridge University Press: Cambridge, 1982.

130 WEISS, A. *Angew. Chem. Int. Ed. Engl.* (1981) *20*, 850–860.

131 CHYBA, C.; SAGAN, C. *Comets as a Source of Prebiotic Organic Molecules for the Early Earth* in *Comets and the Origin and Evolution of Life*; THOMAS, P. J., CHYBA, C. F. and McKAY, C. P., Ed.; Springer: New York, New York, 1997, pp 147–174.

132 SMITH, J. V. *Proc. Nat. Acad. Sci. USA* **1998**, *95*, 3366–3369.

133 SMITH, J. V. *Proc. Natl. Acad. Sci. USA* **1998**, *95*, 3370–3375.

134 PARSONS, I.; LEE, M. R.; SMITH, J. V. *Proc. Natl. Acad. Sci. USA* **1998**, *95*, 15173–15176.

135 VAN OLPHEN, H. *An Introduction to Clay Colloid Chemistry*; Interscience Publishers: New York, New York, 1963.

136 GRIM, R. E. *Clay Mineralogy*; McGraw-Hill: New York, New York, 1968.

137 GLOVER, J. E. *Precambrian Research* **1992**, *56*, 159–166.

138 DEAMER, D. W. *Orig. Life* **1986**, *17*, 3–25.

139 BACHMANN, P. A.; LUISI, P. L.; LANG, J. *Nature* **1992**, *357*, 57–59.

140 BACHMANN, P. A.; LUISI, P. L.; LANG, J. *Chimia* **1991**, *45*, 266–268.

141 SEGRÉ, D.; BEN-ELI, D.; LANCET, D. *Proc. Natl. Acad. Sci. USA* **2000**, *97*, 4112–4117.

142 MOROWITZ, H. J.; HEINZ, B.; DEAMER, D. W. *Origins Life Evol. Biosphere* **1988**, *18*, 281–287.

143 DEAMER, D. W.; BARCHFELD, G. L. *J. Mol. Evol.* **1982**, *18*, 203–206.

144 DEAMER, D. W. *Microbiology and Molecular Biology Reviews* **1997**, *61*, 239–261.

145 DEAMER, D. W.; PASHLEY, R. M. *Origins Life Evol. Biosphere* **1989**, *19*, 21–38.

146 BRANDES, J. A.; BOCTOR, N. Z.; CODY, G. D.; COOPER, B. A.; HAZEN, R. M.; YODER, H. S. *J. Nature* **1998**, *395*, 365–367.

147 CLARK, P. D.; DOWLING, M. I.; HUANG, M. *J. Mol. Evol.* **1998**, *47*, 127–132.

148 McCOLLUM, T. M.; SHOCK, E. L. *Geochim. Cosmochim. Acta* **1997**, *61*, 4375–4391.

149 PFLUMIO, C.; BOULEGUE, J.; TEIRCELIN, J. J. *Chem. Geol.* **1994**, *116*, 85–109.

150 GIOVANNI, G.; TEKLEMARIAM, M. *J. Volcanol. Geotherm. Res.* **1993**, *56*, 429–445.

151 MARSHALL, W. L. *Geochim. Cosmochim. Acta* **1994**, *58*, 2099–2106.

152 HEINEN, W.; LAUWERS, A. *Origins Life Evol. Biosphere* **1996**, *26*, 131–150.

153 HUBER, C.; WÄCHTERSHÄUSER, G. *Science* **1997**, *276*, 245–247.

154 WEBER, A. L. *Origins Life Evol. Biosphere* **1998**, *28*, 259–270.

155 HEINEN, W.; LAUWERS, A. *Proc. Konin. Neder. Akad. Weten.* **1997**, *100*, 11–25.

156 AMEND, J. P.; SHOCK, E. L. *Science* **1998**, *281*, 1659–1662.

157 SHOCK, E. L.; SCHULTE, M. D. *Geophys. Res.-Planets* **1998**, *103*, 28513–28527.

158 SHOCK, E. L. *Geochim. Cosmochim. Acta* **1993**, *57*, 3341–3349.

159 BUJDÁK, J.; LESON, H. L.; YONGYAI, Y.; RODE, B. M. *Catal. Lett.* **1996**, *37*, 267–272.

160 BUJDÁK, J.; RODE, B. M. *J. Mol. Evol.* **1996**, *43*, 326–333.

161 ORGEL, L. E. *Origins Life Evol. Biosphere* **1998**, *28*, 227–234.

162 HILL, R. A. J.; BÖHLER, C.; ORGEL, L. E. *Origins Life Evol. Biosphere* **1998**, *28*, 235–243.

163 LIU, R.; ORGEL, L. E. *Origins Life Evol. Biosphere* **1998**, *28*, 245–257.

164 WHITE, D. H.; KENNEDY, R. M.; MACKLIN, J. *Origins Life* **1984**, *14*, 273–278.

165 SAETIA, S.; LIEDL, K. R.; EDER, A. H.; RODE, B. M. *Origins Life Evol. Biosphere* **1994**, *23*, 167–176.

166 MILLER, S. L.; UREY, H. C. *Science* **1959**, *130*, 245–251.

167 FERRIS, J. P., *Chem. & Eng. News*, 27 Aug. 1984, pp 22–35.

168 KISSEL, J.; KRUEGER, F. R.; ROESSLER, K. *Organic Chemistry in Comets from Remote and In Situ Observations* in *Comets and the Origin and Evolution of Life*; THOMAS, J. P., CHYBA, C. F. and McKAY, C., Ed.; Springer: New York, New York, 1997, pp 69–109.

169 DAVIDSON, A. R.; SAUER, R. T. *Proc. Natl. Acad. Sci. USA* **1994**, *91*, 2146–2150.

170 ROBERTSON, M. P.; MILLER, S. L. *Science* **1995**, *268*, 702–705.

171 ROBERTSON, M. P.; MILLER, S. L. *Nature* **1995**, *375*, 772–773.

172 ERTEM, G.; FERRIS, J. P. *J. Am. Chem. Soc.* **1997**, *119*, 7197–7201.

173 FERRIS, J. P.; HILL, A. R. J.; LIU, R.; ORGEL, L. E. *Nature* **1996**, *381*, 59–61.

174 VISSCHER, J.; BAKKER, C. G.; van der WOERD, R.; SCHWARTZ, A. W. *Science* **1989**, *244*, 329–221.

175 ORÓ, J. *Early chemical stages in the origin of life* IN *Early Life on Earth*; BENGTSON, S., Ed.; Columbia University Press: New York, NY, 1994, pp 48–59.

176 SIGEL, H. *Inorg. Chim. Acta* **1992**, *198–200*, 1–11.

177 PYLE, A. M. *Science* **1993**, *261*, 709–714.

178 RAMOS, A.; GUBSER, C. C.; VARANI, G. *Curr. Opin. Struct. Biol.* **1997**, *7*, 317–323.

179 JAEGER, L. *Curr. Opin. Struct. Biol.* **1997**, *7*, 324–335.

180 PAN, T.; LONG, D. M.; UHLENBACK, O. C. *Divalent Metal Ions in RNA Folding and Catalysis* in *The RNA World*; GESTELAND, R. F. and ATKINS, J. F., Ed.; Cold Spring Harbor Laboratory Press: Cold Spring Harbor, ME, 1993, pp 271–302.

181 BOWRING, S. A.; HOUSH, T. *Science* **1995**, *269*, 1535–1540.

182 MacLEOD, G.; McKEOWN, C.; HALL, A. J.; RUSSELL, M. J. *Origins Life Evol. Biosphere* **1994**, *24*, 19–41.

183 DE WIT, M. *Precambrian Research* **1998**, *91*, 181–226.

184 TAJIKA, E.; MATSUI, T. *Lithos* **1993**, *30*, 267–280.

185 KRUPP, R., OBERTHÜR, T., HIRDES, W. *Economic Geology* **1994**, *89*, 1581–1594.

186 SIGURDSSON, H. *Global Planetary Change* **1990**, *89*, 277–289.

187 TABAZADEH, A.; TURCO, R. P. *Science* **1993**, *260*, 1082–1086.

188 MOJZSIS, S. J.; ARRHENIUS, G.; McKEEGAN, K. D.; HARRISON, T. M.; NUTMAN, A. P.; FRIEND, C. R. L. *Nature* **1996**, *384*, 55–59.

189 SUTHERLAND, J. D.; BLACKBURN, J. M. *Chem. Biol.* **1997**, *4*, 481–488.

190 EIGEN, M.; SCHUSTER, P. *The Hypercycle*; Springer-Verlag: Berlin, 1979.

191 YAO, S.; GHOSH, I.; ZUTSHI, R.; CHMIELEWSKI, J. *Nature* **1998**, *396*, 447–450.

192 KOCH, A. L. *J. Mol. Evol.* **1985**, *21*, 270–277.

193 SEGRÉ, D.; LANCET, D.; KEDEM, O.; PILPEL, Y. *Origins Life Evol. Biosphere* **1998**, *28*, 501–514.

194 NORRIS, V.; RAINE, D. J. *Origins Life Evol. Biosphere* **1998**, *28*, 523–537.

195 WESTHEIMER, F. H. *Science* **1987**, *235*, 1173–1178.

196 MACIA, E.; HERNANDEZ, M. V.; ORÓ, J. *Origins Life Evol. Biosphere* **1997**, *27*, 459–480.

197 MYSEN, B.; HOLTZ, F.; PICHAVANT, M.; BENY, J. M.; MONTEL, J. M. *Geochim. Cosmochim. Acta* **1997**, *61*, 3913–3926.

198 STUMM, W.; MORGAN, J. J. *Aquatic Chemistry*, John Wiley & Sons, Inc.: New York, New York, 1981.

199 SILVER, R. *Form of Phosphate an the Early Earth* in *Proceedings of the Fourth Conference on Origins of Life Chemistry and Radiochemistry*, MARGULIS, L., Ed.; Springer-Verlag: New York, New York, 1973, pp 215–216.

200 MOUNIR, A. M.; FASSAM, R. A.; HENDERSON, J. R. *Chem. Tech. Biotechnol.* **1993**, *58*, 387–389.

201 FROELICH, P. N. *Limnol. Oceanogr.* **1988**, *33*, 649–668.

202 MURASHOV, V. V.; LESZCZYNSKI, J. *J. Phys. Chem. A* **1999**, *103*, 1228–1238.

203 HALMAN, M. *Evolution and Ecology of Phosphorus Metabolism* in *The Origin of Life and Evolutionary Biochemistry*, DOSE, K., FOX, S. W., DEBORIN, G. A. and PAVLOVSKAYA, T. E., Ed.; Plenum: New York, New York, 1974, pp 169–182.

204 SMITH, J. V.; ARNOLD, F. P.; PARSONS, I.; LEE, M. R. *Proc. Natl. Acad. Sci. USA* **1999**, *96*, 3479–3485.

205 YAMAGATA, Y.; WATANABE, H.; SAITOH, M.; NAMBA, T. *Nature* **1991**, *352*, 516–519.

206 VARADACHARI, C. *Ind. Eng. Chem. Res.* **1992**, *31*, 357–364.

207 GEDULIN, B.; ARRHENIUS, G. *Sources and Geochemical Evolution of RNA Precursor Molecules: The role of phosphate* in *Early Life on Earth*; BENGTSON, S., Ed.; Columbia University Press: New York, NY, 1994, pp 91–106.

12

Chemical Reaction Dynamics Looks to the Understanding of Complex Systems*

R. D. Levine

Much of our early conceptual understanding of reaction dynamics [1] was acquired by the study of isolated, simple chemical exchange reactions in which one bond is broken and another bond is formed, in concert. A central feature in these reactions is the high selectivity which can be achieved by the choice of initial conditions and the considerable specificity of the resulting products. Familiar examples [1] include the specificity of energy disposal of exoergic reactions (which gave rise to chemical lasers) or the quite anisotropic distribution of the reaction products in space. More recently similar results were observed in dissociation and/or elimination reactions following photochemical activation. I argue that these themes carry over to the world of more complex reactions [2] (additions, isomerization, substitution) of the kind that are employed in the synthesis of new compounds. Lasers [1, 3, 4] are an already available tool to selectively provide the required free energy of activation, if needed, on the same time scale as that of the chemical event itself. [4–7] Laser control (see, e.g., Ref. [8]) and other modes for ultra fast activation (see, e.g., Ref. [9]) are being explored.

Traditionally, the progress in dynamics has been achieved by a synergetic experimental-theoretical approach where the questions which were being raised by new experimental capabilities stimulated the development of theoretical tools. In a complementary fashion, suggestions by the theory as to hitherto unexplored effects were addressed by new experimental designs.

The rapid growth in both experimental and theoretical capabilities is currently resulting in a progress that is somewhat less coordinated. There is just so much that one can do that experimentalists and theorists alike are right now busy pushing ahead the limits of their respective crafts. It is safe to argue that when this evolutionary stage is over, the two cultures will converge in a new understanding of dynamics of many atom systems.

What drives the current evolution? Overall, it is the general trend in chemistry (certainly motivated by systems of biochemical importance but equally so by material science, catalysis etc.) toward the understanding of weaker interactions. This is sometimes known as "supramolecular" chemistry. [10] The bridges between dynamics and such problems as molecular recognition are already being built. [11] Other examples are the dynamics of clusters of molecules [12] held together by weaker than chemical forces. Experiments and theory of such finite systems [13] are likely to provide for us an understanding of the role of the solvent during the barrier crossing in activated chemical reactions. The reason is that the barrier crossing act is fast – so fast that the reactive system has no time to explore the long range structure of its environment. [14] This is also the reason why an elementary reaction in solution can exhibit the same selectivity in initial conditions and specificity in final conditions that we have come to recognize as typical in the gas phase.

For the theoretician, the new freedom is brought about by the availability of powerful workstations. These provide a benchtop capability to follow the individual motion of a large number (hundreds or even thousands) of atoms. Figure 12.1 is an illustration of this capability for the "burning of air". [9]

The well-recognized limitation of the molecular dynamics (MD) approach [15] is that the results are only as realistic as the assumed forces which are used as an input to generate the motion of the atoms. A less obvious point is that the very strength of the method is also its greatest weakness. It generates so much detailed information that the essence can get lost. I return to this point below.

A major technological innovation that opens up the possibility of novel experiments is the availability of reliable solid state (e.g., TiSapphire) lasers which provide ultra short pulses over much of the spectral range which is of chemical interest. [6] This brings about the practical possibility of exciting molecules in a time interval which is short compared to a vibrational period. The result is the creation of an electronically excited molecule where the nuclei are confined to the, typically quite localized, Franck-Condon region. Such a state is non-stationary and will evolve in time. This is unlike the more familiar continuous-wave (cw) excitation, which creates a stationary but delocalized state. The time evolution of a state prepared by ultra fast excitation can be experimentally demonstrated, [5,7,16] and Fig. 12.2 shows the prin-

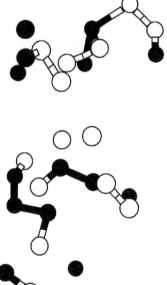

Fig. 12.1 An ultra-fast heating of a cluster containing both N_2 and O_2 molecules upon impact at a surface. [9] Shown is the instantaneous configuration of 14 N atoms (*dark*) and 14 O atoms (*light*) 50 fsec after a cluster of 7 N_2 and 7 O_2 molecules embedded in 97 Ne atoms impacts a surface at a velocity of 20 Mach. The potential used allows for all 125 atoms of the cluster to interact with one another and with the atoms of the surface. In addition, each atom-atom chemical interaction is influenced by the presence or absence of other open-shell atoms nearby.

ciple of one such experiment [16] and the theoretical simulation [17] of the results for an iodine molecule immersed in a rare gas solvent. The optical excitation lasts for about 60 fsec and the dissociation is induced by the solvent.

The methodology used in Ref. [17] overcomes an important limitation of current MD methods, namely that they are limited to a motion on one electronic state. In a problem of the type discussed in Fig. 12.2, where the system can make a transition from a bound to a dissociative state, it is clearly necessary to allow for different dynamics on the different electronic states. In other words, what one needs are methods capable of handling both the intrastate and the interstate dynamics. [18] Quantum mechanical wave packet methods [19] are, in this sense, a very useful tool, but they are currently limited to systems with very few degrees of freedom, and the point about the iodine dissociation is that it is induced by the solvent surrounding the molecule. A traditional route to systems of many particles is a self-consistent field in which each degree of freedom moves in the average field of all the others. [20] However, for many problems of interest there are strong correlations which requires configuration interaction. Extension of such methods to the electronic multistate problems is making good progress. [18]

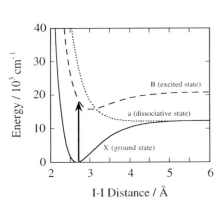
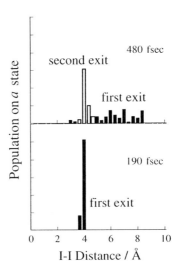

Fig. 12.2 *Left*: The ground (*X, solid line*), excited (*B, dashed line*) and dissociative [*a*1$_g$($^3\Pi$), *dotted line*] electronic state potentials of the iodine molecule. The *arrow* indicates the electronic excitation. The initial excited wave packet is located in the Franck-Condon region near to the inner classical turning point of the *B* state. The transition from the *B* to the *a* state is forbidden by symmetry in the isolated molecule but becomes allowed when the molecule is placed in a solvent.

Right: Histograms of the *a*1$_g$($^3\Pi$) population vs the iodine-iodine separation at 2 time points. Shown are the first two exits for such points in time that the dissociative population has reached an average intramolecular separation of ~4 Å. (This distance is about the upper limit of the experimental probing window.) [16] The localized nature of the dissociative population is a direct result of the vibrational localization on the parent *B* state, as an exit to the *a* state occurs only when there is population in the curve-crossing region.

Another important limitation of current MD methods is that of the conservation of zero point energy (see Ref. [21] and references therein). In a many-atom system the total quantum mechanical zero point energy can be quite high, and, in classical mechanics, this energy is available to the system. Additional sources of reviews of current methodologies are Refs. [22–24].

Two central problems remain. One is that one needs the potential which governs the motion. In many-atom systems, even if the motion is confined to the ground electronic state, this potential is a function of the spatial configuration of all the atoms. It is therefore a function of many variables, so its analytical form is far from obvious, nor do we necessarily want to know it everywhere. Indeed, we really only want it at each point along the actual trajectory of the system (so that the forces can be computed and thereby the next point to which the system will move to can be determined). Such an approach has been implemented [25] and applied to many-atom systems, and an extension to a multi-electronic state dynamics will be important.

The other central problem in a many-atom system is that of the wealth of detail which is provided by the dynamics. One knows that, from one point of view, this detail is definitely necessary, since it is the nature of typical many-body systems that the dynamics is quite chaotic. What this means is that any loss of precision is rapidly (i.e., exponentially) propagated. Indeed, this is the most serious problem in the actual computation of the dynamics. If one really wants to compute a trajectory of the system, one needs an extremely stable method of numerical integration. The resulting time step is then so small that only very short propagations are practical as otherwise the accumulated computer round-off error becomes too high. There are those who adopt a "practical" attitude which is that it does not matter if one is on "the" correct trajectory as long as one is propagating "some" trajectory. (The technical backing for this point is known as the "shadowing" theorem. See Ref. [26] and references therein). The difference between "the" and "some" trajectory is in terms of back integration. The correct trajectory will integrate backward in time to its initial conditions, as it, in principle, should. Some trajectory is a solution of the dynamics. Such a solution need not integrate back to the initial conditions which were specified. The proponents of the "some" approach argue that as long as it is a solution it does not matter. It does not even matter if, during a long integration, a numerical trajectory hops several (or even many) times between possible solutions. Not all MD practitioners care to have this point aired in public. Here I wish to argue that if a pragmatic point is useful, we should ask "why". Why do we have the luxury of not carefully caring that what we compute is the one unique trajectory that corresponds to our choice of initial conditions? (I hasten to add that my students and I do take a purist attitude and when we compute a trajectory, it is "the" trajectory.)

The answer to the problem lies, I believe, in that a single classical trajectory is never the answer to a dynamical question. There are two reasons. One is a reason of principle. To mimic quantal initial conditions it is necessary to generate an entire ensemble of trajectories, each with somewhat different initial conditions. Then one needs to average over these initial conditions. Now comes a wonderful technical point (and a tribute to Stan Ulam [27] who, as far as I know, was the first to realize it. The number of initial conditions that one needs to sample is independent of the number of atoms in the system. It takes just as many initial conditions to mimic the ensemble for an atom-diatom collision as for a cluster collision with hundreds of atoms. The one important role that the number of computed trajectories does play is in determining the numerical accuracy of the sampling. But this is essentially independent of the number of degrees of freedom (as long as it is not unrealistically small). In more technical terms, to answer a simple question the number of trajectories that needs to be computed is independent of the complexity of the system. Hence, to generate an ensemble average it may not matter that the final set of trajectories does not quite correspond to what we think their initial conditions are. It may matter if the actual trajectories correspond to a different ensemble than what we have in mind. This may be alleviated by the second reason for averaging over initial conditions. It is that experiments seldom begin with a well-defined quantum state. This is particularly so for a typical many-atom system. So, in practice, one needs to average over a wider ensemble. (Remember the wonderful theorem of Mr. Ulam: The need to sample a bigger set of initial conditions does not require running more trajectories.) In other words, the less specific is the preparation of the initial state, the less one needs to really worry. A related but not equivalent point is that if the trajectory is ergodic (i.e., it samples all of the available phase space) then one need not worry about the choice of initial conditions. [28]

I do not take these considerations as a license not to compute the correct trajectory in an MD simulation. I do take them as showing that unless one really wants a complete dynamical description, one is over-computing. In other words, if all I want to know is whether a reaction did or did not take place, the precise knowledge of where all the atoms of the system are is not needed.

Figure 12.3 uses a four-center electrocyclic reaction [29] to illustrate the sensitivity to initial conditions. There are two simple yes/no questions. One is: "Did or did not

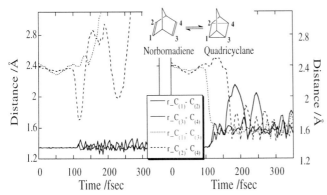

Fig. 12.3 Sequential (*right*) vs a non-reactive four-center ring closure in the isomerization of norbornadiene to quadricyclane (*inset*). The two new C–C bond distances and the two bonds that start as the C=C bonds of norbornadiene are shown vs time in fsec. The two trajectories shown correspond to exactly the same initial conditions for the Ar_{125} cluster. They differ only in the initial conditions for the surface at which the cluster impacts. In a concerted mechanism the two new bonds contract simultaneously.

the impact result in ring closure (cf. Fig. 12.3)?" The other is: "Provided a reaction did take place, what is the branching fraction (i.e., the fraction of trajectories that reacted) in a concerted way (both new bond distances contracting together) vs a sequential way (one new bond contracts first)?" The reaction shown in Fig. 12.3 is the isomerization of norbornadiene to quadricyclane (see inset in the figure) induced by an impact of a rare gas cluster containing one norbornadiene molecule on a surface. [9] The only difference between the two trajectories shown is in the initial conditions of the surface atoms (due to their thermal motion which is at a temperature of 50 K).

How can one make practical use of these observations? For some time I have advocated the use of information theory. (For more details see introductory discussions in Refs. [1] and [3], surprisal analysis in Ref. [23], applications to spectra in Refs. [24] and [30], and a recent prediction of a phase transition induced by cluster impact in Ref. [31]). What this approach seeks to do is to use the minimal dynamical input that is necessary to account for the dynamical observations of interest, the point being that one very rarely has the experimental resolution to probe the individual final quantum states. The information measured is much more coarse grained.

The Monte-Carlo theorem of Stan Ulam tells us that to generate reproducible results for such coarse-grained questions one needs to compute a number of trajectories that is independent of the complexity of the system. Much of the details of the initial conditions are therefore necessarily irrelevant to computing the final answer. Of course, with a finite number of trajectories one only obtains the final answer with a finite accuracy. But to within this accuracy, two sets of computations employing, say, 100 trajectories each for, say, an impact of a cluster containing 125 atoms (as in Fig. 12.1) will yield the same result. This is so even though the number of computed trajectories is smaller than the number of degrees of freedom. (Just the cluster itself has 750, i.e. 6 degrees of freedom per atom × 125 atoms.) It will clearly not be possible for the individual trajectories in the two computations to look alike, yet a coarse-grained result of interest would be the same to within the statistical error. Point one is, therefore, that *to get reproducible results I do not need much of the initial detail that must go into the actual MD computation.* This is central to my argument. To perform an actual MD simulation, each computed trajectory must be specified by 750 initial conditions. Yet the same practical result can be obtained from two different samples of only 100 trajectories each.*

The second point depends on the nature of the dynamics. Say we begin with a limiting case of truly ergodic trajectories. Then I really do not need to compute the actual trajectories. I would get the very same result if I simply postulate that my final set of trajectories is the set of all possible trajectories (where by "possible" I mean that they conserve energy, etc.). Information theory starts from this limit, which we refer to as the "prior" distribution. [1,3,23]

---

*) I should note one technical qualification. The statistical error will depend on the nature of the question, which in a simple way one can describe as "into how many bins do I sort the answer?" "Did a reaction take place in two bins?" (yes or no). "How many NO molecules were formed in eight bins?" (from none up to seven). For a given number of trajectories, the statistical error will be higher in the latter.

The prior distribution is the limiting situation when computing the dynamics will not provide any constraints on the final answer. I must emphasize that while this limit can never be the case, it is a very practical tool. The resolution of the apparent paradox is, as already emphasized, in the nature of the question asked. If I ask for the position and velocities of all atoms at the end of the trajectory, the so-called distribution in phase space, then there is no choice but to accurately compute the trajectory, whether the dynamics is nearly ergodic or not. The reason is that these variables must be highly correlated in order to ensure that, upon time reversal, the trajectory does integrate back to its starting point. Yet one cannot ask for such detailed information because it is more than quantum mechanics allows us, in principle, to measure. Even an ideal experiment can only resolve finite "cells" in phase space ($\equiv$ the space of final positions and momenta). Even for an isolated atom-diatom collision studied in a molecular beam configuration we are not quite yet at such a level of resolution, and for a complex system the size of a typical cell that can be measured is very much larger than the theoretical limit. For such measurements, the Monte Carlo theorem tells us that the detailed fate of an individual trajectory is not required (e.g., Ref. [32]).

The prior distribution is often not what is observed, and there can be extreme deviations from it. [1, 3, 23] By our terminology this means that the dynamics do impose constraints on what can happen. How can one explicitly impose such constraints without a full dynamical computation? At this point I appeal again to the reproducibility of the results of interest as ensured by the Monte Carlo theorem. The very reproducibility implies that much of the computed detail is not relevant to the results of interest. What one therefore seeks is the crudest possible division into cells in phase space that is consistent with the given values of the constraints. This distribution is known as one "of maximal entropy".

There are two ways to implement this program. One is to directly discuss the distribution of final states. This is known as "surprisal analysis". In this simpler procedure one does not ask how this distribution came about. Instead, one seeks the coarsest or "most statistical" ($\equiv$ of maximal entropy) distribution of final states, *subject to constraints*. The last proviso is, of course, essential. If no constraints are imposed, one will obtain as an answer the prior distribution. It is the constraints that generate a distribution which contains just that minimal dynamical detail as is necessary to generate the answers of interest. Few simple and physically obvious constraints are often sufficient [1, 3, 23] to account for even extreme deviations from the prior distribution.

The distribution generated by surprisal analysis is meant to reproduce the results of actual interest, a typical example being the distribution of vibrational energy of products, which is of interest say for chemical laser action. [3] The distribution is not meant to reproduce the fully detailed distribution in classical phase space, which, as already noted, has a to be a highly correlated and complicated distribution.

The other mode of application, known as "surprisal synthesis" is the more fundamental one. It starts from the distribution of initial conditions $\equiv$ the distribution in the initial phase space. As I argued, most of the details of this distribution are irrelevant for obtaining a reproducible final distribution of interest. (Again the caveat: all initial details are essential if what is needed is the final distribution of ultimate res-olution.) One therefore seeks that most coarse-grained initial distribution which is sufficient to get reproducible results for the final distribution. Here too, the most coarse-grained initial distribution is obtained as one of maximal entropy subject to constraints. We now have an initial distribution, but what we need is the final distri-bution. This requires an appeal to the dynamics. One needs a way to phrase the dynamics so as to take advantage of the conclusion that, apart from the constraints, all other details about the initial state are irrelevant for our purpose. Here comes the final theorem that I will appeal to: *In both classical and quantum mechanics, an initial distribution of maximal entropy remains a distribution of maximal entropy throughout its time propagation.* The theorem is exact. [33] It therefore follows that in order to know the final distribution (which, by the theorem, is guaranteed to be one of maximal entropy), it is sufficient to propagate the constraints, and we have shown analytical examples of how this can be implemented. [33]

One can also identify constraints on the basis of physical considerations. One rea-son for selectivity is that different motions often have different time scales. [18] For example, for a chemical barrier-crossing event in solution, it is found that in non-associated solvents it is the motion along the reaction coordinate that is typically faster than the response of the solvent. [14] Figure 12.4 shows the frequency range of the barrier crossing for a symmetric atom exchange reaction relative to the range of frequencies which a typical rare gas solvent can provide. (Negative values of ω^2 are due to unstable motions.) It is evident from the plot that activation by the solvent occurs primarily at the foothills of the barrier to reaction, as is verified by the MD simulations.

The propagation in phase space can also be probed in the frequency domain via the vibrational overtone spectra of polyatomic molecules. [30, 34] The separation of time scales can be demonstrated in this fashion and is useful in representing such spectra by the maximal entropy method. [30] Further details can be found in two recent tutorial reviews in Refs. [23] and [24].

An exact dynamical description requires keeping track of each and every atom in the system. For many problems of chemical interest this represents an enormous

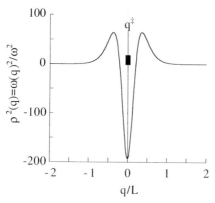

Fig. 12.4 The (squared) frequency of the motion along the reaction coordinate q for a symmetric atom exchange reaction. L is the range of the chemical barrier region, and the frequency is shown in units of the mean frequency of the unperturbed solvent. The range of frequencies in the solvent is indicated as a *solid bar*. The negative values for the solvent correspond to unstable solvent modes.

detail. Experimentalists and theorists both know that, for the kind of questions that they ask, one can obtain reproducible results without the need to specify the initial conditions with such precision. MD simulations mimic this by performing the exact dynamics first and then coarse-graining the result to the required level. Here I discussed a complementary point of view, where the averaging is carried out first, retaining only as much detail as necessary to generate a reproducible result at the desired level of resolution.

Acknowledgement

I thank Michal Ben-Nun, Tamar Raz and Françoise Remacle for their contributions to this review. The work described has been supported by the Air Force Office of Scientific Research (AFOSR).

References

1 R.D. LEVINE and R.B. BERNSTEIN, Molecular Reaction Dynamics and Chemical Reactivity, Oxford University Press, New York (1989).

2 B.K. CARPENTER, *Acc. Chem. Res.* **25**, 520 (1992).

3 A. BEN-SHAUL, Y. HAAS, K.L. KOMPA and R.D. LEVINE, Lasers and Chemical Change, Springer (1981).

4 A.H. ZEWAIL and R.B. BERNSTEIN, *C&E News*, **66**, 24 (1988).

5 F.F. CRIM, *Annu. Rev. Phys. Chem.* **44**, 397 (1993).

6 K.L. KOMPA and R.D. LEVINE, *Acc. Chem. Res.* **27**, 91 (1994).

7 J. MANZ and L. WÎSTE, Femtosecond Chemistry, VCH Weinheim (1994).

8 B. KOHLER et al. *Acc. Chem. Res.* 28, **133** (1995).

9 T. RAZ and R.D. LEVINE, *J. Am. Chem. Soc.* 116, 11167 (1994); J. Phys. Chem. **99**, 7495 (1995).

10 V. BALZANI and F. SCANDOLA, Supramolecular Photochemistry, Harwood, New York (1991).

11 J. JORTNER and R.D. LEVINE, *Adv. Chem. Phys.* **47**, 1 (1981).

12 R.S. BERRY, *J. Phys. Chem.* **98**, 6910 (1994).

13 Reaction Dynamics in Clusters and Condensed Phases, J. JORTNER et al., eds., Kluwer, Dordrecht (1994).

14 M. BEN-NUN and R.D. LEVINE, *Acc. Chem. Res.* **27**, 166 (1994).

15 M.P. ALLEN and D.J. TILDESLEY, Computer Simulations of Liquids, Clarendon Press, Oxford (1987).

16 N.F. SCHERER, D.M. JONAS, and G.R. FLEMING, *J. Chem. Phys.* **99**, 153 (1993).

17 M. BEN-NUN, R.D. LEVINE and G.R. FLEMING, *J. Chem. Phys.* **105**, 3035 (1996).

18 M. BEN-NUN, T, J, MARTINEZ and R.D. LEVINE, *J. Phys. Chem.* **101**, 6389 (1997).

19 R. KOSLOFF, *Annu. Rev. Phys. Chem.* **45**, 145 (1994).

20 J.M. BOWMAN, Acc. Chem. Res. 19, 202 (1986); R.B. GERBER and M. RATNER, *Adv. Chem. Phys.* **70**, 97 (1988).

21 M. BEN-NUN and R.D. LEVINE, *J. Chem. Phys.* **101**, 8768 (1994).

22 Time-Dependent Quantum Molecular Dynamics: Experiment and Theory, J. BROECKHOVE, ed., Plenum Press, New York (1992).

23 Frontiers of Chemical Dynamics, E. YURTSEVER, ed., Plenum Press, New York (1995).

24 Dynamics of Molecules and Chemical Reactions, R.E. WYATT and J.Z.H. ZHANG, eds., M. Dekker, New York (1996).

25 D.K. REMLER and P.A. MADDEN, *Mol. Phys.* **70**, 921 (1990).

26 R.E. GILLIAN and K.R. WILSON, *J. Chem. Phys.* **97**, 1757 (1992).

27 N.G. COOPER, ed., From Cardinals to Chaos, Cambridge University Press (1989).

28 P. BRUMER and M. SHAPIRO, *Adv. Chem. Phys.* **70**, 365 (1985).

29 J.A. BERSON, *Science*, **266**, 1338 (1994).

30 F. REMACLE and R.D. LEVINE, *J. Chem. Phys.* **99**, 2383 (1993).

31 T. KONDOW, U. EVEN, R.D. LEVINE and T. RAZ, U. EVEN, Comments on Modern Physics, 1D , 1 (1999).

32 R. D. LEVINE, *Chem. Phys.* **228**, 255 (1998).

33 Y. ALHASSID and R.D. LEVINE, *J. Chem. Phys.* **67**, 4321 (1977).

34 E.J. HELLER, *Acc. Chem. Res.* 14, 368 (1981).

13
The Past, Present, and Future of Quantum Chemistry

T. Daniel Crawford, Steven S. Wesolowski, Edward F. Valeev, Rollin A. King, Matthew L. Leininger, and Henry F. Schaefer III

Introduction

"Quantum chemistry" refers to the application of quantum mechanics to chemistry. With the rapid development of computers in the middle of the 20th century, the field has blossomed into a major component of the larger realm of chemistry, sometimes superseding traditional experimental approaches to solving chemical problems. W. Graham Richards described the field in his 1979 news article in *Nature*, [1] "Now steady improvements in theoretical calculations have led to the dawning of a third age [of quantum chemistry]: calculations which are more accurate than experiment or at very least sufficiently accurate to be indispensable for interpretation." Now, twenty years later, there exists a plethora of examples for which experimental observations depended on or at least followed quantum chemical predictions. [2–4] Central to all of quantum chemistry is the molecular Schrödinger equation,

$$\left\{ -\frac{1}{2}\sum_i \nabla_i^2 - \frac{1}{2}\sum_A \nabla_A^2 - \sum_i \sum_A \frac{Z_A}{r_{iA}} + \sum_{A>B} \frac{Z_A Z_B}{r_{AB}} + \sum_{i>j} \frac{1}{r_{ij}} \right\} \Psi = E\Psi, \qquad (13.1)$$

where Ψ is the molecular wave function, i and j (A and B) denote electron (nuclear) coordinates, Z_A is the charge on nucleus A, and r_{pq} is the distance between particles p and q. The terms contained within the brackets of the left-hand side of the equation are referred to collectively as the (nonrelativistic) molecular Hamiltonian. For all but the simplest systems (*e.g.* hydrogen-like atoms) the Schrödinger equation cannot be solved exactly, and approximations must be made for systems of chemical interest. One of the first simplifications usually made to Eq. (13.1) is obtained by recognizing that the nuclei are much more massive than the electrons. Hence, one could fix the nuclear coordinates as parameters and solve only the resulting electronic Schrödinger equation,

$$\left\{ -\frac{1}{2}\sum_i \nabla_i^2 - \sum_i \sum_A \frac{Z_A}{r_{iA}} + \sum_{A>B} \frac{Z_A Z_B}{r_{AB}} + \sum_{i>j} \frac{1}{r_{ij}} \right\} \Psi_{\text{elec}} = E_{\text{elec}} \Psi_{\text{elec}}, \qquad (13.2)$$

for which the nuclear-nuclear repulsion (the third term in the electronic Hamiltonian above) is now a constant, and the nuclear kinetic energy is taken to be zero. The Born-Oppenheimer or "clamped-nucleus" approximation is essential to all molecular-level interpretations of chemical phenomena. The electronic energy, E_{elec}, is a functional of the nuclear coordinates, and is frequently referred to as the potential energy surface. The original eigenvalue problem of Eq. (13.1) has now been converted to a series of smaller problems in which the energy, E_{elec}, and the associated wave function, Ψ_{elec}, must be determined for perhaps a large number of nuclear configurations. Unfortunately, even within the Born-Oppenheimer approximation exact solutions to Eq. (13.2) are not feasible for many-electron atoms and molecules, and the search for accurate but reasonably simple approximations to this complicated differential eigenvalue problem represents the bulk of research in quantum chemistry for the last 40 years. The starting point for most models of Ψ_{elec} is the Hartree-Fock approximation, in which the motion of each electron is dictated by an average repulsive field generated by all the other electrons in the system. At the heart of this model is the assumption that Ψ_{elec} is properly described by a single many-electron function (a determinant) constructed from so-called molecular orbitals, each of which is itself a function of the coordinates of a single electron. The primary failing of this approach [often referred to as the self-consistent field (SCF)] is that it neglects the instantaneous repulsive interactions of pairs of electrons in favor of its average field model; that is, it fails to *correlate* the motions of the electrons with respect to one another. This "correlation problem," has occupied the attention of quantum chemists for decades. The focus of the present chapter is on *ab initio*[*)] as opposed to *semiempirical* quantum chemistry. Traditionally, the distinction between these two approaches has lain in the latter method's use of experimental information in the construction of SCF-like wave functions described above. More recently, however, the term "semiempirical" has been applied in a broader sense to the use of any experimental data in the parametrization of Ψ_{elec} or E_{elec}. *Ab initio* techniques, on the other hand, construct wave functions without empirical data via the electronic Schrödinger equation in Eq. (13.2). The most widely used *ab initio* methods include configuration interaction (CI), many-body perturbation theory (MBPT), coupled cluster (CC), density functional theory (DFT) (some variants of which may be considered to be semiempirical), and others. In the first half of the present chapter we examine the rich history of quantum chemistry by focusing on six methodological advances of the last three decades which have had the greatest impact on the field. In the second half of the chapter we discuss several research areas in which quantum chemistry is expected to have a significant impact in the next century.

*) Ab initio (Latin) – from the beginning.

13.2
The History and Present Status of Quantum Chemistry

Prior to the mid-1960s, quantum chemistry served as a tool for providing qualitative interpretation of chemical phenomena and wielded little power to make quantitative predictions of molecular properties and reaction outcomes. However, thanks to developments in theoretical methodology and computational algorithms as well as incredible advances in high-performance computing, the field has evolved to the point that it often provides chemical data more accurate than experiment. In this section we highlight the most important advances in quantum chemistry over the last thirty years. We have organized these topics in order of their importance/impact on the entire field. Needless to say, this ordering represents only the collective opinion of the authors and certainly should not be considered absolute.

13.2.1
The Gaussian Programs

Perhaps the single most important development that has brought computational quantum chemistry to its current level of prominence inscience occurred in 1970 with the first release of the GAUSSIAN suite of programs by John Pople and his research group at Carnegie-Mellon University. In recognition of this achievement, Pople was awarded the Nobel Prize in chemistry in 1998. Indeed, it can be fairly estimated that 90% of all quantum chemical computations carried out today are performed using the descendents of the GAUSSIAN-70 computer program. It is no exaggeration to state that GAUSSIAN is to computational quantum chemistry what IBM is to the computer industry.

Prior to GAUSSIAN-70, only two "general purpose" quantum chemical packages existed for non-linear polyatomic molecules: IBMOL and POLYATOM. The GAUSSIAN programs represented a significant step beyond these predecessors in that they provided new capabilities as well as two orders of magnitude greater efficiency. For example, the GAUSSIAN-70 package was the first to incorporate an automatic geometry optimization algorithm, a feature which is today recognized as essential for any practical chemical application. Another special feature of the GAUSSIAN-70 programs was its capacity to evaluate two-electron repulsion integrals over atomic s- and p-like orbitals built from Gaussian-type functions (hence the name of the package). Such integrals are fundamental to nearly all quantum chemical methodology, and the algorithm developed by Pople and his group was remarkably advanced for its time.

Since the initial version of GAUSSIAN, the package has been in constant development. While the most sophisticated method included in the 1970 version of the program was the Hartree-Fock or self-consistent-field approach, by 1982 the program contained a variety of finite-order many-body perturbation theory techniques, including analytic energy gradients at the second-order Møller-Plesset (MP2) level and full fourth-order (MP4) energies. Later, the GAUSSIAN-92 program was extended to include coupled cluster methods (described in the next section), and the most recent versions, GAUSSIAN-94 and GAUSSIAN-98 [5] include a variety of very popular density-functional theory (DFT) approaches (see Section 13.2.5) as well as new techniques for much larger molecular systems (see the later discussion of this topic). It has been said that while many other program packages have some specialized components that are superior to GAUSSIAN, no program has an overall quantum chemical arsenal that can compare.

13.2.2
Coupled Cluster Theory

The correlation problem alluded to in the Introduction has most often been approached through the use of determinantal wave functions, that is, by writing Ψ_{elec} as a linear combination of Hartree-Fock-like functions, [6]

$$\Psi_{elec} \approx \Psi_{CI} \equiv \hat{C}\Phi_{HF} = \sum_i C_i\Phi_i, \qquad (13.3)$$

where Φ_{HF} represents the Hartree-Fock determinant (or a similarly constructed "reference" wave function) and $\hat{C}$ an operator which produces the list of functions Φ_i from Φ_{HF}. In practice, the form of the operator $\hat{C}$ (and hence the number of Φ_i it produces) is usually limited based on physical arguments as well as the available computational resources. Until the mid-1980s, the expansion coefficients (C_i) were most commonly computed using the configuration interaction (CI) approach: the above expansion for Ψ_{elec} is inserted into the electronic Schrödinger equation and the resulting matrix eigenvalue problem solved for the energy (E_{elec}) and the approximate wavefunction (the C_i). Unfortunately, when Hartree-Fock-like determinants are chosen as the Φ_i, the expansion in Eq. (13.3) converges rather slowly, and frequently high-accuracy wave functions can be obtained for many-electron systems only with the inclusion of billions of terms in the summation. At a 1997 meeting in Cedar Key, Florida entitled "Fifty Years of the Correlation Problem," Isaiah Shavitt, one of the pioneers of quantum chemistry, stated, "if we had realized many years ago that the CI expansion converged so slowly, we may not have been so enthusiastic in our efforts to solve the problem."

However, in the late 1960s, Jiří Čížek and Josef Paldus introduced a some what different approach to the electron correlation problem: instead of using a linear expansion of functions as in Eq. (13.3) they suggested an "exponential ansatz" of the general form [7–9]

$$\Psi_{elec} \approx \Psi_{CC} \equiv e^{\hat{T}}\Phi_{HF}, \qquad (13.4)$$

which had been previously used in nuclear physics. The *cluster* operator, $\hat{T}$, produces the functions Φ_i in the same manner as the operator $\hat{C}$ in the CI approach above, but due to the presence of the exponential, the equations required to obtain the wave function coefficients are necessarily more complicated that those of the linear CI problem. The additional complexity is rewarded as the summation produced by $e^{\hat{T}}$ converges more rapidly than the expansion in Eq. (13.3). Eventually, this approach came to be known as the coupled cluster (CC) method. [10–13]

Unfortunately, the quantum chemical community was slow to accept CC theory, perhaps because the earliest researchers in the field use delegant but unfamiliar mathematical tools such as Feynman-like diagrams and second quantization to derive working equations. By the end of the 1970s, however, computer implementations of a limited form of the theory for realistic molecular systems were reported by the groups of John Pople [14] and Rodney Bartlett [15]. In 1982, Purvis and Bartlett had derived and implemented the now-popular coupled cluster singles and doubles (CCSD) method, [16] where the terms "singles" and "doubles" refer to determinants Φ_i which are related to the Hartree-Fock reference (Φ_{HF}) by the replacement of one (S) or two (D) of its component molecular orbitals. In 1989, another breakthrough was achieved by Raghavachari, Trucks, Pople, and Head-Gordon with the development of an additional correction to CCSD for certain higher-order terms using many-body perturbation theory – the CCSD(T) method. [17]

Thanks to these methodological developments, the coupled cluster approach – and especially its CCSD(T) variant – has become the most reliable yet computationally affordable method for modeling electron correlation effects in small molecules (fewer than twenty atoms) and is now available in a number of widely distributed program packages. [18–21] When implemented with large one-electron basis sets, the results from CC calculations are sometimes considered more accurate than the associated experimental data. As noted in recent performance review articles (see Table 13.1), the CCSD(T) method can be expected in many cases to accurately predict bond lengths to within a few thousandths of an Ångstrom, fundamental vibrational frequencies to within a few wavenumbers (cm^{-1}), and electric dipole moments to better than 0.1 D. This level of accuracy remains unmatched by any other affordable *ab initio* method, and discrepancies between CCSD(T) and experimental results are often attributed to errors in the latter. [12]

In the last ten years, coupled cluster theory has been extended to include excited electronic states via the equation-of-motion (EOM-CC) approach. [22–26] Just as one

Tab. 13.1 Maximum and average errors (relative to experiment) for CCSD(T) using an *spdfg*-type basis set in the determination of molecular properties for a number of small molecules containing first-row atoms and hydrogen. Data taken from Ref. [12].

Molecular Geometries		
Parameter	Max. Error	Average Error
r_{XH}	0.0150 Å	0.0021 Å
r_{XY}	0.0041 Å	0.0024 Å
<	−0.52°	−0.21°

Fundamental Vibrational Frequencies		
Parameter	Max. Error	Average Error
ν(X–H)	17.1 cm^{-1}	1.0 cm^{-1}
ν(X–Y)	−18.0 cm^{-1}	−3.2 cm^{-1}
ν(bend)	−14.0 cm^{-1}	0.5 cm^{-1}

may view the ground state coupled cluster wave function as that in Eq. (13.4), one may equivalently view the associated energy as the lowest eigenvalue of an effective Hamiltonian operator, $e^{-\hat{T}}\hat{H}e^{\hat{T}}$. By diagonalization of this operator within a specified determinantal space (usually all singles and doubles, leading to the EOM-CCSD method), one can obtain energies for excited electronic states. The EOM-CCSD approach is probably the most accurate yet affordable excited state method currently available, providing excitation energies correct to within ca. 0.2 eV. [27] Efforts are currently underway to improve upon this accuracy and to extend the method's applicability to more difficult excited states.

13.2.3
Multireference Approaches

The most common implementations of the computational approaches described thus far make use of a single-reference wave function. That is, *one* Hartree-Fock determinant is used as a starting guess, and correlation is included via linear (CI) or non-linear (e.g., CC) expansions containing excitations built from a *single* set of Hartree-Fock molecular orbitals. Unfortunately, unless an extraordinary number of expansion functions are included in the correlation treatment, these "single-reference" methods often fail when faced with situations where two or more important configurations approach degeneracy. An example of this mathematical finding which may be more palatable to chemists is the familiar notion that more than one "valence-bond" or "resonance" structure may be energetically competitive, and several might be required to qualitatively describe a molecular system. For example, one of the most intensely researched species with such a near-degeneracy (and associated single-reference inadequacies) is ozone, O_3. Its primary resonance structures, including one in which an O–O π-bond is broken in favor of a diradical structure, are depicted in Fig. 13.1. In such a case, a modestly correlated wave function constructed from a single set of Hartree-Fock orbitals may be capable of accurately describing one of these resonance structures, but may fail outright in its attempt to reflect the true overall electronic structure of the molecule. Similar problems often occur in other chemical systems, such as those containing transition metals, and in bond-breaking processes. Only methods which provide a balanced treatment of several configurations can produce even qualitatively accurate electronic wave functions. "Multireference" approaches which employ more than one Hartree-Fock-like reference determinant are currently the most effective means for describing these problematic chemical phenomena.

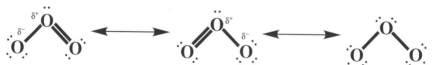

Fig. 13.1 Three important resonance structures for the ground state of the ozone molecule, O_3. The third structure represents a problematic diradical.

When describing multireference methods, it is convenient to divide the correlation energy into two components: dynamic and non-dynamic correlation. The former accounts for the instantaneous repulsion of the electrons and can be treated using coupled-cluster, configuration interaction, or density functional methods described in Sections 13.2.2 and 13.2.4 of this chapter. Non-dynamical correlation, on the other hand, accounts for the existence of near-degeneracy of low-lying electronic-configurations. In the limit that all possible configurations (determinants) are included in the configuration expansion given in Eq. (13.3) or Eq. (13.4), then the dynamical correlation will completely account for the non-dynamical correlation component. However, when the expansions are truncated, as is usually the case, accounting correctly for non-dynamical correlation effects becomes important for difficult multireference cases such as those described above.

An example of a multireference technique is the multiconfigurational SCF (MCSCF) approach, where the wave function is obtained by simultaneously optimizing both the molecular orbitals and the configuration coefficients, thereby blending the different resonance structures together. [28] Historically, the MCSCF approach has been used extensively to provide qualitatively accurate representations of surfaces; however, this method still suffers two primary drawbacks: (1) the ambiguous choice of configurations and (2) the lack of dynamical correlation.

With regard to the former, one would like to include as many important configurations as possible. Unfortunately, the definition of an "important" configuration is often debatable. One popular remedy is the full-valence "complete active space" SCF (CASSCF) approach in which configurations arising from all excitations from valence-occupied to valence-virtual orbitals are chosen. [29] Since this is equivalent to performing a full CI within the valence space, the full-valence CASSCF method is limited to small systems. Nevertheless, the CASSCF approach using a well-chosen (often chemically motivated) subspace of the valence orbitals has been shown to yield a much improved depiction of the wave function at all points on a potential surface. Furthermore, the choice of an active space can be adjusted to describe excited state wave functions.

Dynamical correlation effects in multireference procedures have most often been included through configuration interaction theory using an MCSCF wave function as a starting point. Although computationally intensive, such "multireference CI" or "MRCI" techniques are currently the most accurate yet feasible (for up to 6 atoms) means of providing curvatures of the potential energy surfaces which have played instrumental roles in dynamics (see the discussion of this topic later in the Chapter) as well as rovibrational spectroscopy. A promising technique capable of probing larger molecules employs second-order perturbation theory applied to a CASSCF wave function (CASPT2), developed by Björn Roos and his group at Lund, Sweden. Impressive CASPT2 predictions of electronic spectra of moderately sized aromatic organic compounds have already been produced [30] and provide encouragement for future studies of other larger systems which demand multireference treatments. A second approach toward applications to larger systems is the "internally-contracted MRCI" (IC-MRCI). [31] This approach also reduces the number of variables in a traditional MRCI and provides another tractable alternative approach for studying moderately sized compounds.

13.2.4
Analytic Gradient Techniques

Potential energy surfaces (PESs) govern chemistry. Any molecular-level interpretation of a chemical reaction is dependent upon the notion of such surfaces, which themselves are a consequence of the Born-Oppenheimer or "clamped-nucleus" approximation described in the Introduction. In order to determine reaction mechanisms and rates, for example, one often needs detailed knowledge of the most important regions of the PES, particularly "stationary points" (so called because there exist no forces on the nuclei) such as minimum-energy structures (global and local) and transition states for isomerization or dissociation. Such points may be identified, of course, by constructing the entire PES from repeated energy calculations at all possible nuclear configurations, but such an approach is tedious at best when the molecule contains a large number of degrees of freedom (i.e., any molecule with more than three atoms).

A much more efficient scheme for finding stationary points requires the first derivative of the electronic energy with respect to nuclear motion (i.e., the gradient along the PES). If one wishes to find a minimum-energy structure for the water molecule, for example, the energy gradient with respect to the O–H bond length would indicate whether the bonds should be stretched or compressed in order to lower the energy. Furthermore, once a stationary point has been found, the second derivative of the energy with respect to nuclear motion would characterize the given structure as a (local) minimum ora saddle point on the PES (see Fig. 13.2).

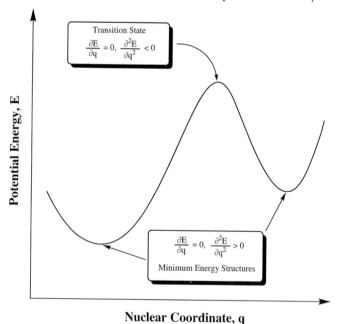

Transition State
$$\frac{\partial E}{\partial q} = 0, \quad \frac{\partial^2 E}{\partial q^2} < 0$$

$$\frac{\partial E}{\partial q} = 0, \quad \frac{\partial^2 E}{\partial q^2} > 0$$
Minimum Energy Structures

Potential Energy, E

Nuclear Coordinate, q

Fig. 3.2 A schematic of a molecular potential energy curve with three stationary points: two local minima ($\partial^2 E/\partial q^2 > 0$) and one transition state ($\partial^2 E/\partial q^2 < 0$).

The importance of analytic derivative techniques in quantum chemistry was probably first recognized by Peter Pulay in 1969 [32]. In his pioneering research, he recognized that, in accord with the well-known Wigner $2n + 1$ rule of perturbation theory, the computational cost of evaluating the analytic derivative of the Hartree-Fock energy should be similar to that of the wave function, if properly implemented. Pulay's "force" method, in which the first derivative of the energy is determined analytically and the second derivative via numerical differentiation of first derivatives, is still in use today with high-level methods such as coupled cluster. In a landmark paper in 1979 [33], John Pople and his colleagues derived analytic second derivatives for Hartree-Fock energies as well as analytic first derivatives for MP2 energies. An important breakthrough was reported in 1984 with the development of the "Z-vector" method by Handy and Schaefer, who showed that the response of the molecular orbitals in analytic gradients of correlated energies such as configuration interaction or MP2 could be included in a particularly efficient manner. [34] Throughout the 1980s, Schaefer and his group at the University of California at Berkeley extended analytic gradient techniques into the realm of configuration interaction and implemented CI energy analytic first and second derivatives [35–38] in their own BERKELEY quantum chemical program package (now superseded by PSI [18]). With the rise of coupled cluster methods, the Schaefer group also implemented analytic first derivatives for closed-shell CCSD energies. [39] Today, thanks primarily to the efforts of Rodney Bartlett and his group at the University of Florida's Quantum Theory Project, analytic energy gradients exist for both open-shell and closed-shell systems for a wide variety of coupled cluster methods including CCSD(T) [40–43].

Today, it is widely accepted that analytic energy derivatives are essential for a quantum chemical theory to be of practical use, and these techniques are available for every standard method. While the most common use of analytic gradients is indeed locating and characterizing stationary points (most often using Pulay's "force" method), derivatives serve an additional important purpose: analytic first, second, and higher-order derivatives of the energy may also be used to compute a variety of spectroscopic properties (see Table 13.2). For example, the permanent electric dipole moment may be determined as the first derivative of the energy with respect to an external electric field. Similarly, polarizabilities and hyperpolarizabilities are second- and higher-order derivatives of the energy with respect to an electric field. Harmonic vibrational frequencies are related to the second derivative of the energy with respect to nuclear motion, while the anharmonic corrections necessary to determine fundamental vibrational frequencies require third- and possibly higher-order derivatives as well. Infrared and Raman vibrational transition intensities are nuclear derivatives of the dipole moment and polarizability, respectively. The extension of derivative methods for computing NMR chemical shifts (the mixed second derivative of the energy with respect to an external magnetic field and a nuclear magnetic moment) has also seen a great deal of activity over the last three decades, including the pioneering work by Ditchfield, [44] Kutzelnigg, [45,46], Pulay, [47] and by Gauss and Stanton. [48–51] The recent work by Stanton, Gauss, and Szalay in efficient coupled cluster analytic second derivative techniques [52–

Tab. 13.2 Relationship between some spectroscopic properties and derivatives of the energy. R represents a nuclear coordinate, F an electric field, B a magnetic field, and m a nuclear magnetic moment.

$\dfrac{\partial E}{\partial R_a}$	Forces on nuclei
$\dfrac{\partial^2 E}{\partial R_a \partial R_b}$	Harmonic vibrational frequencies
$\dfrac{\partial^3 E}{\partial R_a \partial R_b \partial R_c}$	Contributions to fundamental vibrational frequencies
$\dfrac{\partial E}{\partial F_a}$	Dipole moments
$\dfrac{\partial^2 E}{\partial F_a \partial F_b}$	Polarizabilities
$\dfrac{\partial^3 E}{\partial F_a \partial F_b \partial F_c}$	Hyperpolarizabilities
$\dfrac{\partial E}{\partial B_a}$	Magnetic dipole moment
$\dfrac{\partial^2 E}{\partial B_a \partial B_b}$	Magneticsusceptibility
$\dfrac{\partial^2 E}{\partial B_a \partial m_b}$	NMR chemical shielding

54] and their application to systems as large as benzene represent another major step forward in the progress towards high-accuracy prediction of a variety of molecular properties.

13.2.5
Density-Functional Theory

Techniques such as Hartree-Fock, configuration interaction, and coupled cluster all have in common the purpose of approximating the n-electron wave function, Ψ_{elec}. Over the last decade, however, an alternative approach has found increasing success and popularity in the quantum chemical community: density-functional theory (DFT). [55,56] In ground-breaking research in the mid-1960s, Walter Kohn and his group found that all properties of the ground state of a molecular system could be completely determined from only the electron density and the fixed positions of the nuclei. In the first Hohenberg-Kohn theorem, [57] they proved the existence of a unique functional relationship between the density and the Hamiltonian of a given system. Hence, if this exact relationship were known one could (at least formally) compute all the molecular properties from the electron density (which depends on only three coordinates) and thereby avoid explicit construction of the molecular wave function (which depends on $3n$ coordinates). Such an approach may have the potential to streamline quantum chemical calculations, and the challenge of DFT is to answer the question (yet unresolved) of the exact form of the energy-density functional relationship.

A more important step towards the development of *practical* DFT implementations came in 1965, just a year after the initial work by Hohenberg and Kohn. By exploiting the techniques associated with the self-consistent-field (SCF) method within Hartree-Fock theory, Kohn and his colleague Lou Sham developed an SCF-like approach to DFT which avoided explicit construction of the problematic electron kinetic-energy component of the energy functional. The resulting Kohn-Sham density-functional theory (KS-DFT) [58] has many similarities to conventional Hartree-Fock calculations in that a set of one-electron orbitals (Kohn-Sham orbitals) are determined which minimize the DFT energy computed from the associated electron density. The difference between the Hartree-Fock and KS-DFT approaches lies in the so-called exchange-correlation (XC) density functional, which, as the name suggests, accounts for both electron correlation and non-classical exchange effects on the energy (as well as an implicit correction for the use of the approximate electron kinetic energy operator). If the exact form for the XC functional were known, the KS-DFT approach would give the exact energy. Hence, KS-DFT shifts the challenge of quantum chemistry into the construction of accurate XC functionals, but with a computational efficiency rivaling that of simple Hartree-Fock models. In recognition of his seminal contributions to DFT, Walter Kohn shared the 1998 Nobel prize in chemistry with John Pople.

Early attempts at the solution to the DFT problem include the Thomas-Fermi and Thomas-Fermi-Dirac models (both of which predate the KS-DFT formulation), $X\alpha$ (an approximation to Hartree-Fock-Slater theory), and the local density approximation (LDA), all of which rely to varying degrees on uniform and non-uniform, non-interacting electron gas models. While the $X\alpha$ and LDA models, which differ in that the former includes no correlation component, achieved moderate success for molecular systems, the level of accuracy they provide for realistic chemical problems is generally too poor to be reliable. Significant progress was made, however, during the 1980s with the development of so-called generalized gradient approximations (GGAs), which utilize corrections involving the gradient of the density in the construction of exchange and correlation functionals. Examples of GGA functionals include the Becke-88 (B88) exchange functional [59] and the Lee-Yang-Parr (LYP) [60] and Perdew-86 (P86) [61] correlation functionals. Additional progress was made in the early 1990s with the recognition that the purely localized models of the exchange and correlation functionals provided by GGAs fail to reproduce certain types of non-local behavior. In an attempt to correct this problem, hybrid Hartree-Fock/GGA exchange functionals [e.g., the Becke three-parameter exchange functional (B3) [62]] were developed which incorporate varying amounts of the exact (Hartree-Fock-like) exchange contribution to the functional. Composite exchange-correlation functionals such as B3LYP have enjoyed enormous success, particularly in predictions of molecular structure and gas-phase thermochemistry, though significant problems still remain for DFT descriptions of weakly bound systems and many types of bond-breaking processes. More recently, the DFT formalism has been extended to include electronically excited states, an area long thought to be outside the reach of practical DFT implementations. Through established linear-response theories used commonly in traditional, wave-function-based models such as Hartree-Fock and coupled cluster, time-dependent DFT (TD-DFT) methods [63,64] have been implemented which provide encouraging results for the energies of singly excited states. These new methods are also applicable to time-dependent properties such as frequency-dependent polarizabilities and hyperpolarizabilities.

Due to its computational efficiency and reasonable accuracy, DFT has clearly had a dramatic impact on quantum chemistry in the last ten years. However, it is not entirely clear what level of accuracy will be possible in the future using DFT without significant breakthroughs in functional design (e.g., inclusion of higher-order density derivatives and non-local correlation corrections). In its present form, DFT does not allow for a series of systematic improvements toward the exact solution to the electronic Schrödinger equation as offered by wave function models and is therefore sometimes referred to as a "non-convergent" method. New developments in this area have been reported recently, however, based on the design of new functionals using electron densities derived from high-accuracy methods such as coupled cluster theory. [65] Considering the vast research efforts currently dedicated to the task of improving the accuracy and reliability of density functionals, it seems likely that important progress will be reported in the next several years.

13.2.6
Integral-Direct Methods

A common characteristic of all the electronic structure methods described in this chapter is the use of one-electron basis functions, $\chi_p(\mathbf{r})$, usually based on Gaussian-type functions designed to mimic atomic orbitals of varying angular momenta (e.g., s-, p-, or d-type orbitals). Consequentially, it is necessary to evaluate complicated integrals over Hamiltonian operator elements involving these basis functions. The most problematic class of such integrals is the two-electron integral over the electron repulsion operator,

$$(pq|rs) \equiv \int \chi_p^*(\mathbf{r}_1)\chi_q(\mathbf{r}_1)\frac{1}{r_{12}}\chi_r^*(\mathbf{r}_2)\chi_s(\mathbf{r}_2)d\mathbf{r}_1 d\mathbf{r}_2. \tag{13.5}$$

As is apparent from this expression, the number of such integrals grows as N^4 with N being the number of functions, $\chi_r(\mathbf{r})$. Since the original formulations of basis-set schemes for molecular Hartree-Fock and correlated-level calculations, computer implementations have traditionally involved a two-step process in which all two-electron integrals are first computed and stored before construction of Ψ_{elec} commences. Since most methods require iterative algorithms, the latter step necessarily involves repeated retrieval of integrals from disk. For small molecular systems, this two-step scheme is quite reasonable since all the integrals can be conveniently stored on disk. For larger systems, however, the number of integrals to be stored rapidly outstrips the available storage capacity of most modern computers.

The first efforts to overcome this integral-storage bottleneck were reported in the early 1980s by Almlöf, Faergri, and Korsell, [66] who recognized that computer processor speeds were improving much more rapidly than I/O facilities. They developed a one-step Hartree-Fock procedure where the two-electron integrals were recomputed in batches in every iteration of the SCF algorithm. They referred to this algorithm as "integral-direct" and applied it to a number of metal-sandwich compounds such as ferrocene with basis sets having more than 400 functions. The magnitude of this accomplishment is perhaps better understood if one considers Almlöf's own perspective of the computer hardware used for his earlywork: "... calculations with 444 basis functions were carried out in 1982 with computer equipment that would be outperformed in all respects by average wordprocessing equipment ten years later." [67] An obvious disadvantage of direct techniques is the associated increase in CPU requirements; since the two-electron integrals must be frequently recomputed, each iteration of the wave function construction becomes more costly. However, the ever-increasing imbalance between CPU and I/O efficiencies as well as new parallel computer architectures are rapidly making direct techniques more necessary.

Since the pioneering work of Almlöf and coworkers, the quantum chemical community has developed a greater appreciation for integral-direct techniques. Today all competitive SCF, DFT, and MP2 programs commonly use such algorithms for large basis sets, and these same techniques are being extended [68] into high-level wave function theories such as coupled cluster. [69,70] The value of integral-direct methods for the future of quantum chemistry is now well understood since extensions to larger and larger systems would be impossible without them. However, in spite of the computer hardware improvements that have made these techniques practical, it is not reasonable to expect that integral-direct methods alone will be sufficient to allow rapid progress towards molecular systems with hundreds of atoms and thousands of basis functions since the $O(N^4)$ scaling described above would preclude the repeated computation of all the two-electron integrals for such large systems. [67] Much effort is currently being focused not only on generating these integrals more efficiently, but also on schemes for avoiding their evaluation wherever possible. This will be discussed in more detail in the next Section.

13.3
The Future of Quantum Chemistry

Undoubtedly the holy grail of computational quantum chemistry is the ability to determine structural, spectroscopic and dynamic properties of even large molecular systems (e.g., proteins) with high enough accuracy and speed that dangerous or time-consuming experimental analyses may be avoided whenever possible. Thanks to the developments described in the last section, much progress towards this goal has been achieved, and quantum chemistry has evolved into an indispensable tool for understanding and predicting molecular properties and chemical behavior. However, much work remains to be done before the computational tools developed thus far may be applied to problems of a more general chemical interest. In this section, we explore three research directions which are likely to see the most intense activity over the next thirty years.

13.3.1
Extensions to Large Systems

The most important obstacle hindering the progress of quantum chemistry towards its holy grail is the poor computational scaling common to all traditional quantum chemical methods. For example, all the methods described in this chapter rely on the efficient construction of two-electron integrals; hence, one could expect the minimum cost of these quantum chemical calculations to be $O(N^4)$, for N representing the number of one-electron basis functions (roughly equivalent to the size of the molecular system). At the SCF or KS-DFT level of theory, the subsequent construction of the Hartree-Fock or Kohn-Sham molecular orbitals (and the associated density) adds an additional iterative algorithm of $O(N^3)$. For the high-accuracy coupled cluster models, arguably the most reliable yet practical methods developed in the last thirty years, the scalings are even worse; the CCSD method requires the iterative solution of a set of coupled, non-linear, algebraic equations of $O(N^6)$, while CCSD(T) requires an additional, non-iterative correction of $O(N^7)$. Inpractice, these scalings imply that a factor of two increase in the size of the molecular system requires a factor of 128 increase in the cost of the calculation. For example, while a high-accuracy CCSD(T) energy calculation for a molecule of alanine requires only about five days, a comparable calculation for an alanine-alanine amino acid pair would require nearly *two years* to complete. At present, a factor of 128 improvement in computing speed is achieved approximately every ten years. Even considering the superb advances in computing power expected in the next three decades, clearly without dramatic improvements in methods and algorithms, computational quantum chemistry will be unable to make useful contributions to large systems of general chemical interest in the future.

The form of the Hamiltonian in Eq. (13.1) suggests that the addition of m particles (electrons and/or nuclei) to a given molecular system should introduce only $O(m^2)$ new interactions in the molecular wave function. Hence, the high computational scalings described above are strikingly *unphysical*. This artifactual inefficiency is a consequence of the Hartree-Fock-like molecular orbital expansions found at the heart of every many-electron theory. Since these molecular orbitals are delocalized across the entire molecule, even the interactions between electrons from spatially distant parts of the system must be included in the wave function. Fortunately, it is possible to overcome this problem by transforming the orbitals into a more localized representation in which the correlated wave function parameters describing interactions between remote electrons become negligible. [71] This approach, which was pioneered by Peter Pulay and Svein Saebø, has been applied recently to both second-order Møller-Plesset perturbation theory (MP2) [72–75] and to CCSD [76]. The former method, which scales as $O(N^5)$ when implemented with delocalized orbitals, scales as only $O(N^2)$ with localized orbitals. Preliminary studies of CCSD have reduced its scaling to $O(N^4)$ in a localized basis relative to an $O(N^6)$ scaling in the delocalized basis. Unfortunately, further extension of these local correlation techniques to CCSD(T) is problematic because the (T) energy correction may be computed in a non-iterative fashion only in the delocalized basis (a common problem for perturbation-based theories). As a result, even if a localized-orbital (iterative) algorithm for CCSD(T) could reduce the computational scaling to $O(N^5)$, it would so dramatically increase the data storage requirements (perhaps to the same order) that the method would still be applicable to only small molecules. There are a number of possibilities for avoiding these problems, including clever Laplace transform techniques advocated by Almlöf [77] and redundant localized orbital basis sets advocated by Head-Gordon and coworkers, [78] both of which retain the storage advantages of the non-iterative (T) expression.

While localized-orbital methods may eventually succeed in achieving near-linear scaling for the construction of correlated wave functions, the evaluation of the two-electron repulsion integrals still presents an obstacle for all quantum chemical techniques. One can begin to develop strategies for solving this problem, however, by recognizing that each integral of Eq. (13.5) can be considered an overlap between two charge distributions; when this overlap is very small, the integral is essentially zero. Hence, while Eq. (13.5) scales *formally* as $O(N^4)$, the number of *non-negligible* two-electron repulsion integrals actually scales as only $O(N^2)$. In recognition of this fact, much effort has been dedicated in the past several years not only to improved techniques for efficiently computing two-electron integrals, but also to schemes for avoiding their evaluation if possible. In the first category, attention has been focused on techniques for evaluating specific classes of two-electron integrals, especially coulomb interaction integrals which appear prominently in Hartree-Fock and density-functional methods. The most successful algorithms rely on polynomial (multipole) expansions of the $\frac{1}{r_{12}}$ component of the integral. Sophisticated methods for efficient evaluation of these expansions and associated partitioning of the density into interacting components have been developed and applied, including the recursive bisection (RBM) [79,80] and fast multipole methods(FMM). [81–84] For the most efficient implementations, the resulting RBM or FMM algorithms scale linearly $O(N)$ or nearly linearly $O(N\log N)$. It is likely that a coupling of local correlation and linear-scaling multipole techniques will form the foundation of the most efficient future quantum chemical implementations for larger problems.

A vital question which remains unanswered, however, is precisely what sort of "larger problems" quantum chemical research should address. Historically the field has primarily dealt with problems associated with molecular spectroscopy, the methylene problem being a classic example. [4] As a result, the driving force behind nearly all methodological development in quantum chemistry has been applications involving small molecules (fewer than twenty atoms) of rather limited interest to other areas of chemistry or biochemistry. We are currently enjoying a period of symbiotic collaboration between experimentalists and quantum chemists where the fun-

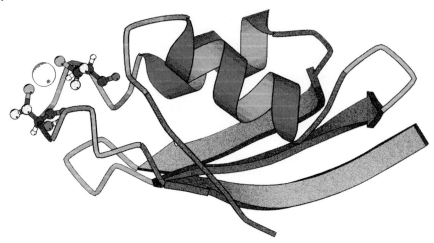

Fig. 13.3 NMR structure of the mercury-bound form of merP, a 72-residue protein from the bacterial mercury detoxification system. The merP protein resides between the inner and outer membranes of the cell in a region known as the periplasm. Mercury(II) entering the cell is bound by merP in the periplasm, passed to the transport protein (merT) which moves it through the inner membrane where it is reduced by mercury reductase (merA) and finally released from the cell as Hg(0) vapor. NMR structures suggest that differences between the mercury-bound and free structures are localized to the metal binding loop where mercury is bicoordinated with sulfur from a pair of cysteine side chains (shown in detail). NMR structures from Ref. [108].

damental questions are often well understood by both groups in terms of a common language. In such spectroscopic problems, quantum chemists are often concerned with locating the most important minimum-energy or transition-state structures on the PES, as well as their dissociation energies, vibrational frequencies, dipole moments, polarizabilities, hyperfine splittings, etc. However, as chemical systems increase in size (see, for example, Fig. 13.3), the nature of the experiments as well as the relevant spectroscopic questions to be addressed by theory shift dramatically as few quantitative properties other than structure are sought. Quantum chemists thirty years from now will require an increased background in experimental areas such as protein crystallography, NMR, molecular biology, surface science, and solvation modeling. In addition, it is likely that other, less accurate theoretical areas such as semiempirical methods and molecular mechanics, both of which have shown tremendous utility thus far in homology modeling and molecular dynamics of proteins, will continue to play key roles in studies of macromolecular structure via hybrid molecular mechanics/quantum mechanics (MM-QM) techniques where active sites are treated quantum mechanically while less important regions are approximated by classical mechanics. [85] As quantum chemists better understand the relevant questions in biology, polymer science, and materials science and find their role in providing appropriate answers, large scale methods will rapidly move beyond their current role as showpieces of the field. Quantum chemists will not only study systems such as the mercury bound form of merP (Fig. 13.3); they will make significant contributions to the underlying science.

13.3.2
Pursuit of Spectroscopic Accuracy

It is ironic that the standard methods of *ab initio* quantum chemistry were traditionally aimed at very simple molecular systems for which the accuracy of available experimental data was too high to be rivaled by theory. Today, advances in theoretical methodology and computer hardware have enabled quantum chemists to study larger systems for which even "chemical accuracy" (~ 1 kcal mol^{-1}) is unnecessary since the corresponding experimental data are often much less accurate. Recently, however, a new generation of techniques has emerged which can finally challenge the accuracy of spectroscopic observations for small molecules (~ 1 cm^{-1}).

The accuracy/cost ratio of conventional multideterminantal methods depends on two factors. The first, the correlation or "*n*-particle" factor (discussed earlier), can be successfully reduced via the rapidly converging coupled-cluster ansatz [see Eq. (13.4)]. The second, the "one-particle" factor, is the primary reason why predictions with spectroscopic accuracy using conventional expansion methods are hardly feasible even for the helium

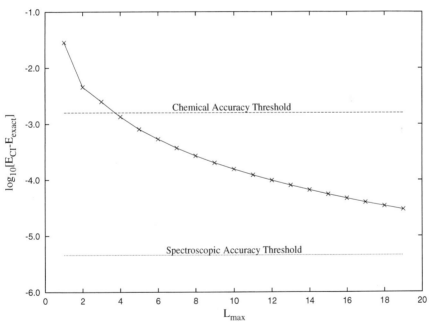

Fig. 13.4 Logarithm of error(E_h) in the configuration interaction energy for the ground state of the helium atom as a function of maximum orbital quantum number, L, of the one-electron basis functions. The data were obtained in an exploratory study of the ground-state He atom problem (E. F. Valeev, W. D. Allen, and H. F. Schaefer, unpublished). Computations were performed as outlined in Ref. [109].

atom. The one-particle factor comes from the fact that most approximate solutions to the molecular Schrödinger equation are given in terms of the *finite* number of one-electron functions of varying orbital angular momentum quantum number L. For example, computing the energy of electron correlation in the ground-state helium atom within 1 kcal mol^{-1} of its complete basis set limit (i.e., the energy computed with a basis set of a necessarily *infinite* size) using conventional multideterminantal methods requires a relatively small basis set with at most f-type functions ($L = 3$) [86] and takes only a few milliseconds to run on a modern workstation (see Fig. 13.4). An analogous attempt to compute the energy with a 1 cm^{-1} accuracy (roughly three orders of magnitude more accurate than the previous example) is far from feasible as functions with $L \simeq 30$ would be needed [86] and would require investments into specialized multiple-precision arithmetic software. Hence, such "brute force" approaches have never been seriously considered.

One obvious way to avoid such a slow convergence rate is to correct energy values computed with smaller basis sets for the basis set incompleteness. This approach is common in atomic calculations, where analytic expressions for asymptotic convergence rates of energies can be derived analytically, [87] and thus a rigorous energy extrapolation is possible. For molecules, a number of approaches have been suggested. The two most widely known schemes are the complete basis set (CBS) family of approaches due to Petersson [88] and coworkers and Gaussian-N (GN) model chemistries developed by Pople and coworkers. [89] Both schemes are computationally feasible for small molecules and have become the *de facto* standard as implemented in the GAUSSIAN suite of programs. Another group of approaches uses asymptotic expressions for convergence rates of molecular electronic energies similar to those used in the atomic case. A carefully constructed series of basis sets is an obvious requirement for such a scheme. Most recent models use the "correlation-consistent" families of basis sets developed by Dunning and coworkers [90] at Pacific Northwest National Laboratories or the atomic natural orbital (ANO) basis sets pioneered by Almlöf and Taylor. [91] These approaches are necessarily more expensive, but normally prove superior to various CBS and GN model chemistries due to their stronger physical motivation.

An obvious problem with these some what empirical approaches is that they are biased toward predictions of energy and therefore do not improve the wave function and electron density. Nevertheless, these types of methods can provide total energies and energy differences of chemical and even spectroscopic quality, and most likely will account for the majority of highly accurate computations during the next 10–20 years.

To improve on the wave function one has to accept that the standard multideterminantal expansion [Eq. (13.3)] is unsuitable for near-exact but practical approximations to the electronic wavefunction. The problem is clear from a simple analysis of the electronic Hamiltonian in Eq. (13.2): singularities in the Coulomb potential at the electron coalescence points necessarily lead to irregularities in first and higher derivatives of the exact wave function with respect to the interparticle coordinate, r_{12}. The mathematical consequences of Coulomb singularities are known as electron-electron (correlation) and electron-nuclear cusp conditions and were derived by

Tosio Kato in 1957. [92] Unfortunately, any trial wave function composed of Slater determinants has smooth first and higher derivatives with respect to the interelectronic coordinates. Thus, even though such expansions are insightful and preserve the concept of orbitals to some extent, from a mathematical point of view they are expected to be slowly convergent.

Fortunately, a now-promising alternative to multideterminantal expansions has been known for more than 70 years. Hylleraas attacked the notorious ground-state helium problem with expansions including terms explicitly dependent on the interelectronic coordinate r_{12}. His three-term trial wave function including a linear r_{12} term variationally optimized by the "paper and pencil" method yields an energy within ~ 0.6 kcal mol^{-1} of the exact non-relativistic value! [93] In contrast, a ~ 30 term conventional configuration interaction expansion (hardly feasible without a computer) is needed to achieve this accuracy. Even higher accuracy is not a problem with the Hylleraas method; the most recent value for the non-relativistic energy of the ground state He atom obtained using a Hylleraas-type expansion is accurate to within $\sim 10^{-12}$ kcal mol^{-1}. [94] The reason for the fast convergence of such expansions is that terms linear in the interelectronic coordinate have a proper cusp behavior near the electron-electron coalescence point and thus approximate the exact wave function in that region very efficiently.

Unfortunately, extending Hylleraas's approach to systems containing three or more electrons leads to very cumbersome mathematics. More practical approaches, known as explicitly correlated methods, are classified into two categories. The first group of approaches uses Boys' Gaussian-type geminal (GTG) functions with the explicit dependence on the interelectronic coordinate built into the exponent: [95]

$$\varphi(i,j) = \exp(-\alpha r_{iA}^2 - \beta r_{jB}^2 - \gamma r_{ij}^2) \tag{13.6}$$

While some of the resulting integrals are not nearly as troublesome as in a Hylleraas-type configuration interaction expansion, a number of technical problems remain to be solved to apply the method to general molecular systems. Note that a Gaussian-type geminal looks just like a product of two Gaussian-type functions enhanced by a correlation factor of the form $\exp(-\gamma r_{ij}^2)$; hence expansions in terms of GTGs do not satisfy Kato's electron-electron cusp condition rigorously. Another group of high accuracy techniques explicitly include terms which are linear in the interelectronic coordinate, following Werner Kutzelnigg's idea that such terms take care of the correlation cusp problem and improve basis set convergence of the expansion. [96] In these "linear R12" methods, potential difficulties with computing many-electron integrals are avoided through a clever application of the resolution of the identity. However, the drawback of this technique is that very large basis sets must be used in order to obtain high quality results. Despite apparently fundamental differences between these two groups of explicitly correlated methods, very impressive results have been obtained for small molecular systems, and even molecules as large as ferrocene have been treated with the linear R12 version of second-order Møller-Plesset perturbation theory.

Truly spectroscopic accuracy for total energies is still difficult to obtain from either explicitly correlated methods or extrapolation approaches for all but the simplest molecular systems, although various energy differences (reaction enthalpies and barriers, ionization potentials, etc.) can be predicted very accurately. High accuracy also means greater attention to effects of special relativity, core correlation, non-adiabaticity, etc., which are often ignored. [97] In the next 30 years we should expect this area to become a part of mainstream quantum chemistry. We do not believe that extrapolation techniques will become obsolete, although explicitly correlated methods as a more complete solution to the problem should enjoy wider applicability in the future, especially in computing highly accurate molecular properties and potential energy surfaces.

13.3.3
Potential Energy Surfaces for Reaction Dynamics

As noted in the Introduction, the electronic Schrödinger equation [Eq. (13.2)] provides the total energy of a molecule while taking the nuclear-nuclear repulsion energy to be constant and neglecting the nuclear kinetic energy. Quantum chemists commonly use this equation to compute equilibrium geometries or relative energies of different geometrical structures. Although in the past computations of vibrational frequencies (often for direct comparison to experiment) rarely extended beyond the simple harmonic oscillator model, quantum chemists are more and more frequently computing spectroscopic properties more directly related to nuclear motion such as fundamental vibrational frequencies or rovibrational sub-bands of electronic transitions. However, there is another vast and fairly distinct research field known as reaction dynamics, wherein theoreticians and experimentalists attempt to understand and to model the mechanisms and rates of reactions at a fundamental molecular level. The potential energy surface (PES) is the convenient nexus between the solution of the Born-Oppenheimer electronic Schrödinger equation and the field of reaction dynamics.

The PES is the electronic energy of a system given as a function of the position of all of the nuclei. A complete potential energy surface is the ideal starting point for any type of accurate dynamical computation. Eugene Wigner noted in 1937, "…every calculation of reaction rates is bound to founder unless we have more detailed energy surfaces than are available at present." [98] Unfortunately, such detailed surfaces were slow in coming to the dynamicists. Thirty-six years later, Nobel Prize winner John Polanyi described the role that accurately computed surfaces had played in the attempt to model chemical reactions by stating, "I can report that *ab initio* surfaces have not figured significantly, and I am surprised." [99] It must be noted that this situation has changed significantly since 1986, a magnificent example being the theoretical work of Werner, Manolopoulos, and coworkers on the $F+H_2$ reaction. [100]

Complete potential energy surfaces have been developed for a few very simple systems. However, they will remain scarce in the future for two primary reasons. First, the geometry of an N-atom molecule is described by $3N-6$ internal coordinates. If the energy of 10 different values along each of these internal coordinates is sufficient to describe the surface, then 10^{3N-6} energy points still must be computed. [101] For benzene, where $N = 12$, this is 10^{30} energy points! Even with the capability to compute first and second derivatives of the energy, exhaustively characterizing such large-dimensionality spaces in a reasonable amount of time will remain generally impossible. Second, for a given chemical system, it is usually extremely difficult to find a computationally practical theoretical method which will accurately treat all portions of a given molecule's PES. For example, Hartree-Fock wave functions (as well as methods based upon them) are often quite poor in regions of the PES corresponding to transition states and to the breaking of bonds, i.e., regions crucial to the dynamics of a reaction. The accurate computation of the properties at only an equilibrium geometry is far more easily accomplished. In fact, quantum chemists often test a theoretical method by computing the energy of the water molecule with its bond lengths stretched to twice their equilibrium values. Ideally, a quantum chemical method is used which parallels the exact PES within the one-electron basis set, or the PES given by the full-configuration interaction method.

Dynamicists have fortunately discovered that most of the global PES is unnecessary for quantitatively accurate computations of reaction rates. In fact, many recent dynamical studies of chemical reactions have been based on very limited information such as the transition state structures and barrier heights on the relevant PES. An example of the reduced requirements by dynamicists was the introduction of the reaction path Hamiltonian by Miller, Handy, and Adams [101] in 1980 which struck a middle ground, requiring a moderate number of computations on the PES. Various dynamical treatments based on the reaction path Hamiltonian model are possible, [102] and the necessary computations are reasonable for many small-molecule reactions. The one-dimensional reaction path is determined by following the steepest-descent paths from a transition state down to energy minima to determine the energy, geometry, and forces along the reaction path. The harmonic force field must also be computed along the reaction path, preferably via analytic second derivatives or analytic first derivatives using finite-displacement techniques. The computed values may be fit to appropriate functional forms, so that values obtained from the parametrized functions are immediately available to the dynamicists. The parametrized force field may be used to compute quantities related to derivatives with respect to the position on the reaction path, such as the curvature coupling matrix which "describes energy transfer between the reaction coordinate and the transverse vibrational modes." [102] Improvements in quantum chemical methodology and computer proficiency will make approaches based on the reaction path Hamiltonian far more feasible for a larger number of small-molecule reactions in the future.

Experiments have also played a critical role in the development of potential energy surfaces and reaction dynamics. In the earliest days of quantum chemistry, experimentally determined thermal rate constants were available to test and improve dynamical theories. Much more detailed information can now be obtained by experimental measurement. Today experimentalists routinely use molecular beam and laser techniques to examine how reaction cross-sections depend upon collision energies, the states of the reactants and products, and scattering angles.

The confluence of improved experimental, dynamical and quantum chemical techniques are making possible the quantitative testing of dynamical rate theories. The ketene molecule (CH_2CO) is a superb example. First, the dissociation of singlet ketene to $^1CH_2 + CO$ is endoergic but barrierless, i.e., there is no potential energy maximum on the reaction path. The quantitative accounting of experimental $k(E)$ measurements [103] for the fragmentation of singlet ketene has already been achieved by variational (Rice-Ramsperger-Kassel-Marcus) RRKM theory. [104] Second, high-resolution measurements of $k(E)$ have also been obtained for the isomerization of ketene, which exchanges its two C atoms via an oxirene intermediate. [105] Interesting step structure observed in $k(E)$ was explained in terms of resonance tunneling via metastable states of oxirene. Using previous computations of relevant parts of the PES [106], a cumulative

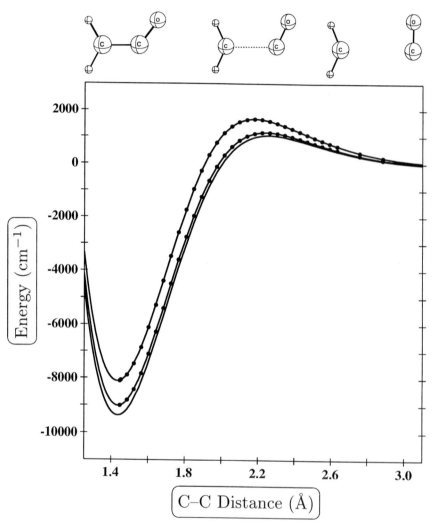

Fig. 13.5 *Ab initio;* potential energy curves for the $C_s^{//}$ (planar) fragmentation of $\tilde{a}\,^3A''$ ketene, represented in terms of the C–C distance along the path. Top: TZ(2d1f,2p) CCSD; middle: cc-pVQZ/TZ UCCSD(T); bottom: cc-pVQZ/TZ UCCSD(T) + cc-pV5Z MP2 basis set and cc-pVDZ UCCSDT correlation corrections. See Ref. [104] for further details.

reaction probability calculation was performed [107] which seems to support the explanation of the structure in $k(E)$, although a more accurate PES is necessary to be able to precisely match up the observed and predicted structures.

Finally, the interpretation of high-resolution experimental measurements [108] of $k(E)$ for the dissociation of triplet ketene over its exit-channel barrier to $^3CH_2 + CO$ has been challenged. The experimentally observed step structure in $k(E)$ was interpreted as confirmation of the RRKM microcanonical form of transition state theory, which predicts that the dissociation rate should increase in a quantized manner as the energetic threshold for each new vibrational level in the transition state is reached. A detailed *ab initio* characterization of the triplet ketene fragmentation surface [109] (see Fig. 13.5), along with quantum reactive scattering computations of the cumulative reaction probability [110], has called this interpretation into question. The efforts to compute the relevant portions of the PES and to reproduce the experimental results by dynamical computation will expose either a false interpretation of the experimental data, a failure in the PES computation, or a failure in the dynamicalt reatment. At this point, it appears possible that the step structure in $k(E)$ for triplet ketene fragmentation is instead due to nonadiabatic surface-hopping dynamics [111] involved in the experimental preparation of triplet ketene.

The discussion of potential energy surfaces thus far has implicitly assumed that gas-phase reactions are in focus. For condensed-phase reaction dynamics, where thermal fluctuations have a significant and intrinsic role, the situation is much more unsettled. The reader is referred to the list of challenges to condensed-phase electronic structure theory recently made by Truhlar, who asks if condensed-phase electronic structure may be "... not only waiting for its Hylleraas, but even waiting for its Schrödinger?" [112] The development of liquid-phase dynamics will surely continue to be an intense area of research through the foreseeable future.

Another major, future advance in the quantum chemical computation of potential energy surfaces for reaction dynamics will be the ability to routinely compute the energies of molecular systems "on the fly". The tedious and time-consuming process of fitting computed quantum chemical values to functional forms could be avoided if it were possible to compute the PES as needed *during* a classical trajectory or quantum dynamics calculation. For many chemical reactions, it should be practical in the near future to prudently select a sufficiently rapid and accurate electronic structure method to facilitate dynamics computations on the fly.

13.4
Conclusions

The field of quantum chemistry has seen tremendous development over the last thirty years. Thanks to high-accuracy models such as coupled-cluster theory and standardized, widely available program packages such as GAUSSIAN 98, what was once merely an esoteric tool of a few specialists has evolved into an indispensable source of knowledge for both the prediction and the interpretation of chemical phenomena. With the development of reduced scaling algorithms for coupled cluster and the rapidly evolving density-functional theory, computational quantum chemistry has the potential to progress even closer to the era described by Professor Robert S. Mulliken in his visionary acceptance speech for the 1966 Nobel Prize in chemistry:

"… I would like to emphasize strongly my belief that the era of computing chemists, when hundreds if not thousands of chemists will go to the computing machine instead of the laboratory for increasingly many facets of chemical information, is already at hand."

13.5
Acknowledgements

This work was supported by the National Science Foundation (grant no. CHE-9815397) and the Department of Energy (grant no. FG02-97-ER14748). The authors thank Meredith Anderson of the University of Georgia for her assistance regarding Fig. 13.3 and the associated discussion of merP, as well as Prof. John F. Stanton of the University of Texas and Drs. Yukio Yamaguchi and Wesley D. Allen of the University of Georgia for their essential role in the research reviewed here.

13.6
Appendix: Nomenclature

A number of non-SI units are commonly used in the literature of quantum chemistry. We therefore provide appropriate conversion factors for certain specialized units used in this chapter:

Unit	SI Value
Wavenumber (cm^{-1})	1.196×10^{-2} kJ/mol
kcal/mol	4.184 kJ/mol
Electron volt (eV)	96.48 kJ/mol
Hartree (E_h)	2625 kJ/mol
Ångstrom (Å)	10^{-10} m
Debye (D)	3.336×10^{-6} C · m

References

1 RICHARDS, W. G. *Nature*, **1979**, *278*, 507.

2 SCHAEFER, H. F. *Science*, **1986**, *231*, 1100.

3 SCHAEFER, H. F. *Chimia*, **1989**, *43*, 1.

4 SCHAEFER, H. F. *J. Mol. Struct. (Theochem)*, **1997**, *398*, 199.

5 FRISCH, M. J.; TRUCKS, G. W.; SCHLEGEL, H. B.; SCUSERIA, G. E.; ROBB, M. A.; CHEESEMAN, J. R.; ZAKRZEWSKI, V. G.; MONTGOMERY, J. A.; STRATMANN, R. E.; BURANT, J. C.; DAPPRICH, S.; MILLAM, J. M.; DANIELS, A. D.; KUDIN, K. N.; STRAIN, M. C.; FARKAS, O.; TOMASI, J.; BARONE, V.; COSSI, M.; CAMMI, R.; MENNUCCI, B.; POMELLI, C.; ADAMO, C.; CLIFFORD, S.; OCHTERSKI, J.; PETERSSON, G. A.; AYALA, P. Y.; CUI, Q.; MOROKUMA, K.; MALICK, D. K.; RABUCK, A. D.; RAGHAVACHARI, K.; FORESMAN, J. B.; CIOSLOWSKI, J.; ORTIZ, J. V.; STEFANOV, B. B.; LIU, G.; LIASHENKO, A.; PISKORZ, P.; KOMAROMI, I.; GOMPERTS, R.; MARTIN, R. L.; FOX, D. J.; KEITH, T.; AL-LAHAM, M. A.; PENG, C. Y.; NANAYAKKARA, A.; GONZALEZ, C.; CHALLACOMBE, M.; GILL, P. M. W.; JOHNSON, B. G.; CHEN, W.; WONG, M. W.; ANDRES, J. L.; HEAD-GORDON, M.; REPLOGLE, E. S.; POPLE, J. A. Gaussian -98. Gaussian, Inc., Pittsburg, PA, U. S. A., 1998.

6 BARTLETT, R. J.; STANTON, J. F. in *Reviews in Computational Chemistry*, Lipkowitz, K. B.; Boyd, D. B., Eds., volume 5, chapter 2, pp 65–169. VCH Publishers, New York, 1994.

7 ČÍŽEK, J. *J. Chem. Phys.*, **1966**, *45*, 4256.

8 ČÍŽEK, J. *Adv. Chem. Phys.*, **1969**, *14*, 35.

9 ČÍŽEK, J.; PALDUS, J. *Int. J. Quantum Chem.*, **1971**, *5*, 359.

10 BARTLETT, R. J. *Annu. Rev. Phys. Chem.*, **1981**, *32*, 359.

11 BARTLETT, R. J. in *Modern Electronic Structure Theory*, YARKONY, D. R., Ed., volume 2 of *Advanced Series in Physical Chemistry*, chapter 16, pp 1047–1131. World Scientific, Singapore, 1995.

12 LEE, T. J.; SCUSERIA, G. E. in *Quantum Mechanical Electronic Structure Calculations with Chemical Accuracy*, LANGHOFF, S. R., Ed., pp 47–108. Kluwer Academic Publishers, Dordrecht, 1995.

13 CRAWFORD, T. D.; SCHAEFER, H. F. in *Reviews in Computational Chemistry*, LIPKOWITZ, K. B.; BOYD, D. B., Eds., volume 14, chapter 2, pp 33–136. VCH Publishers, New York, 2000.

14 POPLE, J. A.; KRISHNAN, R.; SCHLEGEL, H. B.; BINKLEY, J. S. *Int. J. Quantum Chem. Symp.*, **1978**, *14*, 545.

15 BARTLETT, R. J.; PURVIS, G. D. *Int. J. Quantum Chem.*, **1978**, *14*, 561.

16 PURVIS, G. D.; BARTLETT, R. J. *J. Chem. Phys.*, **1982**, *76*, 1910.

17 RAGHAVACHARI, K.; TRUCKS, G. W.; POPLE, J. A.; HEAD-GORDON, M. *Chem. Phys. Lett.*, **1989**, *157*, 479.

18 JANSSEN, C. L.; SEIDL, E. T.; SCUSERIA, G. E.; HAMILTON, T. P.; YAMAGUCHI, Y.; REMINGTON, R. B.; XIE, Y.; VACEK, G.; SHERRILL, C. D.; CRAWFORD, T. D.; FERMANN, J. T.; ALLEN, W. D.; BROOKS, B. R.; FITZGERALD, G. B.; FOX, D. J.; GAW, J. F.; HANDY, N. C.; LAIDIG, W. D.; LEE, T. J.; PITZER, R. M.; RICE, J. E.; SAXE, P.; SCHEINER, A. C.; SCHAEFER, H. F. PSI 2.0.8. PSITECH, Inc., Watkinsville, GA 30677, U. S. A., 1995. This program is generally available without charge.

19 STANTON, J. F.; GAUSS, J.; WATTS, J. D.; LAU-
DERDALE, W. J.; BARTLETT, R. J. ACES II, 1993.
The package also contains modified versions
of the MOLECULE Gaussian integral program of J. ALMLÖF and P. R. TAYLOR, the ABA-
CUS integral derivative program written by T.
U. HELGAKER, H. J. AA. JENSEN, P. JØRGENSEN
and P. R. TAYLOR, and the PROPS property
evaluation integral code of P. R. TAYLOR.

20 FRISCH, M. J.; TRUCKS, G. W.; SCHLEGEL,
H. B.; GILL, P. M. W.; JOHNSON, B. G.; ROBB,
M. A.; CHEESEMAN, J. R.; KEITH, T.; PETERS-
SON, G. A.; MONTGOMERY, J. A.; RAGHAVA-
CHARI, K.; AL-LAHAM, M. A.; ZAKRZEWSKI,
V. G.; ORTIZ, J. V.; FORESMAN, J. B.; CIO-
SLOWSKI, J.; STEFANOV, B. B.; NANAYAKKARA,
A.; CHALLACOMBE, M.; PENG, C. Y.; AYALA,
P. Y.; CHEN, W.; WONG, M. W.; ANDRES, J. L.;
REPLOGLE, E. S.; GOMPERTS, R.; MARTIN, R. L.;
FOX, D. J.; BINKLEY, J. S.; DEFREES, D. J.;
BAKER, J.; STEWART, J. P.; HEAD-GORDON, M.;
GONZALEZ, C.; POPLE, J. A. Gaussian 94, Revi-
sion C.3. Gaussian, Inc., Pittsburg, PA, U. S.
A., 1995.

21 WERNER, H.-J.; KNOWLES, P. J.; AMOS, R. D.;
BERNING, A.; COOPER, D. L.; DEEGAN, M.
J. O.; DOBBYN, A. J.; ECKERT, F.; HAMPEL, C.;
LEININGER, T.; LINDH, R.; LLOYD, A. W.;
MEYER, W.; MURA, M. E.; NICKLASS, A.; PAL-
MIERI, P.; PETERSON, K.; PITZER, R.; PULAY, P.;
RAUHUT, G.; SCHÜTZ, M.; STOLL, H.; STONE,
A. J.; THORSTEINSSON, T. Molpro, 1995.

22 MONKHORST, H. J. *Int. J. Quantum Chem.
Symp.*, **1977**, *11*, 421.

23 MUKHERJEE, D.; MUKHERJEE, P. K. *Chem.
Phys.*, **1979**, *39*, 325.

24 EMRICH, K. *Nucl. Phys. A*, **1981**, *351*, 379.

25 SEKINO, H.; BARTLETT, R. J. *Int. J. Quantum
Chem. Symp.*, **1984**, *18*, 255.

26 STANTON, J. F.; BARTLETT, R. J. *J. Chem. Phys.*,
1993, *98*, 7029.

27 STANTON, J. F.; GAUSS, J.; ISHIKAWA, N.;
HEAD-GORDON, M. *J. Chem. Phys.*, **1995**, *103*,
4160.

28 ROOS, B. O. in *Ab Initio Methods in Quantum
Chemistry–II*, LAWLEY, K. P., Ed., volume 69 of
Advances in Chemical Physics, pp 399–445.
Wiley, New York, 1987.

29 ROOS, B. O.; TAYLOR, P. R.; SIEGBAHN, P. E. M.
Chem. Phys., **1980**, *48*, 157.

30 ANDERSSON, K.; ROOS, B. O. in *Modern Electronic Structure Theory*, YARKONY, D. R., Ed.,
volume 2 of *Advanced Series in Physical Chemistry*, pp 55–109. World Scientific, Singapore,
1995.

31 WERNER, H.-J.; KNOWLES, P. J. *J. Chem. Phys.*,
1988, *88*, 5803.

32 PULAY, P. *Mol. Phys.*, **1969**, *17*, 197.

33 POPLE, J. A.; KRISHNAN, R.; SCHLEGEL, H. B.;
BINKLEY, J. S. *Int. J. Quantum Chem. Symp.*,
1979, *13*, 255.

34 HANDY, N. C.; SCHAEFER, H. F. *J. Chem. Phys.*,
1984, *81*, 5031.

35 BROOKS, B. R.; LAIDIG, W. D.; SAXE, P.; GOD-
DARD, J. D.; YAMAGUCHI, Y.; SCHAEFER, H. F.
J. Chem. Phys., **1980**, *72*, 4652.

36 OSAMURA, Y.; YAMAGUCHI, Y.; SCHAEFER,
H. F. *J. Chem. Phys.*, **1981**, *75*, 2919.

37 LEE, T. J.; HANDY, N. C.; RICE, J. E.; SCHEINER,
A. C.; SCHAEFER, H. F. *J. Chem. Phys.*, **1986**,
85, 3930.

38 YAMAGUCHI, Y.; OSAMURA, Y.; GODDARD, J. D.;
SCHAEFER, H. F. *A New Dimension to Quan-
tum Chemistry: Analytic Derivative Methods in
Ab Initio Molecular Electronic Structure Theory.*
Number 29 in International Series of Mono-
graphs on Chemistry. Oxford Univ. Press,
New York, 1994.

39 SCHEINER, A. C.; SCUSERIA, G. E.; RICE, J. E.;
LEE, T. J.; SCHAEFER, H. F. *J. Chem. Phys.*,
1987, *87*, 5361.

40 ADAMOWICZ, L.; LAIDIG, W. D.; BARTLETT, R. J.
Int. J. Quantum Chem. Symp., **1984**, *18*, 245.

41 SALTER, E. A.; TRUCKS, G. W.; BARTLETT, R. J. *J.
Chem. Phys.*, **1989**, *90*, 1752.

42 GAUSS, J.; STANTON, J. F.; BARTLETT, R. J. *J.
Chem. Phys.*, **1991**, *95*, 2623.

43 GAUSS, J.; LAUDERDALE, W. J.; STANTON, J. F.;
WATTS, J. D.; BARTLETT, R. J. *Chem. Phys. Lett.*,
1991, *182*, 207.

44 DITCHFIELD, R. *Mol. Phys.*, **1974**, *27*, 789.

45 KUTZELNIGG, W. *Israel J. Chem.*, **1980**, *19*,
193.

46 SCHINDLER, M.; KUTZELNIGG, W. *J. Chem.
Phys.*, **1982**, *76*, 1919.

47 WOLINSKI, K.; HINTON, J. F.; PULAY, P. *J. Am.
Chem. Soc.*, **1990**, *112*, 8251.

48 GAUSS, J. *Chem. Phys. Lett.*, **1992**, *191*, 614.

49 GAUSS, J. *J. Chem. Phys.*, **1993**, *99*, 3629.

50 GAUSS, J. *Chem. Phys. Lett.*, **1994**, *229*, 198.

51 GAUSS, J.; STANTON, J. F. *J. Chem. Phys.*, **1995**,
103, 3561.

52 STANTON, J. F.; GAUSS, J. in *Recent Advances in Coupled-Cluster Methods*, BARTLETT, R. J., Ed., pp 49–79. World Scientific Publishing, Singapore, 1997.

53 GAUSS, J.; STANTON, J. F. *Chem. Phys. Lett.*, **1997**, *276*, 70.

54 SZALAY, P. G.; GAUSS, J.; STANTON, J. F. *Theor. Chim. Acta*, **1998**, *100*, 5.

55 PARR, R. G.; YANG, W. *Density-Functional Theory of Atoms and Molecules*. Oxford University, New York, 1989.

56 KOHN, W.; BECKE, A. D.; PARR, R. G. *J. Phys. Chem.*, **1996**, *100*, 12974.

57 HOHENBERG, P.; KOHN, W. *Phys. Rev.*, **1964**, *136*, B864.

58 KOHN, W.; SHAM, L. J. *Phys. Rev. A*, **1965**, *140*, 1133.

59 BECKE, A. D. *Phys. Rev. A.*, **1988**, *38*, 3098.

60 LEE, C.; YANG, W.; PARR, R. G. *Phys. Rev. B.*, **1988**, *37*, 785.

61 PERDEW, J. P. *Phys. Rev. B.*, **1986**, *33*, 8822.

62 BECKE, A. D. *J. Chem. Phys.*, **1993**, *98*, 5648.

63 JAMORSKI, C.; CASIDA, M. E.; SALAHUB, D. R. *J. Chem. Phys.*, **1996**, *104*, 5134.

64 BAUERNSCHMITT, R.; AHLRICHS, R. *Chem. Phys. Lett.*, **1996**, *256*, 454.

65 INGAMELLS, V. E.; HANDY, N. C. *Chem. Phys. Lett.*, **1996**, *248*, 373.

66 ALMLÖF, J.; FAEGRI, K.; KORSELL, K. *J. Comput. Chem.*, **1982**, *3*, 385.

67 ALMLÖF, J. in *Modern Electronic Structure Theory*, YARKONY, D. R., Ed., volume 2 of *Advanced Series in Physical Chemistry*, chapter 3, pp 110–151. World Scientific, Singapore, 1995.

68 TAYLOR, P. R. *Int. J. Quantum Chem.*, **1987**, *31*, 521.

69 KOCH, H.; CHRISTIANSEN, O.; KOBAYASHI, R.; JØRGENSEN, P.; HELGAKER, T. *Chem. Phys. Lett.*, **1994**, *228*, 233.

70 KOCH, H.; DE MER'AS, A. S.; HELGAKER, T.; CHRISTIANSEN, O. *J. Chem. Phys.*, **1996**, *104*, 4157.

71 PULAY, P. *Chem. Phys. Lett.*, **1983**, *100*, 151.

72 SÆBØ, S.; PULAY, P. *Chem. Phys. Lett.*, **1985**, *113*, 13.

73 PULAY, P.; SÆBØ, S. *Theor. Chim. Acta*, **1986**, *69*, 357.

74 SÆBØ, S.; PULAY, P. *J. Chem. Phys.*, **1987**, *86*, 914.

75 SÆBØ, S. *Ann. Rev. Phys. Chem.*, **1993**, *44*, 213.

76 HAMPEL, C.; WERNER, H.-J. *J. Chem. Phys.*, **1996**, *104*, 6286.

77 HÄSER, M.; ALMLÖF, J. *J. Chem. Phys.*, **1992**, *96*, 489.

78 MASLEN, P. E.; HEAD-GORDON, M. *Chem. Phys. Lett.*, **1998**, *283*, 102.

79 PÉREZ-JORDÁ, J. M.; YANG, W. *Chem. Phys. Lett.*, **1995**, *247*, 484.

80 PÉREZ-JORDÁ, J. M.; YANG, W. *J. Chem. Phys.*, **1997**, *107*, 1218.

81 GREENGARD, L. *Science*, **1994**, *265*, 909.

82 PETERSEN, H. G.; SOELVASON, D.; PERRAM, J. W.; SMITH, E. R. *J. Chem. Phys.*, **1994**, *101*, 8870.

83 WHITE, C. A.; JOHNSON, B. G.; GILL, P. M. W.; HEAD-GORDON, M. *Chem. Phys. Lett.*, **1994**, *230*, 8.

84 STRAIN, M. C.; SCUSERIA, G. E.; FRISCH, M. J. *Science*, **1996**, *271*, 51.

85 SVENSSON, M.; HUMBEL, S.; FROESE, R. D. J.; MATSUBARA, T.; SIEBER, S.; MOROKUMA, K. *J. Phys. Chem.*, **1996**, *100*, 19357.

86 CARROLL, D. P.; SILVERSTONE, H. J.; METZGER, R. M. *J. Chem. Phys.*, **1979**, *71*, 4142.

87 KUTZELNIGG, W.; MORGAN, J. D. *J. Chem. Phys.*, **1992**, *96*, 4484.

88 OCHTERSKI, J. W.; PETERSSON, G. A.; MONTOMERY, J. A. *J. Chem. Phys.*, **1996**, *104*, 2598.

89 POPLE, J. A.; HEAD-GORDON, M.; FOX, D. J.; RAGHAVACHARI, K.; CURTISS, L. A. *J. Chem. Phys.*, **1989**, *90*, 5622.

90 DUNNING, T. H. *J. Chem. Phys.*, **1989**, *90*, 1007.

91 ALMLÖF, J.; TAYLOR, P. R. *J. Chem. Phys.*, **1987**, *86*, 4070.

92 KATO, T. *Commun. Pure Appl. Math.*, **1957**, *10*, 151.

93 HYLLERAAS, E. A. *Z. Phys.*, **1929**, *54*, 347.

94 DRAKE, G. W. F.; YAN, Z.-C. *Chem. Phys. Lett.*, **1994**, *229*, 486.

95 BOYS, S. F. *Proc. Roy. Soc. (London) A*, **1960**, *258*, 402.

96 KUTZELNIGG, W. *Theor. Chim. Acta*, **1985**, *68*, 445.

97 CSÁSZÁR, A. G.; ALLEN, W. D.; SCHAEFER, H. F. *J. Chem. Phys.*, **1998**, *108*, 9751.

98 WIGNER, E. *Trans. Faraday Soc.*, **1938**, *34*, 29.

99 POLANYI, J. C. *Faraday Discuss. Chem. Soc.*, **1973**, *55*, 389.

100 MANOLOPOULOS, D. E.; STARK, K.; WERNER, H.-J.; ARNOLD, D. W.; BRADFORTH, S. E.; NEUMARK, D. M. *Science*, **1993**, *262*, 1852.

101 MILLER, W. H.; HANDY, N. C.; ADAMS, J. E.
J. Chem. Phys., **1980**, *72*, 99.

102 MILLER, W. H. in *The Theory of Chemical
Reaction Dynamics*, CLARY, D. C., Ed.,
volume 8, p 27. D. Reidel, Boston, 1986.

103 WADE, E. A.; MELLINGER, A.; HALL, M. A.;
MOORE, C. B. *J. Phys. Chem. A*, **1997**, *101*,
6568.

104 KLIPPENSTEIN, S. J.; EAST, A. L. L.; ALLEN,
W. D. *J. Chem. Phys.*, **1996**, *105*, 118.

105 LOVEJOY, E. R.; MOORE, C. B. *J. Chem. Phys.*,
1993, *98*, 7846.

106 SCOTT, A. P.; NOBES, R. H.; SCHAEFER, H. F.;
RADOM, L. *J. Amer. Chem. Soc.*, **1994**, *116*,
10159.

107 GEZELTER, J. D.; MILLER, W. H. *J. Chem. Phys.*,
1995, *103*, 7868.

108 LOVEJOY, E. R.; KIM, S. K.; MOORE, C. B.
Science, **1992**, *256*, 1541.

109 KING, R. A.; ALLEN, W. D.; MA, B.; SCHAEFER,
H. F. *Faraday Discussions*, **1998**, *110*, 23.

110 GEZELTER, J. D.; MILLER, W. H. *J. Chem. Phys.*,
1996, *104*, 3546.

111 CUI, Q.; MOROKUMA, K. *J. Chem. Phys.*, **1997**,
107, 4951.

112 TRUHLAR, D. G. *Faraday Discussions*, **1998**,
110, 521.

113 STEELE, R. A.; OPELLA, S. J. *Biochemistry*, **1997**,
36, 6885.

114 BROWN, R. T.; FONTANA, P. R. *J. Chem. Phys.*,
1966, *45*, 4248.

14
Quantum Alchemy

Marvin L. Cohen

14.1
From Alchemy to Quantum Theory

Even though the alchemists were unable to turn lead into gold, they did succeed in modifying properties of materials and in laying some of the foundations for what was to become experimental science. In a limited sense, their attempts to explain material properties using a more or less universal approach is a precursor to what became theoretical science. Their conceptual approach involved treating solid-state properties as separate from the base substance or "brute matter". Based on this approach, one might extract the gray color of lead and give it the luster of gold. Color and other properties could be in principle extracted or added to base matter. In contrast, much of modern science is based on the concept that matter is made of atoms, and material properties depend on the properties of atoms. However, this does not mean that a complete understanding of the individual atoms making up a solid necessarily implies that the solid is understood. Atoms interact, and understanding their interactions is critical to understanding properties of many solids. Sometimes the interactions are so important that it is best to view the solid as the basic entity as the alchemist did.

The concept of the atomic nature of matter was refined at the end of the 19th century, and the properties of atoms themselves were "explained in principle" in the 1920s by quantum theory. In fact, "in principle", a large part of physics and chemistry was explained by quantum theory. A famous quote by Dirac [1] could be interpreted as a challenge to researchers trying to understand solids. In 1929, he stated "The underlying physical laws necessary for a large part of physics and the whole of chemistry are thus completely known, and the difficulty is only that the exact application of these laws leads to equations much too complicated to be soluble." It was pointed out often that even if a quantum theory of solids were developed that yielded the positions and momenta of all the individual electrons in a solid, it would be useless as there would not be enough paper to construct a table of the information, and no sense could be made of numbers. Hence, it became important to have a clear goal in order to attempt to answer Dirac's challenge, and this goal became the determination of the solutions of the Schrödinger equation for electrons in a periodic potential caused by the atomic or ionic lattice forming the crystalline solid. Earlier, Bloch [2] had determined the general forms that the electronic wave function in a periodic solid must have. In addition, the adiabatic or Born-Oppenheimer approximation showed that because of the great difference in mass between the electron and an ion, it could be assumed that to a first approximation the electrons were moving through a lattice of stationary ions or atomic cores in a solid at room temperature.

14.2
Applying Quantum Theory

So the model of a crystalline solid became a lattice of periodic atomic cores representing atoms stripped of their valence electrons with these liberated valence electrons flowing through the lattice. Depending on the atom, the electrons might be "free" to wander through the crystal with almost uniform density everywhere, or they might be concentrated in bonds between the cores, or, if different atoms were involved, the electrons might be distributed unevenly allowing some cores to be positively charged and others negatively charged. These three scenarios represent our visualization of the metallic, covalent and ionic bonds of solids. Choosing the second row of the periodic table, it is interesting to examine the first four elements in the row Na, Mg, Al, and Si. For Na, Mg, and Al, the loosely held valence electrons leave their atoms when the solid is formed and contribute to what is called a Fermi sea of free electrons which wander through the entire crystal. In fact, one can ignore the structural arrangement of the cores for many properties and use a "jellium" approximation where the positive cores are smeared out into a positive background jelly in which the electrons move. This is similar to J. J. Thomson's plum pudding model of an atom where the electrons represented by plums exist in a positive pudding. However, for most properties it is critical to treat the electrons quantum mechanically.

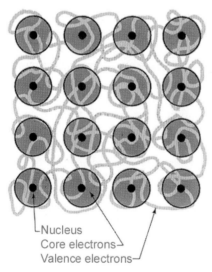

Nucleus
Core electrons
Valence electrons

Fig. 14.1 Model of a solid with cores at fixed lattice positions and valence electrons free to move throughout the crystalline solid.

Hence, the alchemist approach of considering the solid as a whole instead of an atomic view of solids works to a low approximation for free electron metals. However, when we proceed to Si, which is in the fourth column of the periodic table, the situation changes drastically. Here, structure is crucial. The valence electrons concentrate and form bonds between the Si cores which are tetrahedrally coordinated. These are the covalent bonds which hold the crystal together. This picture of bonding is similar to that proposed by Newton where atoms had hooks for bonding. Understanding the atomic nature of matter is critical for explaining bonding in Si. To illustrate ionic bonding, an excellent example is NaCl, where the loose Na valence electron transfers to the Cl and the crystal bonds because of electrostatic attractive forces. Again, electronic charge density is an important factor in the bonding.

Our model of positive atomic cores arranged in a periodic array with valence electrons is shown schematically in Fig. 14.1. The objective is to solve the Schrödinger equation to obtain the electronic wave function $\Psi_{n,\vec{k}}(\vec{r})$ and the electronic energy band structure $E_n(\vec{k})$ where n labels the energy band and $\vec{k}$ the crystal wave vector which labels the electronic state. To explore the bonding properties discussed above, a calculation of the electronic charge density

$$\rho(\vec{r}) = \sum_{n,\vec{k}} e\,|\Psi_{n,\vec{k}}(\vec{r})|^2 \tag{14.1}$$

can be used as a tool for visualizing the metallic, covalent, or ionic nature of the solid.

If we ignore the vibrations of the cores, then the total energy within our model is composed of the kinetic energy of the electrons, the electrostatic interaction between the cores, the electron-electron interaction, and the electron-core interaction. If all these are known for a given structural arrangement of cores, then it is possible to compute the energy as a function of lattice structure and separations between atoms. These structural calculations have been done in recent years, and they have allowed structural determination as a function of pressure, which is related to the change in separation between atoms. This topic will be dealt with later.

If we fix the cores and assume that each electron moves in the average potential generated by the cores and other electrons, this is the so-called Hartree or one-electron approximation. For this model we arrive at a relatively simple expression for the Hamiltonian for determining the electronic band structure and wave functions. The one-electron Hamiltonian is

$$H = \frac{p^2}{2m} + V(\vec{r}) \tag{14.2}$$

where p and m are the electron momentum and mass, and $V(\vec{r})$ is the periodic crystal potential seen by the electron. This potential contains both the electron-core interaction and the electron-electron interaction. The total energy for all the electrons is the sum over the individual electron energies in this model. So the Schrödinger equation for the one-electron wave function becomes

$$\left[\frac{p^2}{2m} + V(\vec{r})\right]\Psi_{n,\vec{k}}(\vec{r}) = E_n(\vec{k})\Psi_{n,\vec{k}}(\vec{r})$$ (14.3)

and the symmetry of $V(\vec{r})$ is reflected in the properties of the wave function $\Psi_{n,\vec{k}}(\vec{r})$ and the band structure $E_n(\vec{k})$. Using group theory and these symmetries, the computational aspects of the problem can be simplified.

The dilemma of electronic structure calculations for solids is that the wave function for a valence electron resembles a free electron between the cores but looks like a localized atomic-like electron near the cores. Around 1940, hybrid wave functions [3, 4] were constructed to account for this schizophrenic nature of the electron, and these approaches had some success. Another approach, which had various origins, is to consider only the outer regions of the wave function since these are the most influential regarding solid-state effects. Fermi [5] introduced this concept when he considered the excited electronic states of alkali atoms and realized that the oscillations of the electronic wave function near the core were not important for many properties. He constructed a pseudopotential and pseudowave function to account for just the outer parts of the valence electrons and did not consider the core electrons. Others came to similar conclusions regarding the advantages of using a weak effective potential and smooth or more free-electron like wave functions. In the hybrid approaches discussed above, it was shown that the effect of the Pauli principle was to keep valence electrons away from the core. Hence, one could arrange terms in the Hamiltonian mathematically [6] so that this repulsive Pauli force would cancel much of the strong attractive ionic potential of the core leaving a net weak effective potential as shown in Fig. 14.2. Hence, the pseudopotential had several

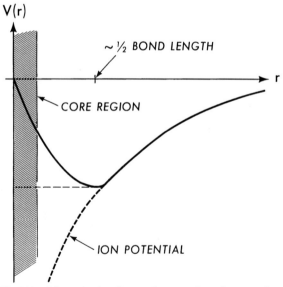

Fig. 14.2 Schematic plot of a pseudopotential as a function of position compared with an ionic Coulomb potential.

beginnings, but the conclusions were the same. It could be argued that the wave function basis set could be plane waves with smoothly varying densities representing free electrons, and the electron-core potential could be calculated or determined empirically.

The empirical approach [7] was by far the most fruitful first attempt. The idea was to fit a few Fourier coefficients or form factors of the potential. This approach assumed that the pseudopotential could be represented accurately with around three Fourier form factors for each element and that the potential contained both the electron-core and electron-electron interactions. The form factors were generally fit to optical properties. This approach, called the Empirical Pseudopotential Method (EPM), gave [7] extremely accurate energy band structures and wave functions, and applications were made to a large number of solids, especially semiconductors. [8] In fact, it is probably fair to say that the electronic band structure problem and optical properties in the visible and UV for the standard semiconductors was solved in the 1960s and 1970s by the EPM. Before the EPM, even the electronic structure of Si, which was and is the prototype semiconductor, was only partially known.

It was surprising to see how well the electronic charge densities based on the pseudowave functions for the semiconductors agreed with experiment. In Fig. 14.3, a ball and stick model for the Si structure is given. The sticks form the bonds. Figure 14.4 shows the calculated and measured electron charge densities with the covalent pileup of electrons in the bonds between the cores. It should be noted that the differences between different experimental data is as large as the differences between theory and experiment. The band structures are still used today and their determination influenced the development of the field of "band-gap engineering", where alloys of semiconductors are made in order to "tune" band gaps for technological applications. Pressure and temperature effects on electronic structures were also calculated. The advent of faster modern computers with larger memories had a

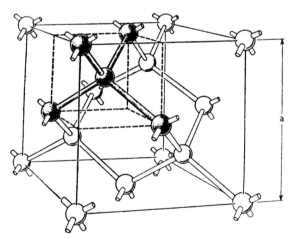

Fig. 14.3. The crystal structure of silicon, germanium and diamond, indicating the tetrahedral coordination.

EXPERIMENT (SILICON)

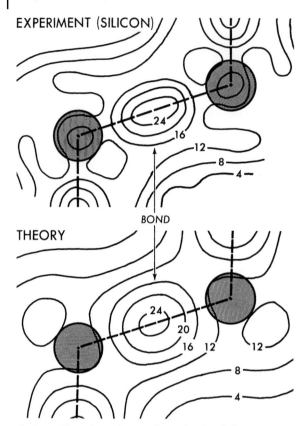

Fig. 14.4. The valence electron charge density of silicon.
Contour spacings are in units of electrons per unit cell volume.
Shaded circles represent atomic cores.

large effect on the productivity of this area. Because some experimental data were used as input and because the output bore on interpreting optical and photoemission experiments, this endeavor resulted in a very effective and active collaboration between theorists and experimentalists.

Returning to Eq. (14:2), the next advance in dealing with the potential was to separate $V(\vec{r})$ into electron-core and electron-electron components. The motivation for this was in part the desire to compute surface properties of semiconductors; it also represented a step in the direction of developing potentials from first principles without experimental input.

The calculation of the electronic structure of surfaces and interfaces requires the notion of self-consistency and a separation of the electron-electron interactions from the electron-core interactions. At interfaces, electronic charge rearranges and differs from the bulk arrangement. This rearrangement changes the electron-electron potential at the surface and hence the changes have to be calculated self-consistently. That is, the change in structure at the surface creates a potential, which changes the

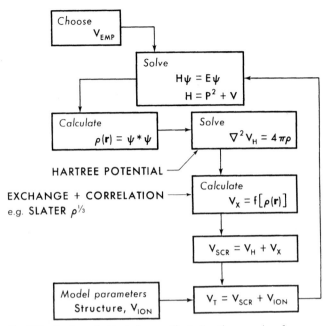

Fig. 14.5   A schematic block diagram illustrating the procedure for calculating self-consistent potentials.

electronic density which in turn changes the potential. A block diagram appears in Fig. 14.5 where the empirical potential V_{EMP} is used to start the calculation. Then the resulting wave function generates a charge density which, when using Poisson's equation, gives an average or Hartree potential V_H for the average Coulomb electron-electron interaction. Because of the Pauli principle for the electrons, a potential V_X, which describes the exchange and correlation contributions to the electron-electron interaction, must be obtained. In early work, a function of $\rho^{1/3}$, as suggested by Slater [9] and Wigner, [10] was used. In more recent work, the density functional approach [11, 12] is used for V_X. So the Hamiltonian now has the form

$$H = \frac{p^2}{2m} + V_{core} + V_H + V_X. \tag{14.4}$$

Returning to the block diagram, the screening potential V_{SCR} is added to the electron-core potential which is now called the pseudopotential, and this new total potential is used together with the new structure to start the loop over again. When input and output agree, the calculation is "self-consistent".

A technical problem occurs when one attempts to apply this approach to study a surface. The calculations described for the bulk crystal assume perfect symmetry and a solid of infinite extent often described in terms of cyclic or periodic boundary conditions. However, for a surface, the translational symmetry is broken, and the usual expansions in Fourier series used for the bulk are not appropriate. For the bulk, a few atoms form a basis which is attached to a lattice cell, and this cell is reproduced throughout the infinite crystal. To retain the same calculational approach and machinery, the concept of a "supercell" was introduced. [13] The supercell used for surfaces is comprised of a slab of atoms and vacuum between the slabs. So typically 6 layers of atoms and 6 layers of vacuum compose the supercell, which is then reproduced infinitely. The interface between the slab with atoms and the vacuum slab becomes the surface. A similar approach is used for interfaces of materials with two connecting slabs of atoms.

A good example of this approach is the application to the Si(111) surface. If one cuts through a model of the Si crystal perpendicular to the (111) crystal direction, the surface exposed has dangling (or half-cut) bonds sticking out of the surface as shown in Fig. 14.6. If no rearrangement of electronic charge were possible, then one would expect to find "half-bonds", which can be pictured by slicing the bonds in Fig. 14.4. However, when the self-consistent calculation [14] of the electronic charge density is performed, as expected, there is a rearrangement of electronic charge density and this has a profound effect. As shown in Fig. 14.7, the cut bonds are "healed" and the redistribution of electronic charge produces a fairly smooth surface. Channels or holes in the surface charge density are evident, and, as shown in Fig. 14.6, these extend into the bulk solid and provide paths for interstitial Si atoms or impurities. The charge density one or two layers down into the crystal looks relatively unaffected by the surface and the bonds resemble the bonds of bulk Si shown in Fig. 14.4. In addition to providing a view of the total charge rearrangement, these calculations [13] yielded energy levels and electron density maps for surface states representing electronic states localized at or near the semiconductor surface. Interfaces [15] between semiconductors and metals and between different semiconductors producing Schottky barriers and heterojunctions were also studied using this

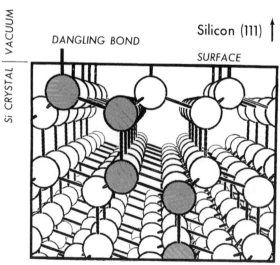

Fig. 14.6 Ball and stick model of the Si(111) surface as viewed in the (110) plane.

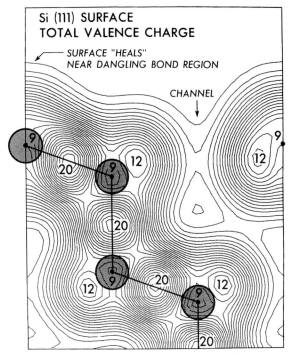

Fig. 14.7   Total valence electron charge density for the ideal Si(111) surface shown in Fig. 14.6. Charge contours are plotted in the (110) plane and normalized to one-electron charge per primitive cell. Shaded circles represent atomic cores.

method. Again, successful interpretations of the data resulted and the research on these interfaces have had wide use in electronics and device physics.

One important aspect not discussed above is the change in atomic structure at a surface. Contrary to the schematic picture of the Si(111) surface shown in Fig. 14.6, a solid surface is usually not just the end of a perfect crystal. Surfaces reconstruct in response to the changes in the electronic distribution caused by the surface itself. Again, all these changes occur self-consistently, and in principle, if the total energy for various configurations of atomic structures at a surface could be evaluated, the shifts in the positions of the atoms and the electronic structures of the surface could be determined theoretically. This approach will be discussed in the next section, but the first calculations for reconstructed surfaces were done using experimental determinations of the atomic positions.

14.3
Total Energy Calculations

The total structural energy of a solid using the model above has the form

$$E_{STR} = E^{cc} + E^{ec} + E_{KIN}^e + E_{COUL}^{ee} + E_X^{ee} \qquad (14.5)$$

where the core-core energy E^{cc} can be calculated from standard point-ion sums of the electrostatic energy, the electron-core energy E^{ec} is evaluated using the electron-core pseudopotential, and the calculated electronic wave functions are used to compute the kinetic energy of the electrons E_{KIN}^e. The Hartree or Coulomb electron-electron term E_{COUL}^{ee} is evaluated as discussed above using the electronic charge density, whereas the exchange-correlation energy E_X^{ee} in most modern calculations is evaluated using the density functional approach [11, 12] in the local density approximations rather than the Slater and Wigner approximations discussed earlier. Using the local density approximation (LDA), [12] the ground state energy can be estimated quite well and E_{STR} can be computed once the appropriate pseudopotential is used.

The EPM required some measured data to determine the Fourier coefficients of the pseudopotential. However, the most modern approaches follow the Fermi [5] concept of developing a pseudopotential to yield a wave function without nodes that coincides with the all-electron atomic wave function outside the core and is still normalized. Several methods were developed [16–19] in the 1970s and 1980s, and new methods for constructing useful pseudopotentials continue to appear in the literature. The applications discussed here are mostly based on the pseudopotentials developed using the approach described in Ref. [19]. The important point to empha-

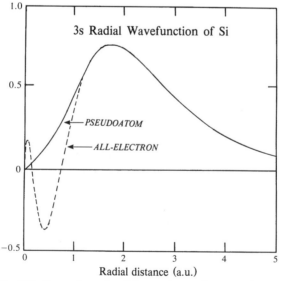

Fig. 14.8 Comparison of an Si 3s pseudowave function with an all-electron wave function.

size is that the only input necessary for generating these "*ab initio*" pseudopotentials is the atomic number. Hence, pseudopotentials can be constructed for the entire periodic table. The pseudo and all-electron wave functions are shown schematically in Fig. 14.8.

Once the pseudopotential is determined and used as input, all the terms in Eq. (14:5) can be evaluated for a given structure and specified lattice constants. A convenient method for computing E_{STR} using reciprocal or Fourier space was developed, [20] and some of the first applications were done [21] for Si. The general approach [22] is to compute the energy of various candidate structures at varying volumes. Results for the energy versus volume curve for Si are given in Fig. 14.9 for several different crystal structures. The lowest energy structure for all volumes is the diamond structure. The volume position of the minimum in the energy curve gives the lattice constant, and the curvature of $E(V)$ near the minimum determines the bulk modulus or its inverse, the compressibility. The results for diamond, Si, and Ge, given in Table 14.1, are typical in that lattice constants are determined to less than 1 % and bulk moduli are evaluated to within around 5 % using this method. Consid-

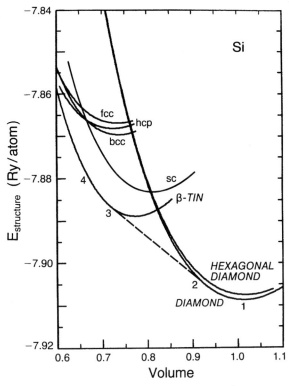

Fig. 14.9 Total energy curves for seven structures of Si as a function of volume normalized to the experimental volume. The dashed line is the common tangent between the diamond and the β-tin phase.

Tab. 14.1 Static structural properties for Si, Ge, and diamond using the atomic number and crystal structure as input.

	Lattice constant (angstroms)	Bulk modulus (GPa)
Si		
calc.	5.45	98
expt.	5.43	99
% diff.	0.4%	−1%
Ge		
calc.	5.66	73
expt.	5.65	77
% diff.	0.1%	−5%
C		
calc.	3.60	441
expt.	3.57	443
% diff.	0.3%	−1%

ering the fact that the only input to these calculations are the atomic numbers and candidate crystal structures, the results are very impressive. Cohesive energies can be estimated by comparing the total energy at the minimum energy volume with that at a large volume where the atoms are separated by sufficient distances so that they can be considered to be non-interacting.

Referring to Fig. 14.9, a comparison between different possible structures for Si can be made. Although the diamond structure is lowest in energy at atmospheric pressure and volume, at smaller volumes the β-tin metallic structure lies lower in energy. Hence, Si will transform from the diamond to the β-tin structure under pressure. The transformation can be viewed as progressing along the dashed line in Fig. 14.9, which is the common tangent between the two structural phases. The slope of the dashed line yields the transition pressure, which is approximately 10 GPa. The calculated transition pressure and transition volumes (volumes at points 1 and 2 on the dashed line) agree well with experiment. Most of the other structural phases shown in Fig. 14.9 have subsequently been found, and the predicted transition pressures have been verified. Two phases, the hexagonal diamond phase in Fig. 14.9, and the BC-8 phase (not shown in Fig. 14.9) are interesting. These phases lie above the diamond phase in energy over the entire range of volume and are realized through non-equilibrium processes. For example, by pressurizing to a higher energy structure, sometimes the barrier back to the higher lying energy phase is smaller than that found in going directly from the diamond structure. Hence, it can be argued that it is likely that many interesting metastable phases exist, but it is difficult to guess their structure at this point.

Surface structure can also be determined in a similar fashion to what is described here for the bulk structure analysis. One begins with an ideal, non-reconstructed surface and calculates the net forces on the atoms or the changes in energy when a surface atom is moved. For example, if there is an energy decrease when an atom is moved in a specific direction, it is assumed that this is a desirable relaxation. All the other atoms near the surface are allowed to relax, and when there are zero residual forces on the atoms, it is concluded that the reconstruction is an appropriate one.

Vibrational properties can also be computed using the total energy formalism. Here, the atomic mass is required as input and the crystal is distorted to mimic a "frozen-in" lattice vibration. The total energy and forces on the atoms can be computed for the distorted configuration, and, through a comparison between the distorted and undistorted crystal, the lattice vibrational (or phonon) spectrum can be computed. Again, the agreement between theory and experiment is excellent. [22]

The above example focusing on Si illustrates how the total energy approach using the standard model and pseudopotentials can yield ground state properties and vibrational spectra with only the atomic number and atomic mass as inputs. Large memory and high speed computers are necessary for detailed calculations based on these physical models; however, shortcuts are possible, particularly if a semiempirical approach is acceptable. Such an approach was developed [23] for calculating bulk moduli for Group IV, III-V and II-VI semiconductors, and there is evidence that the formulae for the bulk modulus can be applied to a wider variety of tetrahedrally coordinated solids. The approach, which is based on a spectral model of semiconductors by Phillips, [23] yields a dependence of the bulk modulus B_0 on the bond length d of the form $B_0 \propto d^{-3.5}$ for the covalent contribution. When one accounts for ionicity, there is a decrease in B_0 which scales like an empirical parameter λ where $\lambda = 0, 1, 2$ for Groups IV, III-V, and II-VI tetrahedral semiconductors and insulators. Coordinations other than 4 can be accommodated by a factor $<N_c>/4$, where $<N_c>$ is the average coordination number. The expression for the bulk modulus becomes

$$B_0 = \frac{\langle N_c \rangle}{4} \frac{(1971 - 220\lambda)}{d^{3.5}} \tag{14.6}$$

where the bond length d is in angstroms and B_0 is given in GPa. Table 14.2, which lists values of B_0 for valence 8 semiconductors illustrates the excellent agreement obtained when the experimental bond length is used as input.

One advantage of having an analytical expression for B_0 is that it enables one to see trends and to consider a broad class of materials such as the valence 8 systems above. By comparing Si and C, it is also possible to gain some insight into the nature of the bonding and its relation to the bulk modulus. For example, the electronic charge density $\rho(\vec{r})$ of C in the diamond structure is compared with that of Si in Fig. 14.10. An obvious difference between the two is the shorter bond length for C and the "two-hump" structure in $\rho(\vec{r})$. Since the core of C has no p-states, the p-valence electrons are not repelled from the core region and they concentrate near their atomic position. For Si, the p-electrons in the core repel their counterparts in the valence electron group and force them to the center of the bond. It is likely that this difference in the C and Si cores is what allows C to bond in sp^2 configurations and to make multiple bonds for organic systems whereas Si does not. This is also likely the reason that C is central to biology, while Si is central to geology.

From Eq. (14:6), it is clear that the shorter bonds and smallest ionicity give the largest values for B_0. This suggests that smaller atoms, such as those in the first row

Tab. 14.2 Bulk Moduli for standard Group IV, III-V, and II-VI semiconductors. The theoretical values are from a semi-empirical formula requiring the measured bond length as input.

Material	d (Å)	Experimental B_0 (GPa)	Calculated B_0 (GPa)
IV			
C	1.54	443	435
Si	2.35	99	99
Ge	2.45	77	85
Sn	2.75	53	57
SiC	1.88	224	213
III-V			
BN	1.57	369	367
BP	1.96	–	166
BAs	2.07	–	138
AlP	2.36	86.0	86.7
AlAs	2.43	77.0	78.3
AlSb	2.66	58.2	57.0
GaP	2.36	88.7	86.7
GaAs	2.45	74.8	76.1
GaSb	2.65	57.0	57.8
InP	2.54	71.0	67.0
InAs	2.61	60.0	61.0
InSb	2.81	47.4	47.1
II-VI			
ZnS	2.34	77.1	78.1
ZnSe	2.45	62.4	66.5
ZnTe	2.64	51.0	51.2
CdS	2.52	62.0	60.3
CdSe	2.62	43.0	52.6
CdTe	2.81	42.4	41.2
HgSe	2.63	50.0	51.9
HgTe	2.78	42.3	42.7

of the periodic table, are particularly interesting for this application. Experimentally, diamond has the highest bulk modulus. To show that it may be possible to produce materials with bulk moduli comparable to diamond, carbon nitride was suggested [23] as a prototype system to explore. Detailed calculations for C_3N_4, assuming the same structure as β-Si_3N_4, gave values for the bulk modulus which were comparable with those of diamond. First-principles calculations [25] of B_0 and values based on Eq. (14:6) were consistent. Although it is not possible at this point to choose the optimum structure for C_3N_4, the values of B_0 computed for the hypothetical β-C_3N_4 structure have motivated experimental searches. For this class of materials, there appears to be a monotonic scaling between B_0 and hardness, and this gives added incentive to search for large B_0 solids based on C and N.

Valence charge density (110 plane)

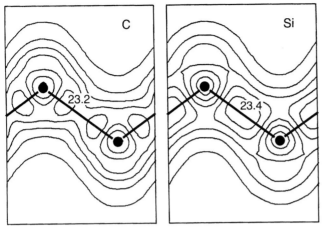

Fig. 14.10 Comparison of the valence electron charge density of
Si and diamond.

Another interesting application of the total energy approach involves superconductivity. For conventional superconductors, the 1957 theory of Bardeen, Cooper and Schrieffer [26] has been subject to extensive tests and has emerged as one of the most successful theories in physics. However, because the superconducting transition temperature T_c depends exponentially on the electron-phonon coupling parameter λ and the electron-electron Coulomb parameter μ^*, it has been difficult to predict new superconductors. The sensitivity is further enhanced because the net attractive electron-electron pairing interaction is proportional to $\lambda-\mu^*$, so when these parameters are comparable, they need to be determined with precision.

Within the formalism described above for determining the phonon spectrum of Si, it is also possible to evaluate the electron-phonon couplings and λ itself. An interesting example is the successful prediction [27] of superconductivity in two high pressure phases of Si. It was shown that the simple hexagonal (sh) and hexagonal close packed (hcp) structures should be stable metals at high pressures. In addition, the calculated values of λ for these phases suggest that they would be good candidates for superconductivity. The values for μ^* were scaled from the dependence of this parameter on the density of states. This is a slowly varying function and reasonable estimates are not difficult to make. The predicted values of T_c were in reasonably good agreement with experiment as was the dependence of the transition temperature on pressure.

The successful prediction of superconductivity in the high pressure Si phases added much credibility to the total energy approach generally. It can be argued that Si is the best understood superconductor since the existence of the phases, their structure and lattice parameters, electronic structure, phonon spectrum, electron-phonon couplings, and superconducting transition temperatures were all predicted from first principles with the atomic number and atomic mass as the main input parameters.

Tab. 14.3 A list of some of the major properties which are determined for real materials using the standard model and employing the pseudopotential total energy method.

For a broad class of solids, clusters, and molecules, the Standard Model describes ground-state and excited state properties such as:

Electronic structure
Crystal structure and structural transitions
Structural and mechanical properties
Vibrational properties
Electron-lattice interactions
Superconductivity
Optical properties
Photoemission properties

Some of the major areas of activity in this field have been the application of the method to more complex materials, molecular dynamics, [28] and the treatment of excited states. [29] We will deal with some of the new materials in the next section. Two major goals of the molecular dynamics calculations are to determine crystal structures from first principles and to include finite temperature effects. By combining molecular dynamics techniques and *ab initio* pseudopotentials within the local density approximation, it becomes possible to consider complex, large, and disordered solids.

For excited state calculations, significant progress has been made based on the "GW" method first introduced by Hybertsen and Louie. [29] By considering quasiparticle and local field effects, this scheme has allowed accurate calculations of band gaps, which are usually underestimated when using the LDA. This "GW" approach has been applied to a variety of crystals, and it yields optical spectra in good agreement with experiment.

There have been a number of improvements in techniques, and more convenient models have been formulated; however, the basic approach of the pseudopotential total energy method has not changed. This general approach or standard model is applicable to a broad spectrum of solid state problems and materials when the electrons are not too localized. Highly correlated electronic materials require more attention, and this is an area of active current research. However, considering the extent of the accomplishments and the range of applications (see Table 14.3) to solids, clusters, and molecules, this approach has had a major impact on condensed matter physics and stands as one of the pillars of the field.

14.4
Novel Materials

The influence of new material systems on the development of condensed matter physics has been substantial. In recent years, the field of nanoscience has evolved and has attracted physicists, chemists, biologists, engineers, and computer scientists. The nanometer scale is of great interest in all of these areas of science and engineering. For example, in cluster science, quantum chemists have been approaching these systems starting with atoms, simple molecules, and then complex molecules. Coming from the larger scales characteristic of solids, physicists have found [30] fruitful models for understanding clusters using approaches developed for solids. Again, the standard model works in this realm.

On a somewhat larger scale, there has been considerable activity in the area of nanocrystals, quantum dots, and systems in the tens of nanometers scale. Interesting questions have arisen regarding electronic properties such as the semiconductor energy band gap dependence on nanocrystal size and the nature of the electronic states in these small systems. Application [31] of the approaches described here, with the appropriate boundary conditions [32] to assure that electron confinement effects are properly addressed, have been successful. Questions regarding excitations, such as excitons and vibrational properties, are among the many that will require considerable scrutiny. It is likely that there will be important input from quantum chemistry as well as condensed matter physics.

At this time, the fastest growing area in the field of nanophysics is in the studies of buckyballs and nanotubes. After the discovery [33] of the C_{60} molecule, many properties of the molecule and solids formed from the molecule were explored. The doped C_{60} crystals showed interesting behavior, including superconductivity. [34] The standard model, including the "GW" quasiparticle theory, was used [35] successfully to explore the energy band structure, and the superconducting properties appear to be consistent with the BCS theory. [36]

Some recent results [37, 38] for C_{36} appear to be suggestive of new possibilities for solids based on molecular fullerenes. In particular, because of the Euler rule, we expect 12 pentagons on the surface of C_{36} as in C_{60}. However, this implies that the C_{36} molecule will be more reactive and more highly curved than C_{60}. Arguments based on these observations suggest that solids composed of C_{36} molecules may be more covalent than those containing C_{60}, which are held together primarily by van der Waals bonding, and higher superconducting transition temperatures are expected for doped solids based on C_{36} molecules. These studies are new, and it will be interesting to see more generally how new molecular structures will influence solid state properties.

The major activity in the field of fullerenes at this time is in the area of nanotube research. The carbon nanotube was discovered by Iijima. [39] A carbon nanotube can be viewed (Fig. 14.11) as a rolled-up graphite-like sheet of carbon atoms all arranged in a hexagonal array. Two lattice vectors are needed to find the points in a 2d hexagonal lattice, and each point can, therefore, be labeled by a pair of integers (m,n) representing the number of unit lengths from the origin $(0,0)$ to the lattice point. The integer pairs can also be used to label tubes since there are different ways of rolling up tubes. For example, an $(8,0)$ tube would have eight hexagons along the circumference of the tube and a line connecting their midpoints would form a circle. However, it is possible to have helicity or a chiral type arrangement when rolling the graphite sheet. Hence, a line connecting the center of the hexagons would trace out a helical path. Figure 14.12 gives a schematic view of such a tube when looking down its axis.

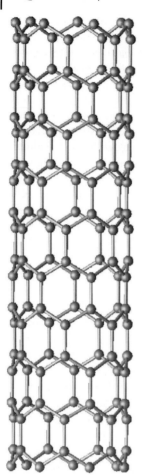

Fig. 14.11 Ball and stick model for a nanotube. The balls represent carbon atoms.

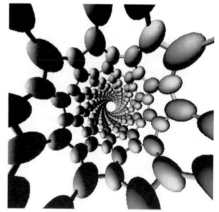

Fig. 14.12 Schematic structural model of a helical nanotube.

An important property of carbon which allows it to form in the tube geometry is the fact that it can exist in an sp^2 or three-fold coordinated bonding arrangement like that found in graphite. If one views graphite as a stacking of sheets of sp^2 covalently bonded atoms with weak van der Waals attraction between the sheets, then, using the arguments of the previous section, the short graphite bonds should result in a material which may be less compressible in the plane (or stronger in some sense) than diamond. Of course, we have the difficulty of comparing two-dimensional sheets with a three-dimensional bulk solid. For graphite, the van der Waals bonds are weak so the overall 3d solid is easily sheared or cleaved. However, once a tube is formed, the bonds are all covalent and the Young's modulus is the highest of any material.

The sp^2 bonding nature of carbon is of great importance for its ability to form the C_{60} soccer ball structure. Because Si does not form graphitic structures, which can be explained with standard model total energy calculations, we do not expect Si buckyballs or nanotubes similar to the carbon variety. However, B and N do form a layer type compound and these first row elements are similar to carbon in that they do not have p-electrons in their cores. Because of the hexagonal and pentagonal rings of bonds on the soccer ball molecule, there is structural frustration if one wants to form an analog of C_{60} with B and N. However, a BN nanotube can be visualized, and a total energy calculation was done, which predicted [40] the BN nanotubes should be stable. These have been found experimentally. [41] In fact, other compound nanotubes have been predicted using the total-energy pseudopotential approach. Two of these, [42, 43] which appear to have been found experimentally, are BC_2N and BC_3. More generally, the theoretical study suggests that $B_xC_yN_z$ compounds should be explored for nanotube formation.

The electronic properties of nanotubes are fascinating. Calculations have been done based on standard band theory suggesting that, depending on the helicity and composition, nanotubes can be metallic, semimetallic, or semiconducting. For example, for carbon nanotubes, the (m,n) values determine these properties; [44] however, BN nanotubes, because of their helicity, are semiconducting irrespective of helicity. [45] The conducting properties of BC_2N depend on structure, and for BC_3 each tube is semiconducting, but bunches of tubes interact if they are in close proximity and behave as metals.

Theoretical calculations predict that it is possible to change the electronic properties of tubes dramatically through doping, defects, and adsorbates. Some of the resulting properties are unusual and very interesting. For example, for BN tubes, the charge density for the lowest unoccupied level in the band structure concentrates electronic charge along the center of the tube. Therefore, it is expected that a BN nanotube, which is doped n-type, will have free-electron-like electronic properties down the center of the tube. For BC_2N, doping is expected to fill a state, which runs along atoms on the tube arranged in a helical form. Hence, BC_2N can be anticipated to be a nanocoil – perhaps the world's smallest. Therefore, because of their structural properties, doping can result in conducting states with unusual properties. Attaching atoms to the outside of tubes, or filling [46] tubes with atoms such as K, can affect their electrical properties substantially.

Defects are of special interest, particularly when one considers connecting tubes with different values of (m,n). A simple defect is to insert a five-fold and a seven-fold ring in place of two six-fold rings. If such a defect were inserted in a graphite sheet, it could be argued that the effect on the electronic properties would be affected only over a distance corresponding to a few bond distances. However, for a tube, the effects are more drastic, particularly when one considers using the 5–7 defect to join two tubes. Figure 14.13 shows such an arrangement where an (8,0) semiconducting tube is joined to a (7,1) metallic tube. The electronic properties of this junction can be shown [47] to be similar to other metal-semiconductor interfaces and to behave like a standard Schottky barrier. Hence, defects in tubes can produce device behavior with junctions only a few atoms thick. Similar geometric arrangements can be formed to join two semiconductor tubes with different gaps and produce a heterojunction. Again, within this extremely small region, it is possible to obtain electronic behavior which is similar to that observed between bulk semiconductors. Metal-

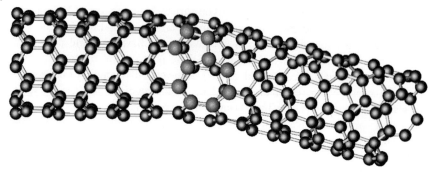

Fig. 14.13 A carbon nanotube Schottky barrier formed from an (8,0) semiconductor nanotube joined by a 5–7 defect to a (7,1) metallic nanotube.

metal junctions for nanotubes are unusual in that one expects conduction when two metals are joined, but this is not always the case for nanotubes. Depending on the geometry of the junction between metal tubes, one can stop the flow of current between the tubes because of a mismatch of wave function symmetry. Bending such a tube usually breaks this symmetry, and the current is then allowed to flow between the tubes.

So it is possible to control the electronic properties of tubes and even do band gaps engineering [48] on the level of what was done for bulk semiconductors. This suggests that new devices may be formed with novel properties. For example, because of their large length to width ratio, nanotubes were expected to be good electron emitters. In fact, the experimental data [49, 50] suggest that this is the case, and nanotubes may play a role in applications focused on producing flat panel displays. Recent experiments have used tunneling microscopes and atomic force microscopes to examine [51–53] electronic properties of single-walled and multi-walled tubes. There is a significant experimental and theoretical effort focused on devices and applications of nanotubes, and progress is being made.

14.5
The Future

A common quote attributed to Yogi Berra is "predicting is difficult – especially about the future". With this statement serving as a caveat, it appears likely that predicting using quantum theory will continue to be an essential tool in the search for new materials and useful properties of materials. Quantum theory is undoubtedly an excellent approximation for dealing with the distances and forces one needs to consider when computing properties of solids. Computers keep getting better and the physical models for how atoms form solids are becoming more robust. One challenge to theory is to determine electronic and structural properties of more complex materials. It is not uncommon for conventional total energy calculation to explore unit cells with an order of 20–50 atoms. This allows studies of surfaces and interfaces with and without adsorbates and defects in addition to restricted geometries and fairly complex structures. Other schemes for considering hundreds or thousands of atoms are in various stages of development allowing interesting models of liquids, amorphous materials, and nanocrystals.

A significant effort is being made to explore highly correlated electronic systems. At this point, much of the emphasis is on formalism. Part of the motivation is to attempt to model the electronic structure of high T_c superconducting oxides and other highly correlated materials where standard band theory fails to give an accurate picture of the electronic states. Lower dimensional systems are also the subjects of great interest. For example, correlation effects can be important for understanding properties of two-dimensional systems which can exhibit unusual electronic properties such as those found in studies of the fractional quantum Hall effect.

Returning to novel materials, several predictions remain to be tested. For example, superhard materials [54] composed of elements in the first row of the periodic table, new nanotubes such as GaSe, [55] and novel carbon-based systems such as pentaheptite. [56] The latter system is based on the 5–7 arrangement of carbon atoms discussed earlier in connection with defects in nanotubes. For pentaheptite, the idea is to have entire planes of atoms composed of 5-fold and 7-fold rings of bonds. This planar arrangement of carbon is analogous to the hexagonal arrangement found in graphite. The major difference is that the 5–7 symmetry changes the electronic properties to produce a metallic sheet. These sheets may be stacked and bonded together as in the graphite case.

Nanotubes have a bright future. Some of the work will undoubtedly be applications driven as the miniaturization of electronics continues. Modern Si chips have geometries ten thousand times smaller than small objects we handle in everyday life. Nanotubes can allow further reduction by another factor of ten thousand. Connecting and fabricating the devices is difficult, but it is being done. Self-assembly is another important technique to employ if nanocrystals and nanotubes will be used in device applications. Another approach would be to attempt to use random access to nanodevices (RAND). Here, one inserts many wires into random arrangements of nanotubes. It is likely that the matrix of tubes contains Schottky barriers, heterojunctions, and other electronic elements necessary for applications, such as making computer chips. Hence, we expect that there will be nonlinear properties when the input and output wires are tested. By keeping track of the responses using conventional computers, it should be possible to produce software to utilize the capabilities of RAND. In other words, given a problem to be solved, the work goes into developing the programs to take advantage of the device properties of the RAND computer. At this stage, the development of these systems has not gone beyond theory and simple prototypes. It is not clear whether systems of this kind will ever be useful, but there is incredible motivation because of the unusually large storage capacity of nanotube arrays. If self-assembly is an alternative path where orderly systems may be fabricated, RAND is still intriguing because systems of this kind may be tolerant of defects.

Theory also has the advantage of not being restricted by the limits of real experimental apparatus. For example, there is a great deal of interest in high pressure science. Metallizing hydrogen is just one example of the goals in this area. Here, theory can do a great service in considering conditions unavailable in the laboratory. Other examples include simulating high temperatures and testing the limits of physical parameters. One can ask about the limits on compressibility, energy band gaps, dielectric constants, mechanical properties, and so on. These studies are particularly useful for materials science investigation. If a system is near the theoretical limit for a given property, it does not make sense to invest a great effort to try for improvement. If it is not, further processing would be recommended.

Hence, it seems safe to conclude that the next century will see considerable research on materials and "alchemy" using quantum theory. If the physical models improve and computations continue to get more accessible, many experimentalists will have access to these tools. In science, all decisions are ultimately made by experiment, and most new discoveries are made by observing physical systems. Perhaps it is not too outrageous to suggest that, if theory continues to improve as it has, theorists may "discover" new states of matter and properties such as superconductivity and magnetism using their computers and analytic modeling.

Acknowledgements

This work was supported by National Science Foundation Grant No. DMR-9520554 and by the Director, Office of Science, Office of Basic Energy Sciences, Materials Sciences Division of the U.S. Department of Energy under Contract no DE-AC03–76SF00098.

References

1 DIRAC, P.A.M. *Proc. Roy. Soc.* (London) **1929**, A123, 714.

2 BLOCH, F. *Z. Physik* **1928**, 52, 555.

3 HERRING, C. *Phys. Rev.* **1940**, 57, 1169.

4 SLATER, J.C. *Phys. Rev.* **1937**, 51, 846.

5 FERMI, E. *Nuovo Cimento* **1934**, 11, 157.

6 PHILLIPS J.C.; and KLEINMAN, L. *Phys. Rev.* **1959**, 116, 287.

7 COHEN, M.L.; BERGSTRESSER, T.K. *Phys. Rev.* **1966**, 141, 789.

8 COHEN, M.L.; CHELIKOWSKY, J.R. *Electronic Structure and Optical Properties of Semiconductors* ; Springer-Verlag, Berlin, 1988.

9 SLATER, J.C. *Phys. Rev.* **1951**, 81, 385.

10 WIGNER, E. *Phys. Rev.* **1934**, 46, 1002.

11 DIRAC, P.A.M. *Proc. Cambridge Philos. Soc.* **1930**, 26, 376.

12 KOHN, W.; SHAM, L.J. *Phys. Rev. A* **1965**, 140, 1133.

13 COHEN, M.L.; SCHLÜTER, M.; CHELIKOWSKY, J.R.; LOUIE, S.G. *Phys. Rev. B* **1975**, 12, 5575.

14 SCHLÜTER, M.; CHELIKOWSKY, J.R.; LOUIE, S.G.; COHEN, M.L. *Phys. Rev. B* **1975**, 12, 4200.

15 COHEN, M.L. *Advances in Electronics and Electron Physics,* Vol. 51; eds. L. MARTON and C. MARTON; Academic Press, New York, 1980; p.1.

16 STARKLOFF, T.; JOANNOPOULOS, J.D. *Phys. Rev. B* **1979**, 19, 1077.

17 KERKER, G.P. *J. Phys. C* **1980**, 13, L189.

18 ZUNGER, A.; COHEN, M.L. *Phys. Rev. B* **1978**, 18, 5449.

19 HAMANN, D.R.; SCHLÜTER, M.; CHIANG, C. *Phys. Rev. Lett.* **1979**, 43, 1494.

20 IHM, J.; ZUNGER, A.; COHEN, M.L. *J. Phys. C* **1979**, 12, 4409. Erratum: *J. Phys. C* **1980**, 13, 3095.

21 YIN M.T.; COHEN, M.L. *Phys. Rev. Lett.* **1980**, 45, 1004.

22 COHEN, M.L. *Physica Scripta* **1982**, T1, 5.

23 COHEN, M.L. *Phys. Rev. B* **1985**, 32, 7988.

24 PHILLIPS, J.C. *Bonds and Bands in Semiconductors*; Academic Press, New York, 1973.

25 LIU, A.Y.; COHEN, M.L. *Science* **1989**, 245, 841.

26 BARDEEN, J.; COOPER, L.N.; SCHRIEFFER, J.R. *Phys. Rev.* **1957**, 106, 82.

27 CHANG, K.J.; DACOROGNA, M.M.; COHEN, M.L.; MIGNOT, J.M.; CHOUTEAU, G.; MARTINEZ, G. *Phys. Rev. Lett.* **1985**, 54, 2375.

28 CAR, R.; PARRINELLO, M. *Phys. Rev. Lett.* **1985**, 55, 2471.

29 HYBERTSEN, M.; LOUIE, S.G. *Phys. Rev. Lett.* **1985**, 55, 1418.

30 DE HEER, W.A.; KNIGHT, W.D.; CHOU, M.Y.; COHEN, M.L. *Solid State Physics,* Vol. 40, ed. H. EHRENREICH and D. TURNBULL; Academic Press, New York, 1987; p.93.

31 MIZEL, A.; COHEN, M.L. *Phys. Rev. B* **1997**, 56, 6737.

32 MIZEL, A.; COHEN, M.L. *Phys. Rev. B* **1998**, 57, 9515.

33 KROTO, H.W.; HEATH, J.R.; O'BRIEN, S.C.; CURL, R.F.; SMALLEY, R.E. *Nature* **1985**, 318, 162.

34 HEBARD, A.F.; RASSEINSKY, M.J.; HADDON, R.C., MURPHY, D.W.; GLARUM, S.H.; PALSTRA, T.T.M.; RAMIREZ, A.P.; KORTAN, A.R. *Nature* **1991**, 350, 600.

35 SHIRLEY, E.L.; LOUIE, S.G. *Phys. Rev. Lett.* **1993**, 71, 133.

36 COHEN, M.L.; CRESPI, V.H. *Buckminsterfullerenes,* eds. W. E. BILLUPS and M. A. CIUFOLINI; VCH Publishers, New York, 1993, p. 197.

37 PISKOTI, C.; YARGER, J.; ZETTL, A. *Nature* **1998**, 393, 771.

38 CÔTÉ, M.; GROSSMAN, J.C.; LOUIE, S.G.; COHEN, M.L. *Phys. Rev. Lett.* **1998**, 81, 697.

39 IIJIMA, S. *Nature* **1991**, 354, 56.

40 RUBIO, A.; CORKILL, J.L.; COHEN, M.L. *Phys. Rev. B* **1994**, 49, 5081.

41 CHOPRA, N.G.; LUYKEN, R.J.; CHERREY, K.; CRESPI, V.H.; COHEN, M.L.; LOUIE, S.G.; ZETTL, A. *Science* **1995**, 269, 966.

42 MIYAMOTO, Y.; RUBIO, A.; COHEN, M.L.; LOUIE, S.G. *Phys. Rev. B* **1994**, 50, 4976.

43 MIYAMOTO, Y.; RUBIO, A.; LOUIE, S.G.; COHEN, M.L. *Phys. Rev. B* **1994**, 50, 18360.

44 HAMADA, N.; SAWADA, S.; OSHIYAMA, A. *Phys. Rev. Lett.* **1992**, 68, 1579. MINTMIRE, J.W.; DUNLAP, B.I.; WHITE, C.T. *Phys. Rev. Lett.* **1992**, 68, 631.

45 BLASE, X.; RUBIO, A.; LOUIE, S.G.; COHEN, M.L. *Europhys. Lett.* **1994**, 28, 335.

46 MIYAMOTO, Y.; RUBIO, A.; BLASE, X.; COHEN, M.L.; LOUIE, S.G. *Phys. Rev. Lett.* **1995**, 74, 2993.

47 CHICO, L.; CRESPI, V.H.; BENEDICT, L.X.; LOUIE, S.G.; COHEN, M.L. *Phys. Rev. Lett.* **1996**, 76, 971.

48 CRESPI, V.H.; COHEN, M L.; RUBIO, A. *Phys. Rev. Lett.* **1997**, 79, 2093.

49 DE HEER, W.A.; CHATELAIN, A.; UGARTE, D. *Science* **1995**, 270, 1179.

50 COLLINS, P.G.; ZETTL, A. *Phys. Rev. B* **1997**, 55, 9391.

51 BOCKRATH, M.; COBDEN, D.H.; MCEUEN, P.L.; CHOPRA, N.G.; ZETTL, A.; THESS, A.; SMALLEY, R.E. *Science* **1997**, 275, 1922.

52 DEKKER, C. *Physics Today* **1999**, 52, 22.

53 COLLINS, P.G.; ZETTL, A.; BANDO, H.; THESS, A.; SMALLEY, R. *Science* **1992**, 278, 100.

54 HAN, S.; IHM, J.; LOUIE, S.G.; COHEN, M.L. *Phys. Rev. Lett.* **1998**, 80, 995.

55 CÔTÉ, M.; COHEN, M.L.; CHADI, D.J. *Phys. Rev. B* **1998**, 58, R4277.

56 CRESPI, V H.; BENEDICT, L.X.; COHEN, M.L.; LOUIE, S.G. *Phys. Rev. B* **1996**, 53, R13303.

15
Quantum Theory Project

Rodney J. Bartlett

15.1
Introduction

With the recent Nobel Prize in Chemistry being awarded to John Pople and Walter Kohn for their contributions to Theoretical and Computational Chemistry, it is especially pertinent to address the question of how our science will develop into the next millennium. There is no doubt that more and more molecular theory, whether we call it chemistry, materials science, or molecular biology, will be done in the future by using quantum and classical mechanical methods coupled to increasingly powerful computers, as this offers a far more effective and cost efficient tool for analysis and design. Such tools are able to investigate regions beyond experimental conditions, and assess processes that are too complex, with too many variables, for normal controlled experimental investigation. However, as we go to more complexity, with vastly different time scales for different processes, we still have to rely upon the fundamental molecular forces described by the quantum theory of matter to be able to relate observed phenomena to basic interactions. So rather than being only a secondary element in the kinds of large scale simulations we see today in molecular biology and materials, the essential, fundamental interactions have to have a much greater role if we are to "get the right answers for the right reason." [1] Consequently, I will direct my remarks in this article to how I believe the quantum chemical description of molecular interactions will evolve from the two viewpoints represented by the Nobel prize: the *ab initio* two-electron, wave function theory (WT) and the one-electron Density Functional Theory (DFT), to obtain a theory which will be far more powerful, and will offer the best of both approaches.

15.2
Background

Molecules are described in terms of a Hamiltonian operator that accounts for the movement of the electrons and the nuclei in a molecule, and the electrostatic interactions among the electrons and the electrons and the nuclei. Unlike the theory of the nucleus, there are no unknown potentials in the Hamiltonian for molecules. Although there are some subtleties, for all practical purposes, this includes relativistic corrections, [2] although for much of light-element chemistry those effects are unimportant. Consequently, obtaining numerical results for most molecular theory is reduced to a problem in applied mathematics, which echoes the famous statement of Dirac, that "the underlying physical laws necessary for the mathematical theory of a large part of physics and the whole of chemistry are thus completely known, and the difficulty is only that the exact application of these laws leads to equations much too complicated to be soluble." [3]

DFT and *ab initio* quantum chemistry, which will be used to mean the wave function theory (WT), with ψ the wave function, which is a complicated function of the coordinates of all *n* electrons, are two conceptually different approaches to the same problem. Each offers a prescription for how to solve the equations that Dirac said were too complicated to solve, and they have different advantages and disadvantages. WT is the approach that leads to Pople's *theoretical model chemistry*, which means for a given atomic orbital basis set representation $\{\chi_1, \chi_2,...\chi_m\}$ for a molecule (today, virtually always contracted Gaussian functions centered on each atom), and a given level of inclusion of electron correlation, meaning the instantaneous electrostatic interactions that keep electrons apart and *correlate* their motions, we obtain numerical results whose calibration against experiment establishes a specific level of accuracy. By extensions of the basis set and the inclusion of electron correlation, we can converge to the exact (non-relativistic) answers. The great power of these methods is that you know how in principle, if not always in practice, to make the next better calculation. No one expects such a series of improved calculations to necessarily converge to a result monotonically, but with some care, the calculations should converge systematically toward the exact value.

The primary characteristic of WT that distinguishes it from DFT, is that two-electron operators are treated explicitly. However, except for a few methods that attempt to use explicit two electron operator ($r_{12} = |r_1 - r_2|$) terms in the wave function, [4] the vast majority of wave function methods attempt to describe the innate correlation effects ultimately in terms of products of basis functions, $[\chi_p(1)\chi_q(2) - \chi_p(2)\chi_q(1)]$, where (1) indicates the space (r_1) and spin (σ) coordinates of electron one (together $x_1 \equiv 1$), while the (2) indicates the coordinates of the second electron. This is the critical approximation in WT. Because of this approximation, all molecular wave functions that do not introduce r_{12} cannot satisfy the electron-electron cusp condition, [5] which is $\partial\Psi/\partial r_{12} = \eta\Psi(r_{12})|_{r_{12}=0}$, where η is a non-zero constant that depends upon the particular state of the atom or molecule. How important this weakness is for a given quantum chemical calculation (even most explicit r_{12} methods like Gaussian geminals [6] do not satisfy this condition) depends upon a lot of factors, but it is certainly true that rather than correlation (see Fig. 15.1), basis set limitations are the greatest present weakness in *ab initio* electronic structure theory. The only current solution for non-r_{12} methods is to perform a series of highly correlated calculations with increasingly large Gaussian basis sets, such as the correlation-consistent, polarized *N* zeta (cc-pVNZ) sets, [7] where *N* is 2–6, and extrapolate to the basis set limit. The Gaussian 2 (G2) approach [8] attempts to extrapolate correlation and basis, as does the CBS (complete basis set extrapolation) approach. [9] Even r_{12} methods suffer from basis set limitations, since the complicated integrals

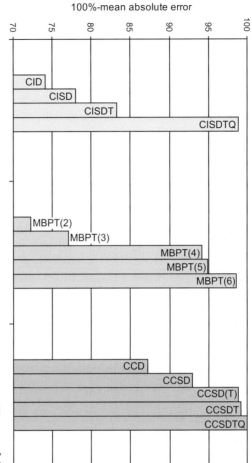

Fig. 15.1 Comparison of CI, MBPT and CC energies with respect to full CI. The data set consists of the BH, HF and H_2O molecules at bond distances of R_e, 1.5 R_e and 2.0 R_e within a DZP basis set.

that arise in such a calculation typically require some type of *resolution of the identity* approximation, which assumes the basis is complete.

The DFT approach, on the other hand, focuses on the density, which is

$$\rho(1) = \int \Psi^*(123\ldots n)\Psi(123\ldots n)d\tau_2 d\tau_3 \ldots d\tau_n \tag{15.1}$$

$$n = \int \rho(1)d\tau_1 \tag{15.2}$$

where $d\tau_n$ is the space-spin volume element and n is the number of electrons. Unlike the wave function, the density is a quantum mechanical observable, meaning that one can devise an experiment to measure it. It is routinely plotted in molecular studies today. Hohenberg and Kohn [10] posed the question of whether the density alone could be used to determine the ground state energy, and proved the *existence* theorem that the energy is a functional of the density, $E[\rho]$, and eventually implemented the Kohn-Sham DFT method. [11] This HK theorem was generalized to non-degenerate ground states by the constrained search formalism of Levy. [12] The theorem does not say what $E[\rho]$ is, however. Unlike the wave function, which is a function of n space and spin variables for n electrons, the density is a function of only one electron's coordinates. This has the consequence that DFT, discussed in more detail below, is a one-particle theory, which has great computational advantages. One is that one-particle equations can be solved in a space of basis functions much more easily than can two-particle equations. In particular, the basis product approximation above does not have to be made, making DFT results much less sensitive to basis set limitations. [13, 14] The primary negative element in DFT is that all results depend upon a knowledge of an unknown $E[\rho]$, and, doubtless, very complicated exchange-correlation functional, $E_{XC}[\rho]$; and eventually its functional derivatives with respect to ρ, which define potentials (first derivatives) and kernels (second derivatives). Both frequently even have the wrong signs compared to the exact results for model systems. [15, 16] In practice, an $E_{XC}[\rho]$ is defined for a model system like the electron gas (LDA) [17]or semi-empirically, now even with many parameters, [18]and the potentials and kernels defined from differentiation. Though DFT is an exact theory in principle, in practice DFT is semiempirical, so the result of a calculation is what the particular choice of functional gives, and there is no way to do the next better calculation to systematically converge to the right answer as there is in WT partly illustrated by Fig. 15.1 for the energy, though it will apply to essentially any property. [19]

In the near future, quantum chemistry will inevitably forge a union between the WT and DFT approaches, as these are two different approaches to the same problem; and this union will, hopefully, offer the best of both worlds. In the following, I will attempt to point the way that this will possibly be accomplished. However, to do so, we need to pay more attention to some of the pertinent mathematical structure of the theories.

15.3
Wave Function Theory

Treating the electronic Schrödinger equation in the usual clamped nuclei (Born-Oppenheimer) approximation, [20] we have (in atomic units) the Hamiltonian, H, and the spectrum of eigenvalues and eigenvectors, E_k and Ψ_k,

$$H \, \Psi_k = E_k \Psi_k \tag{15.3}$$

$$H = \sum_i^n \left[\hat{t}(i) + \hat{v}(i) \right] + \frac{1}{2} \sum_{i,j}^n \frac{1}{|r_i - r_j|} + \frac{1}{2} \sum_{\alpha,\beta} \frac{Z_\alpha Z_\beta}{|R_\alpha - R_\beta|} \tag{15.4}$$

$$H = \hat{T} + \hat{v} + \hat{V}_{ee} + \hat{V}_{NN} \tag{15.5}$$

where the last term that depends solely upon the positions of the nuclei α and β and their atomic numbers Z_α and Z_β is a constant, and need not be considered any further. The first term accounts for the electronic kinetic energy, $\hat{t}(i) = -\frac{1}{2}\nabla^2(i)$ and the second the potential, $v(i) = -\sum_\alpha Z_\alpha/|r_i - R_\alpha|$ which accounts for the electron-nuclear attraction. The n indicates the number of electrons. These operators depend upon the coordinates of only one electron, so they are one-particle operators. The third term, the electron-electron repulsion, $\hat{V}_{ee}$, depends upon the positions of two electrons at a time, so we refer to it as a two-electron operator. Whereas we can exactly solve the hydrogen atom one-particle equation, the existence of the two-particle operator is why we cannot offer an *exact*, as opposed to a very accurate, numerical result for He or H_2. The latter has been solved [21] to an accuracy that exceeds spectroscopic for all of its properties, but such solutions for more than a very few electrons are not obtainable today. Instead, we hope to obtain results short of spectroscopic, but on the order of ~1 kcal/mol, which is usually considered to be chemical accuracy.

The wave function theory can be reduced to knowing about only one and two particles at a time (we only consider the ground state in the following), since the energy is given by

$$E = \frac{\langle \Psi | H | \Psi \rangle}{\langle \Psi | \Psi \rangle} \tag{15.6}$$

which may be rewritten in terms of density matrices,

$$\rho(1, 1') = \int d\tau_2 d\tau_3 ... d\tau_n \Psi^*(1, 2, ...) \Psi(1, 2, ...) \tag{15.7}$$

$$\rho(1, 2; 1', 2') = \frac{1}{2!} \int d\tau_3 d\tau_4 ... \Psi^*(1', 2', 3'...) \Psi(1, 2, 3...) \tag{15.8}$$

where by virtue of integrating over the space and spin coordinates of all electrons but one, we define the one-particle density matrix, and in the case of all electrons but two, the two-particle density matrix. [22] Then we have for the energy,

$$E = \int d\tau_1 \hat{h}(1) \rho(1, 1') + \int \int d\tau_1 d\tau_2 \hat{V}_{ee}(1, 2) \rho(1, 2; 1, 2) \tag{15.9}$$

where I have grouped the kinetic energy and one-particle potential energy operator together in $\hat{h}$. Because of the differential operator nature of the kinetic energy, the $1'$ guarantees that the operator works only on the coordinates of electron 1, while $1'$ is then replaced by 1 for integration. The density itself is $\rho(1) = \rho(1,1)$, which corresponds to the diagonal elements of the one-particle density matrix. Since the $\hat{v}(1)$ operator is multiplicative, $\rho(1,1')$ reduces to just the density, $\rho(1)$, there. Similarly, because of the multiplicative nature of the $\hat{V}_{ee}(1, 2)$ operator, the last quantity has no primes on the electron coordinates, 1 and 2, so it can be replaced by its diagonal form, often designated $\rho(1,2)$.

The usual first *ab initio* approximation to the wave function leads to the Hartree-Fock theory, where $\Psi \approx \Phi_o = A(\varphi_1(1)\varphi_2(2)...\varphi_n(n))$ is an antisymmetrized (i.e. determinant) of molecular spin orbitals $\{\varphi_k\}$ with one for each electron. Then, asking the question what is the single determinant solution with the lowest possible energy, we obtain the Hartree-Fock equations and density,

$$f(1)\varphi_k(1) = \varepsilon_k \varphi_k(1) \tag{15.10}$$

$$\rho(1) = \sum_i^n \varphi_i(1)\varphi_i^*(1) \tag{15.11}$$

$$f(1) = \hat{h}(1) + \hat{J}(1) + \hat{K}(1) \tag{15.12}$$

$$J(1) = \sum_k^n \int d\tau_2 \frac{\varphi_k^*(2)\varphi_k(2)}{|r_1 - r_2|} = \int d\tau_2 \frac{\rho(2)}{r_{12}} \tag{15.13}$$

$$K(1) = -\sum_k^n \int d\tau_2 \frac{\varphi_k^*(2)P(12)\varphi_k(2)}{|r_1 - r_2|} \tag{15.14}$$

The operator $\hat{J}$ corresponds to the classical Coulomb repulsion, while the quantum mechanical exchange operator, $\hat{K}$, contains the permutation operator $P(12)$, which has the effect of interchanging the coordinates of electron 1 and 2. This causes the exchange operator to be non-local and difficult to plot, unlike the local, multiplicative Coulomb operator, $\hat{J}$. However, the exchange energy is

$$E_X = -\frac{1}{2}\sum_{k,l}^n \iint d\tau_1 d\tau_2 \frac{\varphi_l^*(1)\varphi_k^*(2)\varphi_l(2)\varphi_k(1)}{|r_1 - r_2|} \tag{15.15}$$

which should be contrasted to the Coulomb repulsion energy

$$E_J = \frac{1}{2}\sum_{k,l}^n \iint d\tau_1 d\tau_2 \frac{\varphi_l^*(1)\varphi_k^*(2)\varphi_l(2)\varphi_k(1)}{|r_1 - r_2|} = \frac{1}{2}\iint \frac{\rho(1)\rho(2)}{r_{12}} \tag{15.16}$$

which can be written entirely in terms of the density. For one electron, $E_J + E_X = 0$, which eliminates the self-interaction of one electron with itself. The great weakness in HF theory is that in the above equations the two-electron interactions are treated in an averaged way, which means that no consideration is made of the instantaneous interactions of electrons among themselves which tends to keep them apart. That is, the motions of electrons are *correlated*, causing the correlation error in an HF calculation. To attain anywhere near a ~1 kcal/mol accuracy in chemistry, electron correlation must be included.

The basic idea for including electron correlation in wave function theory is to introduce additional functions orthogonal to the occupied orbitals in HF theory, and allow these functions to mix into the wave function in such a way that they offer additional space for the electrons to move, enabling the electrons to be better able to avoid each other. In practice, this is most frequently accomplished by solving the HF equations above in a basis set (usually contracted Gaussian functions centered on each atom) whose dimension, m, significantly exceeds the number of electrons in the problem. Then, in addition to the occupied orbitals, we have solutions for the unoccupied (virtual) orbitals $\{\varphi_{n+1}, \varphi_{n+2},...\varphi_m\}$, which we will refer to generically as $\varphi_a, \varphi_b,...$ to distinguish them from the occupied orbitals, $\varphi_i, \varphi_j,...$ The correlated methods in quantum chemistry, configuration interaction (CI), [23] multi-configuration self-consistent field (MCSCF), [24] many-body perturbation theory (MBPT), [25] and coupled-cluster (CC) [26, 27] theory, all offer prescriptions for how to do this.

For our purposes, CC theory and its finite order MBPT approximations offer a convenient, compact description of the correlation problem and give rapid convergence to the basis set (i.e. full CI) limit with different categories of correlation operators (see Fig. 15.1). The coupled-cluster wave function is

$$\Psi = \exp(T)|\Phi_o\rangle = (1 + T + \frac{T^2}{2!} + \frac{T^3}{3!} + ...)|0\rangle \tag{15.17}$$

where the operator

$$T = T_1 + T_2 + T_3 + ... \tag{15.18}$$

is composed of different cluster excitations, single, T_1, doubles, T_2, etc. The effect of each T_p operator is to create p-fold excited determinants by promoting electrons from an occupied orbital into unoccupied orbitals to introduce effects of electron correlation. That is, $T_1\Phi_0 = \sum\limits_{i,a} t_i^a \Phi_i^a$, $T_2\Phi_0 = \sum\limits_{i<j,a<b} t_{ij}^{ab} \Phi_{ij}^{ab}$, etc., where $\Phi_{ij...}^{ab...}$ is the excited determinant. In particular, we can write the CC functional for the correlation energy conveniently in terms of the CC wave function and a de-excitation operator, Λ, which give E_C

$$E_C = \langle\Phi_o|(1 + \Lambda)\exp(-T)[H - \langle\Phi_o|H|\Phi_o\rangle]\exp(T)|\Phi_o\rangle$$
$$= \langle\Phi_o|(1 + \Lambda)\bar{H}|\Phi_o\rangle \tag{15.19}$$

once the CC equations satisfy the decoupled equations

$$\left\langle\Phi_{ij...}^{ab...}\middle|\bar{H}\middle|0\right\rangle = 0 \tag{15.20}$$

$$\left\langle\Phi_o\middle|(1 + \Lambda)\bar{H} - E_C\Lambda\middle|\Phi_{ij...}^{ab...}\right\rangle = 0 \tag{15.21}$$

The excitation level matches that for T_p. The simplest possible approximation to E_C is given by MBPT(2) or MP2,

$$E_C = \frac{1}{4} \sum_{\substack{i,j \\ a,b}} \frac{|\langle ij\|ab\rangle|^2}{\varepsilon_i + \varepsilon_j - \varepsilon_a - \varepsilon_b} \tag{15.22}$$

where

$$\langle ij\|ab\rangle = \iint d\tau_1 d\tau_2 \varphi_i^*(1)\varphi_j^*(2) \frac{(1 - P(12))}{r_{12}} \varphi_a(1)\varphi_b(2) \tag{15.23}$$

and a canonical HF reference function is assumed.

Due to the work of John Pople and many others, today we recognize a paradigm in electronic structure theory for correlated methods, MBPT(2) ~ MBPT(3) < CCD < CCSD ~ MBPT(4) < CCSD(T) < CCSDT < CCSDT(Q_f) < CCSDTQ < Full CI, [19] (see Fig. 15.1) where MBPT(n) means perturbation theory through order n, and CCD means coupled-cluster limited to double excitation cluster operators, T_2; CCSD, includes singles(T_1) and doubles, CCSDT adds triples, (T_3); and CCSDTQ adds quadruple clusters, T_4. The meaning of these cluster operators is like that in statistical mechanics, except here we refer to clusters of electrons. The ultimate result in a basis set is the full CI, which, because of its ~m^n computational dependence, where m is the number of basis functions (typically ~$3n$), is seldom possible to obtain. Since the Hamiltonian only has one- and two-particle operators, and correlation means keeping two electrons apart, the most important cluster operator is T_2, and the CCD wave function is written as $\exp(T_2)|\Phi\rangle = (1 + T_2 + T_2^2/2 + T_2^3/3 + ...)|\Phi\rangle$ so all doubles, plus the simultaneous double excitations (disconnected quadruples), plus their higher products are introduced. CCD, or CCSD are said to be m^6 methods, which is their asymptotic computational dependence on the number of basis functions. Once we consider three-electron (~m^8) and four-particle clusters (~m^{10}), the extreme non-linear dependence on the number of functions is apparent. The methods CCSD(T) [28] (~m^7) and CCSDT(Q_f) [29] combine perturbation ideas with CC ideas to offer non-iterative estimates of higher-order clusters in a more tractable computational scheme. The Q_f indicates that the *connected* T_4 operator is further factored to give a very attractive computational procedure which is a non-iterative, ~m^7 procedure for the quadruples, though ~m^8 for the underlying CCSDT. In principle, this non-linear dependence can be drastically reduced if we are able to exploit the fact that chemical bonds are really fairly short-sighted, as is electron correlation. In fact, it has been shown that the classical electrostatic many-body problem can be reduced to just linear scaling with proper handling, [30, 31] and the quantum many-body problem can be similarly simplified, [32–34] particularly for DFT. However, in WT, this has only been demonstrated so far for HF [35] and MBPT(2). [36]

15.4
Density Functional Theory

DFT can be developed from the constrained search formulation of Levy, [12] since we know that

$$E_o = \mathop{Min}_{\Psi}\langle\Psi|\hat{T} + \hat{V}_{ee} + \sum_i \hat{v}(r_i)|\Psi\rangle \tag{15.24}$$

$$E_o = \mathop{Min}_{\rho}\left\{ \mathop{Min}_{\Psi\to\rho}\left[\langle\Psi|\hat{T} + \hat{V}_{ee}|\Psi\rangle + \int v(r_i)\rho(1)d\tau_1\right]\right\} \tag{15.25}$$

$$E_o = \mathop{Min}_{\rho}\left\{ F[\rho] + \int v(r_i)\rho(1)d\tau_1 \right\} = \mathop{Min}_{\rho} E[\rho] \tag{15.26}$$

where $F[\rho]$, composed of the kinetic energy and two-particle term, is the *universal* functional of DFT theory. The kinetic energy and electron-electron repulsion pertain to any group of electrons, whether they be in a molecule or an electron gas, while the distinction among molecules arises due to the different $\hat{v}(r_i)$. Then, from the usual expectation value and variational principle of WT, we know that the exact energy must be obtained by the antisymmetric wave function that minimizes the expectation value. However, many wave functions can give the same density. Hence, among all the wave functions that give a particular density, we then search all such densities until the minimum energy is obtained.

The universal functional in Kohn-Sham theory is chosen to be

$$F[\rho] = \langle\Psi_\rho|\hat{T} + \hat{V}_{ee}|\Psi_\rho\rangle = T_s[\rho] + E_J[\rho] + E_X[\rho] + E_C[\rho] \tag{15.27}$$

where $T_S[\rho]$ is the kinetic energy of the single determinant wave function composed of KS orbitals, $\Phi_{KS} = A(\varphi_1(1)\varphi_2(2)...)$, subject to the condition that the exact density is $\rho(1) = \sum_i^n \varphi_i(1)\varphi_i^*(1)$. The two-electron interaction is decomposed into the classical Coulomb repulsion energy, E_J, the electron exchange energy, E_X, and the electron correlation energy, E_C. At this point, we recognize that all appropriate densities can be represented in terms of orbitals, $\{\varphi_i\}$, so variation of the energy functional (not usually an expectation value) with respect to the density, or with respect to orbital variation, leads us to the same kind of one-particle equations as in the HF theory above,

$$h_{KS}(1) = \hat{h}(1) + \hat{J}(1) + \hat{V}_X(1) + \hat{V}_C(1) = \hat{t}(1) + \hat{v}_s(1) \tag{15.28}$$

($\hat{v}_S(1)$ is the Sham effective potential)

$$h_{KS}(1)\varphi_i(1) = \varepsilon_i\varphi_i(1) \tag{15.29}$$

$$\rho(1) = \sum_{i=1}^n \varphi_i(1)\varphi_i^*(1) \tag{15.30}$$

except that we now have a local, multiplicative exchange potential, $\hat{V}_X(1) = \delta E_X/\delta\rho(1)$, and a correlation potential, $\hat{V}_C(1) = \delta E_C/\delta\rho(1)$, which are the functional derivatives with respect to the density. By virtue of using the density as an intermediate, we are able to eliminate explicit consideration of the two-particle character of the electron correlation. Hence, we have *correlation* in DFT unlike HF theory. Furthermore, the Kohn-Sham (KS) orbitals, [11, 37] being the solution of the exact one-particle equations, give the *exact density*. There is an underlying wave function in KS theory, namely Φ_{KS}, which as the determinant of KS orbitals has the property that

$$\rho(1) = \int \Phi_{KS}(1, 2...)\Phi_{KS}^*(1, 2...)d\tau_2 d\tau_3...d\tau_n = \sum_{i=1}^{n} \varphi_i(1)\varphi_i^*(1). \tag{15.31}$$

Faced with the basis product approximation in WT because of its explicit treatment of two-particle interactions, plus the high non-linear dependence upon m, there are great potential computational advantages in DFT. First and foremost, it is a one-particle theory. Second, it is more conceptual, as you can plot and look at exchange and correlation potentials. In fact, in this way something as complicated as electron correlation can be explained to freshmen chemistry students. Thirdly, it offers a more immediate way to derive conceptually useful quantities like the Mulliken electronegativity, [37, 38] i.e. $(I–A)/2$ where I and A are the ionization potential and electron affinity of some element. Fourth, linear scaling algorithms are more immediate, readily permitting very large-scale applications to be made. [32–34]

The weakness, of course, is that $E_X[\rho]$ and $E_C[\rho]$ are not known, nor their derivatives. Consequently, either exchange and correlation potentials are taken from a soluble model like the homogeneous electron gas, which gives the local density approximation (LDA), or they are improved via gradient corrections, [39, 40] with various justifications. Yet none of these are entirely satisfactory, because other *exact* conditions on the functionals and potentials are not satisfied. Hence, in DFT we do not know how to systematically improve upon a given calculation. For example, gradient corrected functionals like BLYP, [41] PW92, [42] and PBE [43] might be better for molecules, but for most crystals and polymers, LDA actually performs better. Today it is generally thought that the best current method is B3LYP, [44] where some percent of Hartree-Fock non-local exchange, evaluated for the KS orbitals, is incorporated to rectify failures of BLYP. Furthermore, what works well for ground states and properties can fail dramatically for excited states obtained from the time-dependent DFT, where the second derivative

$$\hat{f}(1, 2) = \frac{\delta^2 E_{XC}}{\delta\rho(1)\delta\rho(2)} = \frac{\delta V_{XC}(2)}{\delta\rho(1)} \tag{15.32}$$

is required, and, as we have shown, [16] can be expected to frequently even have the wrong sign. I propose that one solution to not knowing $\hat{V}_X[\rho]$ and $\hat{V}_C[\rho]$ and $\hat{f}(1, 2)$ is **ab initio DFT**.

15.5
Ab Initio Density Functional Theory

15.5.1
Exact Exchange

For example, the exact exchange in DFT is defined as

$$E_X = -\frac{1}{2}\sum_{i,j}^{n} \langle ij\,|\,ji\rangle = \langle \Phi_{KS}|V_{ee}|\Phi_{KS}\rangle - E_J \tag{15.33}$$

provided that the orbitals i, j are obtained from the exact KS equations, Eq. (15.29). Consequently, this exact relation can be exploited to define $\hat{V}_X[\rho]$ from functional differentiation. However, we do not immediately know $E_X[\rho]$. We can circumvent this problem, however, by using a functional derivative chain rule to permit this derivative to be taken through the KS orbitals and the Sham $\hat{v}_s$ potential of Eq. (15.28). Thus,

$$\hat{V}_X(1) = \frac{\delta E_X}{\delta\rho(1)} = \sum_{i}^{n} \iint \frac{\delta E_X}{\delta\varphi_i(2)}\frac{\delta\varphi_i(2)}{\delta\hat{v}_s(3)}\frac{\delta\hat{v}_s(3)}{\delta\rho(1)}d\tau_2\delta\tau_3 \tag{15.34}$$

The first derivative in Eq. (15.34) is straightforward, while the second requires only standard one-particle perturbation theory,

$$(\varepsilon_i - h_{KS}(1))\delta\varphi_i = (\delta v_s - \delta\varepsilon_i)\varphi_i \tag{15.35}$$

whose solution is

$$\frac{\delta\varphi_i(1)}{\delta v_s(2)} = \sum_{p\neq i}\frac{\varphi_p^*(2)\varphi_i(2)\varphi_p(1)}{\varepsilon_i - \varepsilon_p} \tag{15.36}$$

The third term derives from the response of the density,

$$\delta\rho(3) = \int d\tau_2 X(2,3)\delta\hat{v}_s(2) \tag{15.37}$$

whose solution is

$$X(2,3) = \frac{\delta\rho(2)}{\delta v_s(3)} = \sum_{i,a}\left(\frac{\varphi_i^*(2)\varphi_a(2)\varphi_i(3)\varphi_a^*(3)}{\varepsilon_i - \varepsilon_a} + c.c.\right) \tag{15.38}$$

However, we require the inverse of $X(2,3)$ in the chain rule, so we expand $X(2,3)$ in a Gaussian basis set using inner projection theory, [45]

$$X^{-\frac{1}{2}}O_hX^{-\frac{1}{2}} = X^{-\frac{1}{2}}|h\rangle\langle h \mid h\rangle^{-1}\langle h|X^{-\frac{1}{2}} = |g\rangle\langle g|X|g\rangle^{-1}\langle g|$$
$$= \sum_{p,q}X_{pq}^{-1}g_p(2)g_q(3) \approx X^{-1} \tag{15.39}$$

where

$$X_{pq} = \sum_{i,a}\left(\frac{(i^*a|p)(ia^*|q)}{\varepsilon_i - \varepsilon_a} + c.c.\right) \tag{15.40}$$

with $(ia|p)$ being the overlap of the occupied orbital i with the virtual a and the Gaussians p and q introduced. These are simple integrals to evaluate. Putting everything together, we have the exchange potential,

$$\hat{V}_X(1) = \sum_{i,j,a}\sum_{p,q}\left(\frac{\langle ij \mid ja\rangle(ia|q)X_{pq}^{-1}g_p(1)}{\varepsilon_i - \varepsilon_a} + c.c.\right) \tag{15.41}$$

which depends upon the exchange integral and the overlaps. In practice, it is actually the matrix elements $\langle\mu|\hat{V}_X|\nu\rangle$ which introduce another overlap $(\mu\nu|p)$ that are used. We must evaluate the above in each SCF cycle, and use a singular value decomposition to ensure that any singularity is excluded from the $\mathbf{X}^{-1}$ evaluation. The inverse of the response function does not in general exist, but in a basis set with singularities eliminated it does. These singularities are partly due to $\hat{v}_s$ only being defined up to constant functions. [46]

Using the above *exact exchange* procedure, there is no arbitrariness in $\hat{V}_X$. Unlike the Hartree-Fock exchange, it is a local, multiplicative operator, yet unlike any other exchange potential used in DFT, it satisfies **all** the conditions of the exact exchange. It correctly accounts for self-interaction cancellation. It has the correct (-1/r) long-range behavior. Representing it as above in a Gaussian basis set eliminates any need for numerical integration. It obviously satisfies the exchange hole sum rule [37] and formally satisfies Koopmans' theorem for the highest occupied orbital, homo (see below). Furthermore, since the exchange energy is an order of magnitude larger than the correlation energy, it is essential that we first get the exchange right before concentrating on the correlation.

The above procedure is a generalization for molecules of the Optimized Effective Potential method used for atoms by Talman and Shadwick [47] and by Li, Iafrate, and Krieger. [48] It has also been used for crystal applications by Görling. [49] Our work [50] and that of Görling [51] are the first for molecules. For the total energy and the ionization potentials for the occupied orbitals, the results are very similar to those of Hartree-Fock theory. However, unlike HF theory, the unoccupied orbitals for exact exchange are moving in the same $\hat{V}_J + \hat{V}_X$ as the occupied orbitals, making their orbital energies lower, and the orbitals more localized than the corresponding HF virtual orbitals.

Another interesting feature of an exact exchange only KS result is that

$$E[\rho] = \langle \Phi_{KS}|\hat{T}|\Phi_{KS}\rangle + v[\rho] + E_J[\rho] + \int d\tau \; \hat{V}_X(1)[3\rho(1) + r \cdot \nabla\rho(1)]$$

$$= \langle \Phi_{KS}|H|\Phi_{KS}\rangle \tag{15.42}$$

which establishes a connection between the exact exchange functional and the expectation value of the Hamiltonian. The fourth term is the exchange functional expressed in terms of the exact potential. [52] Note the functional derivative of the exchange functional above is $\hat{V}_X(1)$. Because of the equivalence of the functional and the expectation value, for this case WT and DFT become equivalent, allowing for the power of wave function theory to be exploited. For example, from the right hand side of Eq. (15.42), we can consider a $\Phi_{KS}(n-1)$ wave function where the jth electron has been removed to give a cation without allowing the orbitals to relax. Hence, $\langle \Phi_{KS}|H|\Phi_{KS}\rangle - \langle \Phi_{KS}(n-1)|H|\Phi_{KS}(n-1)\rangle \approx \varepsilon_j \approx -I_j$, where the KS orbital energy is an approximation to the ionization potential of the jth electron. When j is the homo, we know that the difference above gives exactly ε_j, since $\langle j|\hat{K}|j\rangle = \langle j|\hat{V}_X|j\rangle$ is an exact condition. [48] For other orbitals, this tends to offer a reasonable approximation to most ionization potentials. (In practise, we actually improve this exact condition by shifting all orbital energies to make it true. [50]) Note that this KS version of Koopmans' theorem holds for all principal ionizations in a molecule instead of just for the highest occupied orbital as in the usual KS considerations, where that, alone, determines the long range behavior of the density. This also ascribes a meaning to the other KS orbital energies. Without a reasonable description of the other ionization potentials, many excited state descriptions will fail because most excited states for molecules are of Rydberg character, or require some mixture of Rydberg character in their accurate description.

15.5.2
Exact Correlation

The next step is the correlation. Similar to the exchange, we can formally express the correlation energy as

$$E_C = \langle \Psi[\rho]|T + V_{ee}|\Psi[\rho]\rangle - \langle \Phi_{KS}[\rho]|T + V_{ee}|\Phi_{KS}[\rho]\rangle \tag{15.43}$$

where the first term on the right side is for the exact correlated wave function. This expectation value can be equally well represented in the biorthogonal CC way used in Eq. (15.19), which has many advantages compared to CI descriptions.

In orbital form, the second order correlation expression [MBPT(2)] is

$$E_2 = \sum_{i>j,a>b} \frac{|\langle ij||ab\rangle|^2}{\varepsilon_i + \varepsilon_j - \varepsilon_a - \varepsilon_b} + \sum_{i,a} \frac{|f_{ia}|^2}{\varepsilon_i - \varepsilon_a} \tag{15.44}$$

Adiabatic perturbation theory is useful in general to keep the density fixed, [53] but it can assume slightly different forms depending upon chirals that are made. E_2's functional derivative can be obtained similarly to that for E_X, but it actually has been obtained from a different viewpoint as presented elsewhere. [55] To use the same chain rule procedure we used with exact exchange, in addition to the occupied orbitals we must consider changes in the unoccupied orbitals too, plus the orbital energies. These exact second-order correlation formulas are now being implemented to complement our treatment of exact exchange. From adiabatic perturbation theory, [53] only MBPT(2) is required in the high-density case. Further extensions of this procedure to higher orders of perturbation theory, or eventually to CC theory, will provide a series of better and better correlation potentials not unlike the paradigm of WT from MBPT(2) to CCSD, CCSD(T), CCSDT, and full CI discussed above and illustrated in Fig. 15.1, but now within in a one-particle theory. We can envision that this set of *ab initio* potentials will provide a solution to the criticism of DFT that there is no way to ensure what would be a higher level calculation to assist convergence to the exact solution. I suspect that these *ab initio* potentials will be difficult to evaluate compared to those usually used today in DFT. In fact, since they depend upon derivatives with respect to orbitals, the orbital-dependent expressions which emerge will be more like those encountered in WT, and can likely be systematized diagrammatically as in many-body theory. However, these potentials will have the overwhelming advantage of being those from an exact, converging theory. Once the exact exchange is incorporated, we have shown that all the current correlation potentials used in DFT hurt the agreement of the homo orbital energy compared to the exact ionization potential for the molecule. [50] Quantum Monte Carlo calculations have shown that the exact potentials frequently have the wrong sign, [15] as do the kernels. [16] Hence, only an exact correlation potential can be recommended, and only with such potentials can we expect TDDFT [55–60] to provide realistic descriptions of excited states and associated properties. From the wave function viewpoint, the TDDFT equations are only the RPA equations with a modified potential, but up to the adiabatic approximation that replaces the time-dependent exchange-correlation potential with the usual ground state choices, it is an exact theory of excited states. [56]

In summary, the original Thomas-Fermi-Dirac DFT was unable to give binding in molecules. This was corrected by Kohn-Sham, [11] who chose to use an orbital rather than density evaluation of the kinetic energy. By the virial theorem, $\langle \hat{T} \rangle = -E$, so this was a necessity to obtain realistic results for energies. Next, it was shown that the exact exchange requires an orbital-dependent form, too. [47, 48] The future seems to demand an orbital-dependent form for the correlation.

Acknowledgements

The new work on exact exchange for molecules was done in collaboration with Dr. S. Ivanov and Dr. S. Hirata, who have also taught the author much about DFT. I would like to also acknowledge useful discussions with Prof. Mel Levy. This work was supported by the U.S. Air Force Office of Scientific Research under grant number F49620-98-0116.

References

1 This phrase commonly used in quantum chemical circles is associated with Prof. E. Davidson.

2 Pykko, P.; *Chem. Rev.* **1988**, 88, 563–94.

3 Dirac, P.A.M.; *The Principles of Quantum Mechanics;* Clarendon: Oxford, 1930.

4 Kutzelnigg, W.; *J. Chem. Phys.* **1996**, 105, 5878–85.

5 Kato, T.; *Commun. Pure Appl. Math.* **1957**, 10, 151–77.

6 Bukowski, R.; Jeziorski, B.; Szalewicz, K.; *J. Chem. Phys.* **1999**, 110, 4165–83.

7 Wilson, A.K.; Woon, D.E.; Peterson, K.A.; Dunning, T.H., Jr.; *J. Chem. Phys.* **1999**, 110, 7667–76.

8 Curtiss, L.A.; Raghavachari, K.; Pople, J.A.; *J. Chem. Phys.* **1995**, 103, 4192–200.

9 Montgomery, J.A., Jr.; Frisch, M.J.; Ochterski, J.W.; Petersson, G.A.; *J. Chem. Phys.* **1999**, 110, 2822–27.

10 Hohenberg, P.; Kohn, W.; *Phys. Rev.* **1964**, 136, B864–71.

11 Kohn, W.; Sham, L.J.; *Phys. Rev. A* **1965**, 140, A1133–38.

12 Levy, M.; Perdew, J.P.; In *Density Functional Methods in Physics,* eds. Dreizler, R.M.; Providencia, J.; Plenum: New York, 1985; 11–30.

13 Oliphant, N.; Bartlett, R.J.; *J. Chem. Phys.* **1994**, 100, 6550–61.

14 Ivanov, S.; Levy, M.; Bartlett, R.J.; to be published.

15 Umrigar, C.J.; Gonze, X.; *Phys. Rev. A: At. Mol., Opt. Phys.* **1994**, 50, 3827–37.

16 Ivanov, S.; Bartlett, R.J.; *Chem. Phys. Lett.* **1999**, 308, 449–55.

17 Perdew, J.P.; In *Density Functional Methods in Physics,* eds. Dreizler, R.M.; Providencia, J.; Plenum: New York, 1985; 265–308.

18 Van Voorhis, T.; Scuseria, G.E.; *J. Chem. Phys.* **1998**, 109, 400–10.

19 Bartlett, R.J.; In *Modern Electronic Structure Theory,* ed. Yarkony, D.R.; World Scientific: Singapore, 1995, 1047–1131.

20 Slater, J.C.; *The Quantum Theory of Molecules and Solids,* vol. 1; McGraw-Hill: New York, 1963.

21 Kolos, W.; *J. Chem. Phys.* **1994**, 101, 1330–32.

22 Löwdin, P.O.; *Phys. Rev.* **1955**, 97, 1474–89.

23 Shavitt, I.; *Mol. Phys.* **1998**, 94, 3–17.

24 Roos, B.O.; *Adv. Chem. Phys.* **1987**, 69, 399–445.

25 Bartlett, R.J.; *Annual Rev. of Phys. Chem.* **1981**, 32, 359–401.

26 Èížek, J.; Paldus, J.; *Physica Scripta* **1980**, 21, 251–54.

27 Bartlett, R.J.; *J. Phys. Chem.* **1989**, 93, 1697–1708.

28 Raghavachari, K.; Trucks, G.W.; Head-Gordon, M.; Pople, J.A.; *Chem. Phys. Lett.* **1989**, 157, 479–83.

29 Kucharski, S.A.; Bartlett, R.J.; *J. Chem. Phys.* **1998**, 108, 5243–54.

30 Greengard, L.; Rokhlin, V.; *J. Comput. Phys.* **1987**, 73, 325–48.

31 Greengard, L.F.; *The Rapid Evolution of Potential Fields in Particle Systems;* MIT: Cambrige, 1988.

32 Yang, W.; *Theochem* **1992**, 87, 469–79.

33 White, C.A.; Johnson, B.G.; Gill, P.M.W.; Head-Gordon, M.; *Chem. Phys. Lett.* **1994**, 230, 8–16.

34 Strain, M.C.; Scuseria, G.E.; Frisch, M.J.; *Science* **1996**, 271, 51–53.

35 Challacombe, M.; *J. Chem. Phys.* **1999**, 110, 2332–42.

36 Ayala, P.Y.; Scuseria, G.E.; *J. Chem. Phys.* **1999**, 110, 3660–71.

37 Parr, R.G.; Yang, W.; *Density-Functional Theory of Atoms and Molecules;* Oxford University Press: Oxford, 1989.

38 Mulliken, R.S.; *J. Chem. Phys.* **1934**, 2, 782–93.

39 Perdew, J.P.; Langreth, D.C.; Sahni, U.; *Phys. Rev. Lett.* **1977**, 38, 1030–33.

40 Becke, A.D.; *J. Chem. Phys.* **1992**, 96, 2155–60.

41 Lee, C.; Yang, W.; Parr, R.G.; *Phys. Rev. B* **1988**, 37, 785–89.

42 Perdew, J.P.; Wang, Y.; *Phys. Rev. B: Conds. Matter* **1992**, 45, 13244–49.

43 Perdew, J.P.; Burke, K.; Ernzerhof, M.; *ACS Symp. Ser.* **1996**, 629, 453–62.

44 Becke, A.D.; *J. Chem. Phys.* **1993**, 98, 5648–52.

45 Löwdin, P.O.; *Int. J. Quant. Chem.* **1971**, 4S, 231–37.

46 Görling, A.; Levy, M.; *Phys. Rev. A* **1994**, 69, 763–69.

47 Talman, J.D.; Shadwick, W.F.; *Phys. Rev. A* **1976**, 14, 36–40.

48 Krieger, J.B.; Li, Y.; Iafrate, G.J.; *Phys. Rev. A* **1992**, 45, 101–26.

49 Görling, A.; *Phys. Rev. A* **1996**, 53, 7024–29.

50 Ivanov, S.; Hirata, S.; Bartlett, R.J.; *Phys. Rev. Lett.* **1999**, 83, 5455–5458.

51 Görling, A.; *Phys. Rev. Lett.* **1999**, 83, 5459–5462.

52 Engel, E.; Vosko, S.H.; *Phys. Rev. A* **1993**, 47, 2800–11.

53 Görling, A.; Levy, M.; *Phys. Rev. A* **1992**, 45, 1509–17.

54 Ivanov, S.; Burke, K. Levy, M.; *J. Chem. Phys.* **1999**, 110, 10262–68.

55 Burke, K.; Gross, E.K.U.; In *Density Functionals: Theory and Applications. Proceedings of the Tenth Chris Engelbrecht Summer School in Theoretical Physics*, ed. Joubert, J.; Springer: Berlin, 1998.

56 Runge, E.; Gross, E.K.U.; *Phys. Rev. Lett.* **1984**, 52, 997–1000.

57 Gross, E.K.U.; Kohn, W.; *Adv. Quantum Chem.* **1991**, 21, 255–91.

58 Jamorski, C.; Casida, M.; Salahub, D.R.; *J. Chem. Phys.* **1996**, 104, 5134–47.

59 Bauernschmitt, R.; Ahlrichs, R.; *Chem. Phys. Lett.* **1996**, 256, 454–64.

60 Tozer, D.J.; Handy, N.C.; *J. Chem. Phys.* **1998**, 109, 10180–89.

Subject Index